NATIONAL ACADEMIES

Sciences
Engineering
Medicine

NATIONAL
ACADEMIES
PRESS
Washington, DC

The Potential Impacts of Gold Mining in Virginia

Committee on Potential Impacts of Gold Mining in Virginia

Committee on Earth Resources

Board on Earth Sciences and Resources

Division on Earth and Life Studies

Consensus Study Report

NATIONAL ACADEMIES PRESS 500 Fifth Street, NW Washington, DC 20001

This activity was supported by a contract between the National Academy of Sciences and the Virginia Department of Energy and the National Academy of Sciences' Arthur L. Day Fund. Any opinions, findings, conclusions, or recommendations expressed in this publication do not necessarily reflect the views of any organization or agency that provided support for the project.

International Standard Book Number-13: 978-0-309-69121-5
International Standard Book Number-10: 0-309-69121-4
Digital Object Identifier: https://doi.org/10.17226/26643

This publication is available from the National Academies Press, 500 Fifth Street, NW, Keck 360, Washington, DC 20001; (800) 624-6242 or (202) 334-3313; http://www.nap.edu.

Suggested citation: National Academies of Sciences, Engineering, and Medicine. 2023. *The Potential Impacts of Gold Mining in Virginia*. Washington, DC: The National Academies Press. https://doi.org/10.17226/26643.

The **National Academy of Sciences** was established in 1863 by an Act of Congress, signed by President Lincoln, as a private, nongovernmental institution to advise the nation on issues related to science and technology. Members are elected by their peers for outstanding contributions to research. Dr. Marcia McNutt is president.

The **National Academy of Engineering** was established in 1964 under the charter of the National Academy of Sciences to bring the practices of engineering to advising the nation. Members are elected by their peers for extraordinary contributions to engineering. Dr. John L. Anderson is president.

The **National Academy of Medicine** (formerly the Institute of Medicine) was established in 1970 under the charter of the National Academy of Sciences to advise the nation on medical and health issues. Members are elected by their peers for distinguished contributions to medicine and health. Dr. Victor J. Dzau is president.

The three Academies work together as the **National Academies of Sciences, Engineering, and Medicine** to provide independent, objective analysis and advice to the nation and conduct other activities to solve complex problems and inform public policy decisions. The National Academies also encourage education and research, recognize outstanding contributions to knowledge, and increase public understanding in matters of science, engineering, and medicine.

Learn more about the National Academies of Sciences, Engineering, and Medicine at **www.nationalacademies.org**.

Consensus Study Reports published by the National Academies of Sciences, Engineering, and Medicine document the evidence-based consensus on the study's statement of task by an authoring committee of experts. Reports typically include findings, conclusions, and recommendations based on information gathered by the committee and the committee's deliberations. Each report has been subjected to a rigorous and independent peer-review process and it represents the position of the National Academies on the statement of task.

Proceedings published by the National Academies of Sciences, Engineering, and Medicine chronicle the presentations and discussions at a workshop, symposium, or other event convened by the National Academies. The statements and opinions contained in proceedings are those of the participants and are not endorsed by other participants, the planning committee, or the National Academies.

Rapid Expert Consultations published by the National Academies of Sciences, Engineering, and Medicine are authored by subject-matter experts on narrowly focused topics that can be supported by a body of evidence. The discussions contained in rapid expert consultations are considered those of the authors and do not contain policy recommendations. Rapid expert consultations are reviewed by the institution before release.

For information about other products and activities of the National Academies, please visit www.nationalacademies.org/about/whatwedo.

COMMITTEE ON POTENTIAL IMPACTS OF GOLD MINING IN VIRGINIA

WILLIAM A. HOPKINS, *Chair*, Virginia Polytechnic Institute and State University
KWAME AWUAH-OFFEI, Missouri University of Science and Technology
JOEL D. BLUM, University of Michigan
ROBERT J. BODNAR, Virginia Polytechnic Institute and State University
THOMAS CRAFFORD, U.S. Geological Survey (retired)
FIONA M. DOYLE, University of California, Berkeley
JAMI DWYER, Barr Engineering Company (retired)
ELIZABETH HOLLEY, Colorado School of Mines
PAUL A. LOCKE, Johns Hopkins Bloomberg School of Public Health
SCOTT M. OLSON, University of Illinois
BRIAN S. SCHWARTZ, John Hopkins Bloomberg School of Public Health
M. GARRETT SMITH, Montana Department of Environmental Quality
SHILIANG WU, Michigan Technological University

Staff

STEPHANIE JOHNSON, Study Director
MARGO REGIER, Study Director
CLARA PHIPPS, Senior Program Assistant (until June 2022)
MILES LANSING, Program Assistant (starting July 2022)
CHIOMA ONWUMELU, Christine Mirzayan Science and Technology Policy Fellow (March–May 2022)

NOTE: See Appendix B, Disclosure of Unavoidable Conflicts of Interest.

Reviewers

This Consensus Study Report was reviewed in draft form by individuals chosen for their diverse perspectives and technical expertise. The purpose of this independent review is to provide candid and critical comments that will assist the National Academies of Sciences, Engineering, and Medicine in making each published report as sound as possible and to ensure that it meets the institutional standards for quality, objectivity, evidence, and responsiveness to the study charge. The review comments and draft manuscript remain confidential to protect the integrity of the deliberative process.

We thank the following individuals for their review of this report:

STEPHEN BUCKLEY, Alaska Department of Natural Resources
BARTHOLOMEW CROES, California Air Resources Board
JACLYN GOODRICH, University of Michigan School of Public Health
GEORGE HORNBERGER, Vanderbilt University
STEVE KESLER, University of Michigan
MOLLY KILE, Oregon State University College of Public Health and Human Sciences
STEPHEN LANG, Hudbay Minerals and Hycroft Mining Holding Corporation
JORGE MACEDO, Georgia Institute of Technology
ANN MAEST, Buka Environmental
JOHN MARSDEN, Metallugium and John O. Marsden LLC
FRANK McCORMICK, U.S. Forest Service and Rocky Mountain Research Station
PATRICIA McGRATH, U.S. Environmental Protection Agency
JOHN PRICE, University of Nevada, Reno
NORMAN SLEEP, Stanford University

Although the reviewers listed above provided many constructive comments and suggestions, they were not asked to endorse the conclusions or recommendations of this report nor did they see the final draft before its release. The review of this report was overseen by **HERMAN GIBB**, Gibb Epidemiology Consulting, LLC, and **CORALE L. BRIERLEY**, Brierley Consultancy LLC. They were responsible for making certain that an independent examination of this report was carried out in accordance with the standards of the National Academies and that all review comments were carefully considered. Responsibility for the final content rests entirely with the authoring committee and the National Academies.

Acknowledgments

Many individuals assisted the committee and the National Academies of Sciences, Engineering, and Medicine staff in their task to create this report. The committee would like to thank the following people who gave presentations, participated in panel discussions, provided public comments to the committee, or served as field trip guides.

Jeeva Abbate
Byron Amick, South Carolina Department of Health and Environmental Control
Nelson Bailey
David Ball
Betsy Baxter, Chair, Good Neighbor Agreement Task Force
Heidi Berthoud
Stephen Buckley, Alaska Department of Natural Resources
Kevin Cook, OceanaGold
Carolyn Curley, Alaska Department of Natural Resources
Kristin Davis, Southern Environmental Law Center
Jeremy Eddy, South Carolina Department of Health and Environmental Control
Damien Fehrer, Virginia Department of Energy
Lakshmi Fjord
Marie Flowers
Paul Fry, California State Mining and Geology Board
Susan Fulmer, South Carolina Department of Health and Environmental Control
Robert Ghiglieri, Nevada Division of Minerals
James Golden, Virginia Department of Environmental Quality
Shawn Gooch, Nevada Department of Environmental Protection
Debbie Gordon
Christine Green
William Groom, Alaska Department of Natural Resources
Elizabeth Guzman, Virginia House of Delegates
Kenda Hanuman, Friends of Buckingham
Chinsuk Henshaw
Claire Horan, Southern Environmental Law Center

Karen Hudson-Edwards, University of Exeter
Larry Innes, Townshend LLP
Wayne Jepson, Montana Department of Environmental Quality
Suzanne Keller
Karen Kreps
Robery Kuczynski, Nevada Department of Environmental Protection
Jim Kuipers, Kuipers and Associates
Irene Leech
Natathanial Leies
Kelly Lowry, Attorney, Lowry and Associates
Sara MacDonald, South Carolina Department of Health and Environmental Control
Ann Maest, Buka Environmental
Karl McCrea, Nevada Department of Environmental Protection
Jim Mclain, Gemini Consulting
Theresa McMannis
Thomas Jordan Miles III
Chris Miller
Chris Mitchell
Erica Morrell, St. Lawrence University
Robby Morrow, South Carolina Geological Survey
Kyle Moselle, Alaska Department of Natural Resources
Allan Nakanishi, Alaska Department of Environmental Conservation
James Nooten, Jr.
Chad Oba
Deacon William Perkins and members of the Warminster Baptist and Union Hill Church
Timothy Pilon, Division of Water, Alaska Department of Environmental Conservation
Todd Process, Nevada Department of Environmental Protection
Manuel Rauhut, Idaho Department of Water Resources
Abhaya Rhiek
Stephanie Rinaldi, Press Pause Coalition
Quin Robinson
Ella Rose
Paul Saunders, Virginia Department of Energy
Erica Schoenberger, Johns Hopkins Environmental Health and Engineering
Eric Schwamberger, International Cyanide Management Institute
Jennifer Serafin, U.S. Army Corps of Engineers
Jessica Sims, Appalachian Voices
Mareesa Singleton, South Carolina Department of Health and Environmental Control
Michael Skiffington, Virginia Department of Energy
J. Phil Skorupa, Virginia Department of Energy
Christopher Slocum, South Carolina Department of Health and Environmental Control
Maggie Snoddy
David Spears, Virginia Department of Energy
Brianna St. Pierre, California State Water Resources Control Board
Janice Thore
Lukas Tyree, Bureau of Indian Affairs
Thomas Ullrich, Aston Bay Holdings, Ltd.
Joshua Vana

Michael Visher, Nevada Division of Minerals
David Waters
Katie Whitehead
Eric Wilson, Idaho Department of Lands
Mindy Zlotnick

Contents

Summary

Gold is a high-value metal utilized primarily in electrical devices, in jewelry, and for investments. Western states, especially Nevada, have dominated recent gold production in the United States, but as current deposits become depleted and gold prices rise, mining companies are increasingly exploring for lower-grade gold deposits or those that are deeper in the Earth. The Commonwealth of Virginia was one of the first major gold-producing states in the nation, but only intermittent exploration activity and small operations have occurred in the past 70 years. Recently, there has been renewed attention to the potential for gold exploration and mining in Virginia, both at new sites and at historical sites where advances in mining and processing techniques might allow for the profitable production or "remining" of gold from deposits that were previously uneconomic.

The National Academies of Sciences, Engineering, and Medicine's Committee on Potential Impacts of Gold Mining in Virginia was formed following House Bill 2213, which passed in response to stakeholder concerns regarding gold exploration in central Virginia. The committee's task focused on the technical aspects of potential gold mining in Virginia—including a review of the geologic characteristics of the main gold deposits and probable modern mining techniques that could be used in such deposits, an evaluation of the potential impacts of those activities, and an assessment of the sufficiency of existing regulations in the Commonwealth to protect air and water quality (see Box 1-3 for the full Statement of Task). A parallel committee (the "state agency committee") formed by the Virginia Department of Energy, the Virginia Department of Environmental Quality, and the Virginia Department of Health focused on local equity and environmental justice issues and environmental and human health concerns of the local community.

Overall, the committee found that the regulatory framework of Virginia appears to have been designed for operations like crushed stone quarrying and sand and gravel operations, not gold mining. As such, Virginia's current regulatory framework is not adequate to address the potential impacts of commercial gold mining.[1] More specifically, Virginia's regulatory framework lacks an adequate financial assurance system, which poses a fiscal and environmental risk to the Commonwealth. Additionally, Virginia lacks opportunities for a diverse public to be engaged in permitting processes and a modern system for review of environmental impacts from potential gold mining projects. These and other portions of Virginia's regulatory framework fell short in comparison to other states, the federal government, and modern best practices.

---

[1] In this report, commercial gold mines refers to larger and more technologically complex operations than small-scale gold mines. Small-scale gold mines are typically low-tech, labor-intensive mineral extraction and processing carried out mostly by local people (Hilson and Maconachie, 2020).

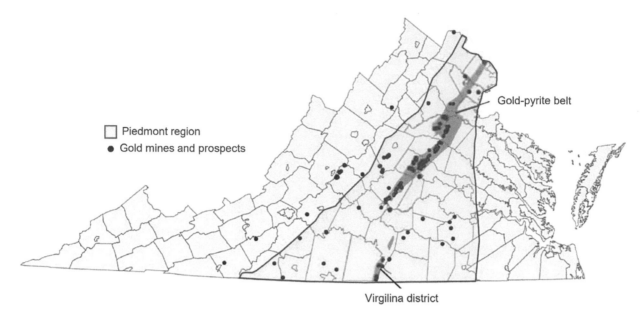

FIGURE S-1 Locations of historic gold mines and prospects in Virginia. The majority (>95 percent) are found in the Piedmont region. Two major gold districts occur in the Piedmont: the gold-pyrite belt in the north-central part of the state, and the Virgilina district in south-central Virginia. A few deposits are located to the west of the Piedmont, in the Blue Ridge region. SOURCE: Modified from Sweet (2007).

GEOLOGIC CHARACTERISTICS AND PROBABLE MINING OPERATIONS

Most known gold occurrences in Virginia are associated with metamorphic and igneous rocks in the Piedmont physiographic province, except for a few small occurrences in the Blue Ridge province (see Figure S-1). These deposits in Virginia occur in lens-shaped, low-sulfide, gold-quartz veins (1–5 percent pyrite) that dip at steep angles, making shallow open pit and underground mining the most likely excavation methods. As demonstrated by the historic London and Virginia, Buckingham, and Williams mines, massive sulfide bodies can occur in close proximity to the low-sulfide, gold-quartz vein deposits in Virginia (see Chapter 2), and could release acid rock drainage (ARD) and metals if disturbed during mining. All available evidence indicates that Virginia gold deposits are generally smaller than those in other gold-producing states, which suggests that it may be more economical for companies to ship ore or pyrite concentrates off-site for the later stages of processing. This is significant because the magnitude of the potential impacts of gold mining can scale with the size of the operations and whether processing occurs on- or off-site.

THE POTENTIAL IMPACTS OF GOLD MINING

The potential human health and ecological impacts identified in this report are based on a review of the impacts of gold mining at U.S. and international sites and on the concerns expressed by community members during the information-gathering activities in this study. Given the statewide focus of the Statement of Task, the committee could not predict site-specific impacts from gold mining. Instead, the committee evaluated the impacts reported at other gold mining sites in the context of the environmental, geologic, and social conditions of the gold-bearing regions of the Commonwealth. The major potential impacts of concern are related to surface water and groundwater contamination, groundwater table drawdown, remobilization of legacy mercury from past uses, rare but catastrophic events such as dam failures and spills, and cumulative health effects due to interacting stressors. All of these factors are likely to affect some communities more than others, particularly those with lower socioeconomic status and higher proportions of racial and ethnic minorities, which could further exacerbate environmental injustice and health disparities. A robust regulatory framework and modern best practices can significantly reduce many of the impacts associated with gold mining, but the risk of adverse impacts cannot be completely eliminated. The largest potential impacts, and factors that could mitigate or exacerbate those impacts, are discussed below.

Remobilization of Legacy Contaminants

Remobilization of legacy mercury from mining operations that take place at historically mined sites poses a significant risk to human health and the environment. Mercury is no longer used for the processing of gold in the United States, but it was used at historical gold mines in Virginia. As a result, considerable legacy mercury may exist in surface waters, soil, and mine waste at previously mined sites. These areas may still harbor unmined gold deposits and unrecovered gold in historic waste material, and future gold mining operations could remobilize this legacy mercury unless appropriate extraction and processing circuits are implemented to capture the mercury. Because of mercury's high toxicity, careful characterization for mercury is essential at all potential mine sites in order to protect environmental and human health.

Impacts to Water Quality

ARD is among the most important potential environmental impacts of concern and poses a substantial risk if massive sulfides are disturbed during gold mining operations and if proper engineering controls are not in place. ARD can persist long after mining has ended and can cause acidity, high salinity, and elevated concentrations of toxic metals in surface water and groundwater if appropriate engineering controls are not in place. Many gold deposits in Virginia are not directly associated with large quantities of sulfide-containing minerals, reducing the likelihood of extensive ARD associated with mining. However, if adjacent massive sulfide deposits or sulfide-bearing country rock are disturbed and if appropriate engineering controls are not applied, ARD could adversely impact sensitive freshwater fauna in nearby streams and wetlands, resulting in substantial remediation costs. Site-specific characterization, engineering controls, and monitoring throughout the life cycle of gold mines are important to minimize and mitigate ARD that could negatively impact surface water and ecological communities.

Site-specific geologic conditions determine whether metals could be released from gold mining operations in sufficient quantities to pose human health threats to surrounding communities. The primary elements of concern for human health that could be released from Virginia gold deposits or from nearby rocks disturbed during mining include antimony, arsenic, cadmium, lead, mercury, and thallium. Most Virginia gold deposits occur in low-sulfide, gold-quartz veins and the few reliable geochemical data that are available for these deposits show low concentrations of metals of concern in discharge waters. However, some gold deposits in Virginia are located in close proximity to massive sulfide deposits, which have higher concentrations of pyrite and higher risk of toxic metal discharge, leaving considerable uncertainty in predicting risk across the state. Therefore, any future efforts to mine gold deposits in Virginia should be accompanied by detailed studies to characterize the mineralogy, metal content, and geochemistry of each deposit and its surrounding rock. Site-specific characterization, water quality management, and monitoring throughout the life cycle of gold mines will be important to minimize and mitigate the release of metals that could negatively impact surface water and groundwater quality.

Mining can increase nitrate loading to local waterways, which can contribute to eutrophication of local surface waters. Although best practices for blasting activities can limit nitrogen loading of surface water and groundwater (see Chapter 3), incomplete combustion of ammonium nitrate and fuel oil explosives under wet, nonideal conditions may result in nitrate-laden, mine-influenced water that can exceed water quality criteria. If this water is not appropriately managed and it reaches local surface waters without significant dilution, depleted dissolved oxygen and reduced pH due to eutrophication may result, which can be lethal to invertebrates and fish. Mining could also contribute to the total loading of nitrogen to more distant habitats (e.g., the Chesapeake Bay), although the relative contributions to the total loads are expected to be small. Elevated nitrate in drinking water can also be harmful to human populations, but these higher concentrations are likely only possible in groundwater in the immediate vicinity of the mine site and can be prevented with best practices for blasting activities.

Open impoundments that contain cyanide pose acute toxicity risks to wildlife unless proper management and deterrents are in place. Wildlife species are attracted to virtually any kind of surface water body, natural or constructed, including waste and treatment impoundments. In the arid western United States, there have been numerous acute toxicity events affecting wildlife (especially birds) at cyanide impoundments in gold mining sites, although there have been fewer reports documenting these toxicity events following the establishment of modern best practices for cyanide management. Although surface water is plentiful in Virginia, the Commonwealth hosts diverse and abundant wildlife species that are dependent on access to open surface water. Unless best practices (e.g., deterrent systems, cyanide destruct systems) or alternative methods (e.g., enclosed tank leaching) are used, wildlife acute toxicity events could occur at open impoundments containing cyanide.

Rare But Catastrophic Events

Catastrophic failures of gold mine tailings dams and cyanide solution containment structures are low-likelihood but high-consequence events that have caused significant impacts where they have occurred. Tailings dam failures can lead to acute danger (e.g., fatalities, injury, destruction of property) as well as long-term ecological effects that are caused by the dispersal of toxic metal-containing mine wastes in rivers and floodplains. The magnitude of the long-term ecological effects depends on the scale of the spill, bioavailability of the contaminants, and effectiveness of cleanup efforts. In contrast, cyanide spill events do not pose long-term risks because cyanide degrades in the surface environment relatively quickly. However, because of cyanide's high acute toxicity, accidental spills have caused mass mortality events of aquatic life and pose an acute human health risk where water affected by the spill is used as a drinking water supply. If tailings and cyanide containment structures are not designed to accommodate seismic, high-precipitation, and flooding events, then the likelihood of these potential high-consequence events will increase. This is especially pertinent in light of the potential for increased frequency and severity of precipitation events due to climate change.

Impacts to Air Quality

The committee did not find evidence to indicate that gold mining in Virginia would significantly degrade air quality if appropriate engineering controls were in place. Fugitive dust produced from excavation activities, heavy equipment, and mine road traffic can be a nuisance that impacts the quality of life of affected neighbors. In addition, toxic fine particles and gaseous pollutants generated from fuel combustion and gold processing can be hazardous if released, because of their greater respiratory impacts and longer atmospheric transport distance. Given the likely small scale of future commercial gold mining in Virginia that would lead to limited heavy equipment operation and traffic, and the technological advancements in recent decades that allow for effective dust suppression and control of hazardous air pollutants, the impacts of air pollutants on surrounding communities are expected to be limited.

Impacts to Water Quantity

Drawdown of the water table associated with the dewatering of an open pit or underground mine could impact local groundwater users, depending on aquifer conditions and the proximity of wells to the mine site. Unless appropriately mitigated, drawdown of the water table could significantly affect the quality of life and the cost of living for residents near the mine site who rely on groundwater supplies. Rigorous site characterization and modeling is needed to estimate the level and geographic span of groundwater impacts and to evaluate whether alternative sources of water or new wells need to be provided to local citizens. Public engagement and participation during permitting is essential if alternative sources of water or new wells may need to be provided.

Cumulative Risk

Robust analyses of the potential impacts of mining consider cumulative health risks. Human populations are exposed to multiple hazard types, including biological, physical, chemical, psychological, and social (e.g., poverty, discrimination, unemployment, limited access to health care). These hazards can occur through different exposure settings (e.g., environmental, occupational) and multiple media (e.g., air, water, soil). Different hazard types, especially chemical and nonchemical stressors, can interact to affect human health in complex and dynamic ways. These multiple, sometimes synergistic, stressors can lead to asymmetric impacts within and between communities, and historically underresourced and underrepresented populations are often most affected.

***

The above conclusions outline the potential impacts of gold mining across the Commonwealth of Virginia, but only robust site- and project-specific analyses can assess the potential impacts of a particular project on human and ecological health.

> **RECOMMENDATION: To minimize impacts to human health and the environment, the Virginia General Assembly and state agencies should ensure that robust site- and project-specific analyses of impacts are completed prior to the permitting of a gold mining project.**

VIRGINIA'S REGULATORY FRAMEWORK

Gold mining has a long history in Virginia, dating from the 1800s. At present, however, there are few metal mining activities in the state and no active commercial gold mines. Given the current lack of metal mining activities in the Commonwealth, it is not surprising that the present regulatory framework appears geared toward projects such as sand and gravel mining and not gold mining. Although most of Virginia's mineral mining laws and regulations seem suitable for the types of mines now operating in the state, the current regulatory framework is not adequate to address the potential impacts from commercial gold mining. Gold mining raises a number of environmental and public health issues that merit additional attention and suggest a need for changes in laws, regulation, and guidance.

Review of Impacts

Virginia's current regulatory system lacks an effective and consistent process for review of environmental impacts from potential gold mining projects. As a result, it is unlikely that a robust collection, evaluation, and review of site-specific data regarding potential impacts of gold mining activities and their impact on the public health and welfare of surrounding communities will take place. The National Environmental Policy Act (NEPA) requires federal agencies to consider the potential environmental effects on natural resources, as well as social, cultural, and economic resources, before permitting. Virginia law does not require a NEPA-like review of environmental impacts for private lands, where gold mining is most likely to occur. Additionally, while baseline studies in Virginia appear to be recommended, they are not required. This means that in the absence of a major federal action that triggers the federal NEPA process, there may be limited collection of baseline information and no formal documentation of the regulatory program's analysis, disclosure of impacts, or decision making for a range of environmental resources or factors. Some states have a state-specific NEPA-like process that allows for a consistent approach to collecting and considering baseline information and other material relevant to environmental impacts (e.g., Montana and California). Other states have regulation, code, and guidance documents that emphasize the importance of baseline studies (e.g., Colorado, Nevada, Montana, California). The protection of air and water quality would be strengthened if Virginia adopted laws and promulgated regulations that required up-front, robust data collection and a NEPA-like analysis that discusses and evaluates reasonable alternatives.

Exemptions

Virginia provides exemptions from regulatory oversight for off-site processing and exploratory drilling which are not commensurate with the potential impacts from those operations.

- **Off-site processing:** Gold processing facilities in Virginia that are not located on site with active mining or extraction ("toll mills") would not require a permit from the Mineral Mining Program for the operation and reclamation of the site. Toll mills may look very similar to permitted on-site processing facilities and similar environmental impacts may result from toll mills. In fact, the waste materials at toll mills may contain a broader range of potential contaminants if the source materials come from different locations. While toll mills may be required to obtain permits from other agencies to protect air quality and water quality, the lack of regulatory oversight by the Mineral Mining Program means that site characterization, project plans and designs, and the implementation of best practices for operations, reclamation, and long-term stewardship may not be adequately addressed.
- **Exploratory drilling:** Virginia's current laws and regulations exempt exploratory drilling for mineral resources. Impacts on the environment during initial exploration are generally minor, localized, and easily reclaimed. However, advanced exploration methods may be associated with greater impacts (see Chapter 3). While surface impacts including erosion and runoff may be regulated by the Virginia Department of Environmental Quality and the Virginia Department of Conservation and Recreation, there are currently no mineral mining regulations for exploration in Virginia that mandate the plugging of drill holes or the covering of drill cuttings from the hole. If best practices are not utilized for these closure activities, pollution of the local groundwater and surface water could occur. This exemption for exploratory drilling also means that public notice to citizens and local communities is not required. Greater oversight of exploration drilling would ensure community participation starting at the earliest appropriate stage and continuing throughout the life cycle of a potential gold mine, and would lessen the likelihood of these localized impacts, especially in regard to more advanced and intensive drilling programs. This oversight could include requirements to file plans for drilling, closure, and reclamation, and a requirement to provide notice to those around the exploration site.

Underground gold mining without significant surface effects is also currently exempt from regulations under Virginia's mineral mining codes and regulations. While significant surface effects related to disturbances and facilities would require a permit, the exemption for underground gold mining could cause important aspects of underground mines to be excluded from operations and closure plans of the surface permit. Additionally, the level of technical assessment and oversight for underground gold mines by Virginia Energy is not clear.

Financial Assurance

Virginia's bonding requirements are insufficient to cover the costs of reclamation and long-term stewardship of gold mining and processing operations, which poses a fiscal and environmental risk to the Commonwealth in the case of the bankruptcy of mining enterprises or abandonment of their mining sites.

- **Bonding rates:** Virginia's bonding rates are based solely on disturbed acreage. This type of bond calculation often leads to undercollection of bonds for gold mining and processing operations because it focuses only on aspects of land reclamation and does not account for additional costs like postclosure water management. Additionally, Virginia offers a bond pool, called the Minerals Reclamation Fund, with even lower per-acre rates and pooled risk. The complex reclamation and long-term stewardship activities that might be necessary for some gold mining projects could greatly deplete or potentially exhaust the Minerals Reclamation Fund used by the Commonwealth to guarantee reclamation. The regular recalculation of potential costs using verifiable engineering estimates would constitute an improved model for determining bonding rates. This model would estimate the costs for reclamation and long-term stewardship for all aspects of the operation over the project's life, including any postclosure water management, treatment, and monitoring that may be required to achieve long-term hydrologic, physical, and chemical stability. The integrity of the Minerals Reclamation Fund could be maintained using a similar bond calculation model, or by establishing membership criteria that are based on the operation's characteristics and its potential impacts.

- **Exemptions from bonding:** Virginia's exemptions from bonding for underground gold mining (without significant surface effects), small-scale gold mining, and toll mills do not reflect the costs necessary to conduct reclamation and long-term stewardship at those operations. No financial assurance is provided to the Commonwealth for these exempt operations, which poses a fiscal and environmental risk to the Commonwealth and its citizens.
- **Bond release:** Virginia does not have clear guidance regarding the criteria for bond release for projects that require complex closure and reclamation. To ensure successful mine reclamation, bonds should only be released following the demonstration that performance standards for reclamation have been achieved over a sufficient period of time. These performance standards may include requirements for slope stability, vegetation establishment, water quality, and hydrologic balance. Incremental bond release for areas at which successful reclamation has been demonstrated can encourage the timely completion of reclamation.

Standards and Their Enforcement

To incorporate best practices, build a mutual understanding among permittees and regulators, and better support protection of human health and the environment, Virginia agencies will need to review the regulatory performance standards pertinent to gold mining and update guidance documents. Virginia's performance-based laws and regulations provide flexibility for the site-specific designs of each project, but do not provide sufficient guidance for operators to achieve objectives and do not offer sufficient metrics for regulators to evaluate during the review of applications and inspection of activities. Fiscal and environmental risks to the Commonwealth would be reduced with improved guidance and performance standards on best practices for the collection of baseline information, geochemical characterization, water management, waste rock management, tailings management, and impoundment design. Specifically, performance standards for impoundment designs could recommend a probabilistic framework for designing for seismic events and a consideration of the predicted increased frequency and magnitude of major storm events due to climate change. Performance standards would also be improved with conservative recommendations for slope angles and safety factors that reflect best geotechnical practices and incorporate the potential for undrained loading and liquefaction in saturated tailings. Additionally, decision makers may want to reconsider the current practice of using incremental damage assessments to calculate design flood requirements for impoundments.

The capacity to regulate is as important as a strong regulatory framework and is a concern for Virginia given the limited experience with the regulation of metal and gold mining. The capacity to regulate requires robust funding of the regulatory entities, as well as diverse and appropriate technical expertise of the regulators, supplemented by periodic reviews of evolving best practices. In addition, effective coordination between multiple regulatory entities is critical for protecting air quality and water quality, particularly when evaluating, permitting, and monitoring compliance for stormwater and process water management, treatment technologies, and methods for discharge. Given the lack of experience of Virginia regulatory entities in regulating metal and gold mining, regulators' current expertise and familiarity with best practices may be limited. There are also key gaps in Virginia's capacity to implement and enforce some of its laws and regulations, such as the inability to directly issue penalties or fines for noncompliance without lengthy adjudication, and the lack of requirements for impoundment inspections by the associated Engineer of Record. Higher-level technical reviews, third-party reviews, or audits would enhance the evaluations of Mineral Mining Plans and inspections of individual permits.

Public Engagement and Environmental Justice

The current requirements for public engagement in Virginia are inadequate and compare unfavorably with other states, the federal government, and modern best practices because they require the provision of limited information, place the burden of public notification on the mine permit applicant, and apply only to a limited scope of recipients. Industry best practices are adopting a greater emphasis on public engagement, consultation, and partnership with communities before and after mining activities are initiated, as well as free, prior, informed consent to govern interactions with tribes. In Virginia, there is a scarcity of project details in the new permit notifications, a short deadline provided for filing objections or a request for hearing, and a limited number of area residents that are required to be notified, with no specific inclusion of tribal communities. In addition, Virginia Energy does not make technical reports, designs, and other components of the permit application package readily available for public review. Finally, there are no requirements in Virginia for public notice or opportunity for public input for exploratory drilling or when an application is renewed, a permitted project is expanded, or a bond is released. These permitting actions are critical milestones for the mining operation, and they warrant meaningful engagement with nearby landowners, communities, and other stakeholders.

Current Virginia regulations that are applicable to mineral mining will need to be amended to reach the goals set out in the Environmental Justice Act. In 2020, the Virginia legislature passed the Virginia Environmental Justice Act to better incorporate environmental justice into regulatory decision making in the Commonwealth. In the context of potential gold mining projects, an emphasis on environmental justice requires a regulatory structure that recognizes existing environmental injustice, population vulnerabilities, and economic and health disparities, and aims to reduce existing disparities and prevent future disparate impacts. This regulatory structure should ensure that those experiencing existing environmental injustice and health disparities are notified in a timely fashion about potential gold mining projects, are able to consult meaningfully with potential gold mining project proponents, and can contribute to decision making.

***

As detailed above, Virginia's present regulatory structure is not adequate to protect against the potential environmental degradations that could accompany gold mining activities. Stronger requirements for bonding, public engagement, and the review of environmental impacts are necessary; as well as updated regulatory capabilities, exemptions, performance standards, and guidance documents in order to protect human health and the environment.

RECOMMENDATION: To protect against the potential impacts of gold mining, the General Assembly and state agencies should update Virginia's laws and its regulatory framework.

1

Introduction

Gold is a precious metal that is mined for investments, jewelry, and technology (see Figure 1-1). The primary use of gold is for investing purposes, by individuals who buy gold and gold-backed exchange-traded funds as a hedge against inflation and market volatility, or by national central banks that hold gold reserves to support the value of the national currency and stabilize financial markets. In contrast, the most important technological use of gold is the production of corrosion-free microcircuitry in electrical devices—every cell phone, for example, contains an average of 0.03 g of gold (EPA, 2022d; USGS, 2006). Despite its significance to modern society, gold is not considered a critical mineral because its supply chains are not particularly vulnerable to disruption given current levels of domestic production, resources in reserve, and the number of reliable foreign sources (USGS, 2022c).

Gold is one of the least abundant of the nonradioactive elements and is estimated to occur in the Earth's crust at a concentration of only about 4 parts per billion (Lide, 1999). However, it is not uniformly distributed. Geologic processes have concentrated gold into "lode" deposits within bedrock, and the gold in these lodes can be released from the rock by weathering and then transported and concentrated in streams to form "placer" deposits. Grades for lode gold mines, which reflect the amount of gold obtained relative to the amount of ore needed to obtain it, currently range from approximately 0.2 to 30 g/t (grams/metric ton[1]) with an average of 1.5 g/t for ores mined in the United States (Butterman and Amey, 2005). The high unit value of gold—$1,500 to $2,100 per troy[2] ounce since 2021 (GoldPrice.Org, 2022)—can make it economically profitable to mine and process at grades that would be uneconomic for most other metals. The methods employed to mine and process gold depend on the type of deposit. Placer deposits often only require minimal processing that is based on the gravity-based separation of gold, because gold is much denser than most other rock-forming minerals. Historically, these efforts were aided with a process called amalgamation, which involved the addition of mercury to form an alloy with gold. This process was used in Virginia gold mines in the 1800s and early 1900s, but is no longer used in the United States because of mercury's high toxicity and persistence in the environment. Modern, more complex, processes used for commercial gold production in lode deposits are described in Chapter 3.

The United States currently ranks fourth among the world's top gold-producing countries (see Figure 1-2). Approximately 74 percent of gold produced from the United States in 2021 was mined in Nevada (USGS, 2022d), though several states have long histories of gold mining.

---

[1] A metric ton or tonne is equivalent to 2,204 lbs and a ton, short ton, or avoirdupois ton is equivalent to 2,000 lbs. In this report, metric ton is abbreviated as t (e.g., grams per metric ton or g/t), whereas ton is not abbreviated (e.g., ounces per ton or oz/ton).

[2] Gold and other precious metals are measured in troy ounces. One troy ounce is approximately 1.097 avoirdupois ounces.

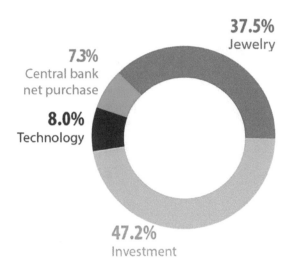

FIGURE 1-1 Global utilization of gold is primarily for investments, jewelry, technology, and purchases by central banks (figures as of 2020).
SOURCE: Image modified from Natural Resources Canada (2022).

GOLD MINING IN VIRGINIA

Virginia was one of the first gold-producing states in the nation and it claims a long history of exploration and mining of gold (Laney, 1917; Lonsdale, 1927; Pardee and Park, 1948; Park, 1936; Spears and Upchurch, 1997; Sweet, 1971, 1980, 1995; Sweet and Lovett, 1985; Sweet et al., 2016; Taber, 1913). Most of the gold deposits in Virginia occur in the Piedmont region, which runs northeast to southwest along much of the state from just south of Washington, DC, to midway between the James and Roanoke Rivers. Other gold deposits occur in the Virgilina district in the south-central counties of Halifax, Charlotte, and Mecklenburg (see Figure S-1 and Chapter 2).

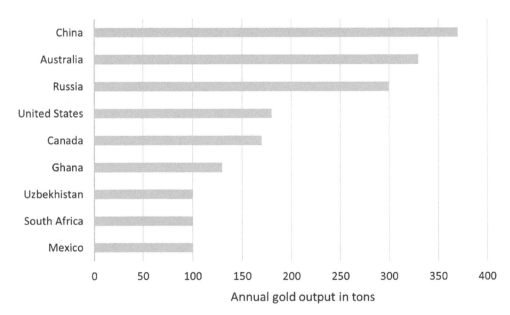

FIGURE 1-2 A ranking of the top 10 gold-producing countries in 2021 places the United States as the fourth largest producer.
SOURCE: Data from USGS (2022d).

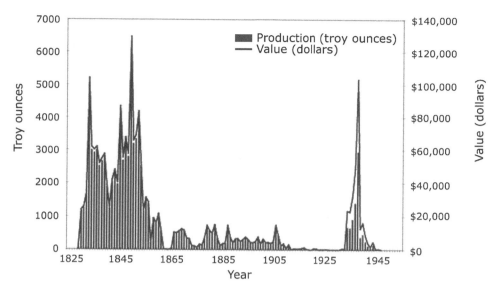

FIGURE 1-3 Gold production in Virginia in ounces and dollars at the time of mining from 1825 to 1960. SOURCE: Figure modified from Virginia Energy (2021).

The earliest known reference to a gold discovery in Virginia was in 1782, when Thomas Jefferson described a 4-pound gold-bearing rock that was found along the Rappahannock River near Fredericksburg (Sweet, 1971). Then, in 1806, the first lode gold occurrence was reported at a deposit in western Spotsylvania County that would later become the Whitehall Mine (Sweet et al., 2016). Several small gold mines were operating in Amherst County and elsewhere along the James River as early as 1825 (Sweet et al., 2016), and by 1836 gold mines were being operated in Spotsylvania, Orange, Louisa, Fluvanna, and Buckingham Counties (Park, 1936; Sweet et al., 2016). A total of approximately 100,000 ounces of gold was produced from 1804 to 1947, and peak gold production occurred between the mid-1830s and the mid-1850s (see Figure 1-3). The large majority of these gold mines were first operated as small open pits or trenches (see Figure 1-4) that exploited the near-surface oxidized ore (Sweet, 1980) containing free gold. The free gold in oxidized ore was more easily mined and processed than the deeper unoxidized ore, which often required underground mining and more intensive processing because the gold is bound to other minerals. Virginia gold production began to decline with the onset of the California Gold Rush in 1849 (Sweet, 2007), but another peak occurred in the 1930s, when gold was produced in Virginia as a by-product of sulfur mining. Following the last commercial gold production in 1947, when gold was produced as a by-product of lead and zinc mining in Spotsylvania County (Sweet, 2007), Virginia has seen only intermittent exploration activity and attempts to mine placers, including operations at Brush Creek and Laurel Creek in Floyd County in the 1980s (Sweet et al., 2016).

Mineral mining activities in Virginia are primarily regulated by the Virginia Department of Energy (Virginia Energy), which prior to October 2021 was known as the Department of Mines, Minerals and Energy (DMME). The Division of Mineral Mining was established in 1987 as part of DMME, by merging two preexisting regulatory programs that were responsible for mine safety and reclamation (Virginia Energy, 2022a). Today, the Virginia Energy Mineral Mining Program administers the health, safety, and land reclamation programs for all noncoal mineral mining operations, which include quarries, sand and gravel pits, and surface and underground mines producing metals or industrial minerals (Virginia Energy, 2022a).

There were approximately 440 active permits for nonfuel mines in Virginia as of 2021 (Virginia Energy, 2022a). These permitted areas produce more than 80 million tons of nonfuel commodities annually (Virginia Energy, 2022a). Most of these permits are for extraction of sand, limestone, granite, shale, and clay (see Table 1-1), and about 93 percent consist of open pit and quarry operations (see Table 1-2). Only one permit—the Moss Mine in Goochland County—lists gold as the primary produced metal (see Box 1-1). Approximately half of all permits

FIGURE 1-4 Photographs of the abandoned Bondurant Mine site in Buckingham County in January 2022. (A) Shallow surface depressions that are the remains of shallow trenches and open cuts used to mine oxidized near-surface ore at the site. (B) The surface expression of a shaft used to access underground ore.
SOURCE: Photos by Robert J. Bodnar.

are less than 50 acres in total permit area, and one-quarter are less than 10 acres (2020 data; see Figure 1-6). Only 2 percent of permit areas exceed 1,000 acres and these large operations produce limestone, granite, titanium, or kyanite.

Although Virginia has seen very little gold production in the past 70 years, the rising price of gold (see Figure 1-7) has brought renewed attention to the potential for gold exploration and mining in the Commonwealth. Additionally, in recent decades, the efficiency of valuable minerals extraction has increased, which means that lower-grade materials can be profitably mined and processed. Increasing gold prices and extraction efficiencies mean that deposits that at one time were previously not economic may become economic. This has often resulted in mining companies returning to sites previously considered to be "mined out" (i.e., all the mineral reserves at the time of active mining had been removed), and restarting operations to mine material that had previously been considered to be noneconomic or waste (see Box 1-1). According to Heylmun (2001), conservative estimates of Virginia's lode gold reserves and placer gold reserves are 378,000 and 274,000 ounces, respectively (see Box 1-2), although methods used to determine these values and uncertainties associated with the estimates are not discussed. Heylmun (2001) notes that "the Piedmont region became sort of a 'forgotten province' after the American West opened up . . . [and that] there are good possibilities for commercial gold production in Virginia, as well as to the south in the Carolinas, Georgia, and Alabama. New exploration methods might uncover gold deposits in areas which have never been exploited."

In the fall of 2020, it became widely known that Aston Bay Holdings, Ltd., was performing exploration drilling on privately owned forestland in Virginia's Buckingham County with a goal of assessing the area's potential for gold mining. County officials were reportedly unaware of the activity and, when informed, believed the local zoning ordinance did not permit mineral exploration drilling. However, Buckingham County's Board of Supervisors soon voted to retroactively allow exploratory drilling (Vogelsong, 2021). This and other renewed interest in mineral exploration and mining in Virginia has raised concerns among stakeholders regarding the potential impacts of such operations on the Commonwealth's people and environment.

TABLE 1-1 Permits Issued for Mining Activities in Virginia in 2020, Listed by Commodity

Commodity	Number of Permits	Permitted Area (acres)	Disturbed Area (acres)
Aggregate	2	677.8	438.4
Amphibolite	1	76.9	22.0
Aplite	1	223.5	119.8
Basalt	1	143.7	129.2
Clay	9	1,637.8	836.4
Diabase	2	427.6	303.3
Diorite	1	160.3	77.6
Dolomite	2	556.8	282.0
Fullers Earth	2	97.7	84.3
Gemstones	2	27.8	27.8
Gneiss	4	969.0	505.8
Gold	1	1.0	1.0
Granite	41	12,837.5	8,080.5
Gravel	5	121.6	82.8
Greenstone	2	859.2	205.9
Iron oxide	1	25.2	9.3
Kyanite	1	2,746.9	737.0
Limestone	71	16,825.9	8,917.9
Marble	1	14.7	14.7
Marl	2	39.6	34.3
Quartz	2	299.8	88.2
Quartzite	2	1,409.4	534.0
Salt	1	184.3	83.7
Sand	219	21,866.3	11,591.5
Sandstone	9	1,000.7	510.0
Shale	27	937.0	167.4
Silica	1	821.3	530.2
Slate	3	924.5	456.9
Soapstone	1	507.7	72.4
Titanium	2	6,168.6	3,893.8
Traprock	8	2,733.9	1,929.4
Unknown	1	17.0	17.0
Uranium	1	194.0	2.0
Vermiculite	1	199.7	132.2

NOTE: Only the primary commodity is listed for operations at which multiple resources are mined.
SOURCE: Virginia Energy (2022d).

TABLE 1-2 Mining Methods Reported for All Mineral Mine Permits in Virginia That Were Active in 2020

Reported Method	Number of Sites	Notes
Dragline	14	All sand, 1 quartz
Dredge	12	10 sand, others gravel + clay
Exploratory	1	Uranium
Open Pit	253	Many commodities
Other	3	2 sand + clay, 1 salt
Quarry	147	
Underground	2	1 gemstone, 1 limestone

SOURCE: Virginia Energy (2022d).

BOX 1-1
Moss Mine

The Moss Mine in Goochland County is currently the only permitted active mining operation in the Commonwealth of Virginia whose primary purpose is to recover gold. This operation provides an example of some reprocessing methods that may be implemented at other locations in Virginia to recover gold from historically mined materials and to possibly reclaim contaminated land. Moss mine was originally operated intermittently by several different owners between 1836 and 1936 (Mindat, 2022c). During an earlier phase of reclamation at the site, waste material was spread across a field. This waste material includes mercury as a result of the amalgamation method used to extract gold prior to 1936. Today, a ~1 acre operation site is being operated by Big Dawg Resources, LLC, to reprocess surface material to remove metallic mercury and extract gold. The material is collected and stockpiled using several large excavators and track loaders (see Figure 1-5A).

FIGURE 1-5 Photographs of Moss Mine reclamation site, Goochland County, Virginia. (A) Photograph of some of the equipment used to collect rock and sediment materials for reprocessing. (B) Photograph of the crusher being used to process waste material at the Moss Mine. (C) Three jigs used to capture mercury, gold, and lead. (D) Recovered gold and mercury from the Moss Mine site.
SOURCES: Photograph A by Robert J. Bodnar; photographs B–D by Paul Busch.

The material is then fed through a crusher (see Figure 1-5B) and screened, with the fine material sent to jigs or gravity separators (see Figure 1-5C). Jigs retain the denser materials, including gold, lead, and mercury. The lighter fraction of feed leaves the bottom of the jigs and is pumped up to a wave table that further concentrates and removes any other waste, with gold and mercury collected at this point (see Figure 1-5D). The amalgam is retorted off-site to recover gold and mercury, and the mercury is shipped to a vendor for further reprocessing. The lighter fraction emerging from treatment is returned to the site. Although committee members were informed that mercury was not detectable in the material being returned to the site, the committee did not assess any analytical data on the chemical or mineralogical composition of the waste material before or after processing, or quantitative information on mass balances. Accordingly, the committee is unable to comment on the efficacy of the processing approach, or its long-term impact on the site and local environment.

CHARACTERISTICS OF VIRGINIA RELEVANT TO GOLD MINING IMPACTS

The following sections outline key characteristics of gold-bearing regions of Virginia with potential relevance to gold mining activities.

Land Ownership and Mineral Rights in Virginia

Approximately 84 percent of all land in Virginia is privately owned (Virgina DCR, 2022b), which means that in Virginia, exploration and potential development of gold resources is most likely to occur on private lands. The remaining land in the Commonwealth is owned by private organizations and local, state, federal, and tribal governments. Approximately 5 percent of Virginia's land (Virgina DCR, 2022b) is owned by the state government and is comprised of Wildlife Management Areas, State Forests, and State Parks. Some of these lands could be mined for gold if the project proponent were granted a lease by the state, but it is unclear which state lands in Virginia might be open to such mineral extraction. The federal government owns approximately 2.4 million acres

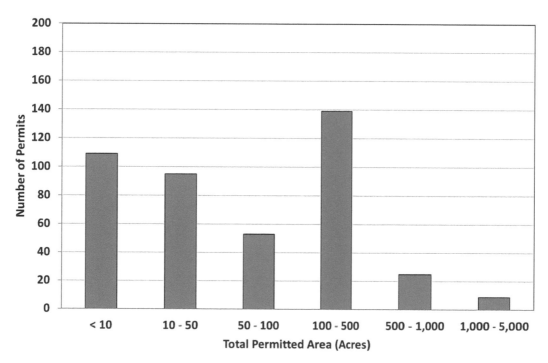

FIGURE 1-6 Distribution of mining permits in Virginia by total permitted area.
SOURCE: Data from Virginia Energy (2022d).

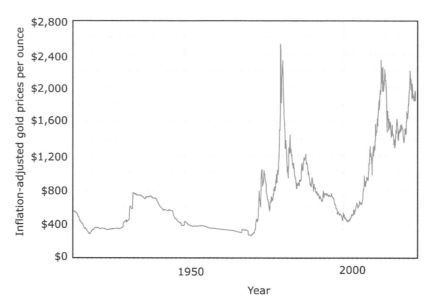

FIGURE 1-7 Price of gold from 1915 to 2021 (dollars per ounce), adjusted for inflation levels in 2022. Note the significant increase in the price of gold in 1971 after the United States abandoned the gold standard.
SOURCE: Graph downloaded from Macrotrends (2022).

BOX 1-2
Gold Mining Definitions

The meaning of the term *gold mining* might appear at first glance to be obvious, but a more rigorous definition is necessary for the purposes of this report. When mining companies explore for minerals, the eventual goal is to locate, delineate, and estimate *mineral reserves* that can currently be mined at a profit. A mineral reserve is "an estimate of tonnage and grade or quality of indicated and measured mineral resources that, in the opinion of the qualified person, can be the basis of an economically viable project" (17 CFR § 229.1300). To achieve this goal, exploration must generate enough data to allow mining companies to estimate *mineral resources*. A mineral resource is "a concentration or occurrence of material of economic interest in or on the Earth's crust in such form, grade or quality, and quantity that there are reasonable prospects for economic extraction" (17 CFR § 229.1300).

It is important to note that not every accumulation of gold qualifies as a mineral resource, but only those that have the "form, concentrations, and quantity" to have "prospects of economic extraction." Mining companies typically commission technical studies (prefeasibility or final feasibility studies) to evaluate technical, economic, and legal factors to determine whether mineral reserves exist and, if so, to estimate the quantity and value of such mineral reserves. In this report, the committee considers gold occurrences that, even if they are not considered mineral resources or reserves today, could be deemed mineral resources or reserves with more exploration or changes in economic conditions. One example of this is the properties of Aston Bay Holdings, Ltd., who have not yet reported mineral resources or reserves at their exploration project in Buckingham County.

Mineral deposits can be described based on the ore mineralogy and elements being recovered. In a very general sense, metal mineral deposits in the gold-pyrite belt and in the Virgilina district of Virginia may be classified as "gold deposits," "base metal[a] deposits," or "pyrite deposits" (see Chapter 2) and, depending on factors outlined above, gold could be mined from any of these deposit types. It is important to note that there is a continuum between different types of deposits. In fact, in some individual deposits, such as the London and Virginia Mine in Buckingham County and the Cofer Mine in Louisa County, mining companies produced both gold and base metals. In this report, the committee considers gold deposits to be those deposits where gold is a primary metal of material economic interest, where "material economic interest" refers to an economic interest that could, or might reasonably be thought to, influence judgment or action.

---

[a] A base metal is a metal (e.g., zinc, lead) of comparatively low value compared to precious metals (e.g., gold, silver).

in Virginia, or approximately 9 percent of the Commonwealth (CRS, 2020), but most of these lands are managed for conservation and recreation purposes. Other large tracts, like military reservations, are closed to mineral exploration and development. There are currently no tribal reservations in the areas of Virginia that are most favorable for gold mineralization, although historic tribal lands cover the state and tribal communities maintain strong ties to a number of areas (see Chapter 5). As a result, the committee focused its deliberations on private and state-owned lands and did not focus on considerations specific to gold mining in tribal reservations or federal lands in Virginia.

In general, the owner of the surface estate in Virginia also owns the underground minerals. However, because the ownership of the underground minerals can be sold or disposed of separately from the surface estate, there can be situations where the surface and subsurface resources are owned by different parties. Such "split estate" scenarios can lead to conflict between the surface estate owners and the owners of the subsurface mineral rights.

Population Demographics

Population density and demographic characteristics vary widely in the gold-bearing areas of Virginia. Metropolitan communities in northeast Virginia have relatively higher population densities (e.g., Fairfax County has approximately 2,941 people per square mile). In contrast, many of the state's rural areas have relatively lower population densities (see Figure 1-8)—for example, Buckingham County has fewer than 30 people per square mile (U.S. Census Bureau, 2020). Future mining and exploration activities are likely to be concentrated in the more rural areas of the Commonwealth, which also have different demographic characteristics from the state overall. As discussed in Chapter 4, approximately 1 in 5 Virginia residents are African American, 1 in 10 are Hispanic, and around 10 percent live below the poverty line (U.S. Census Bureau, 2020), but certain regions have higher populations of minorities and underserved populations (see Chapter 5). For example, in Buckingham County, where recent exploration activity raised concerns, 32 percent of residents are African American and 21 percent are living below the poverty line (U.S. Census Bureau, 2020).

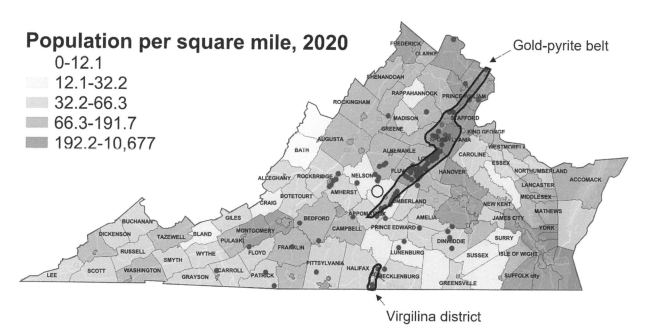

FIGURE 1-8 Population density in 2020 for all counties in Virginia.
NOTES: Overlain on the map are historic gold mines (red dots) and the gold-pyrite belt and Virgilina district outlined in black. The large yellow circle denotes the location of Aston Bay's exploration property in Buckingham County.
SOURCES: Population density map from 2020 Census data and locations of gold deposits from Sweet (2007).

Percentage of population on private wells

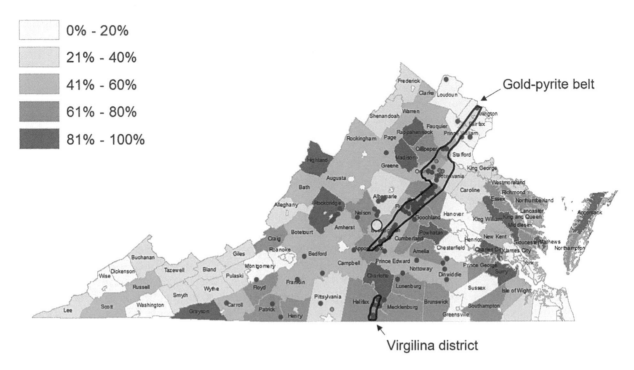

FIGURE 1-9 Percentage of population on private wells by county.
NOTES: Overlain on the map are historic gold mines (red dots) and the gold-pyrite belt and the Virgilina district outlined in black. The large yellow circle denotes the location of Aston Bay's exploration property in Buckingham County.
SOURCE: Map modified from Virginia Department of Health (2022b) and locations of gold deposits from Sweet (2007).

More than 3 million Virginians rely on drinking water intakes downstream of the gold-bearing regions in Virginia (Maest, 2022). Additionally, whereas Virginians who live in more urban areas are generally served by public water utilities, ~20 percent of the Commonwealth's population and a majority of households in some rural areas use private wells (VanDerwerker et al., 2018; see Figure 1-9). Private well owners generally do not receive the same water treatment and monitoring services compared to those provided by public utilities (EPA, 2022k). Instead, private well owners are responsible for ensuring the safety of their own water. Therefore, people who utilize private wells often have higher risks of waterborne contaminant exposure than those who receive water from regulated public utilities (MacDonald Gibson and Pieper, 2017).

Climate

Virginia has a humid, subtropical climate, receiving an average of more than 42 inches of precipitation annually (NCEI, 2022). Many of its counties—including those in the gold-pyrite belt and Virgilina district—are subject to extreme weather events including hurricanes and tropical storms, thunderstorms, and heavy rain and snow events. From 1851 to 2021, 27 hurricanes and tropical storms made landfall in Virginia (AOML, 2022a,b), including Hurricane Camille in 1969, which brought 27 inches of rain to the central part of the state, and Hurricane Irene in 2011, which brought approximately 12 inches of rain to central and eastern Virginia (NWS, 2016).

Future changes in climate may affect overall temperature and precipitation trends as well as the frequency and magnitude of extreme weather events in Virginia. The Sixth Assessment Report from the Intergovernmental

Panel on Climate Change shows that Eastern North America will very likely experience increased average and extreme precipitation in the future (IPCC, 2021), leading to more severe flooding and flooding outside of current floodplains. Some studies estimate that before the end of the century, extreme summer thunderstorms in the southeastern United States will result in between 40 and 80 percent more rain than the same storms do today (Prein et al., 2017). To account for these changes, the Virginia Department of Transportation updated its guidance for bridge construction in 2021 (VDOT, 2020), requiring engineers to anticipate a 20 percent increase in rainfall intensity and a 25 percent increase in discharge.

Seismicity

Earthquakes and seismic activity can damage mining infrastructure, including open pits, underground workings, processing facilities, tailings storage facilities, and waste dumps (Lenhardt, 2009). While the East Coast of the United States experiences far less seismic activity and fewer large earthquakes than the West Coast, significant earthquakes have occurred historically. Around 6 to 12 earthquakes per year originate in Virginia and one or two of those is large enough it can be felt (Virginia Energy, 2022b; VTSO, 2021). Most earthquakes in Virginia occur in three zones: the Central Virginia seismic zone, the Eastern Tennessee seismic zone, and the Giles County seismic zone (see Figure 1-10). The Central Virginia seismic zone, between Charlottesville and Richmond, overlaps with much of the gold-pyrite belt. Earthquakes documented in this zone include a magnitude 5.8 earthquake in the town of Mineral in 2011, a magnitude 4.8 earthquake northwest of Goochland in 1875, and a magnitude 4.5 earthquake in Powhatan County in 2003 (Tarr and Wheeler, 2006). There is also evidence for earthquakes of magnitude 6 or higher between 350 and 2,800 years ago (Tuttle et al., 2015, 2021).

Seismic activity in this region is relatively frequent, but generally low magnitude. However, because seismic waves propagate much farther in the eastern United States (Frankel et al., 1990; Pollitz and Mooney, 2015) than

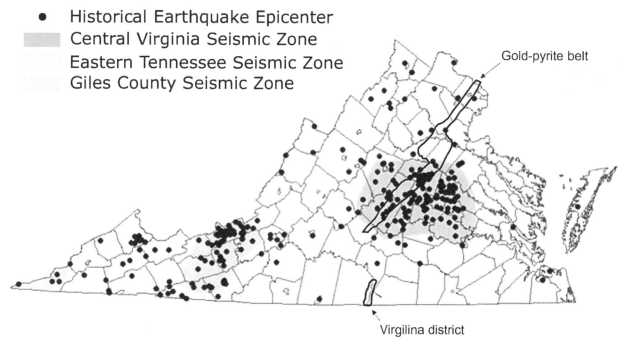

FIGURE 1-10 Locations of the Central Virginia, Eastern Tennessee, and Giles County seismic zones showing historical earthquake epicenters.
NOTES: Overlain on the map are the gold-pyrite belt and Virgilina district outlined in black. The Central Virginia seismic zone overlaps with much of the gold-pyrite belt.
SOURCE: Image modified from Virginia Energy (2022b).

in the western United States (Nishenko and Bollinger, 1990), earthquake damage can occur at some distance from the epicenter. As an example, the magnitude 5.8 earthquake that occurred in Mineral in 2011 caused significant damage approximately 80 miles away in Washington, DC, including $15 million in damage to the Washington Monument and $34 million in damage to the National Cathedral.

STUDY TASK AND APPROACH

In response to citizen concerns over potential impacts of gold mining, the Virginia State Assembly passed House Bill (HB) 2213, which mandates a study of gold mining in Virginia. The workgroup to address HB2213 consists of two components: (1) a committee appointed by the National Academies of Sciences, Engineering, and Medicine (the National Academies) to evaluate the technical aspects and potential impacts of gold mining in Virginia (see Box 1-3) and (2) a committee formed by the Department of Mines, Minerals and Energy, the Department of Environmental Quality, and the Virginia Department of Health ("state agency committee") to focus on local equity and environmental justice issues and environmental and human health concerns of the local community. This report reflects the work of the first committee.

To address the task, the National Academies appointed a committee of 13 experts from academia, industry, and state government with expertise in hydrogeology, geology, geochemistry, ecotoxicology, epidemiology, public health, environmental medicine, environmental policy, environmental law, mining regulations, environmental engineering, mining engineering, and geotechnical engineering (see Appendixes A and B). The committee held six virtual information-gathering sessions focused on the geology, mining, potential environmental and health impacts of mining, public participation in mine permitting processes, and the regulations of Virginia and other states. These sessions included presentations and discussions with representatives from industry, academia, community-based organizations, and governments. Information-gathering meetings featured case studies from a variety of places including South Carolina, Montana, Alaska, California, Canada, and Spain.

Committee members also visited several sites where they received tours and presentations relevant to gold mining activities in Virginia and elsewhere in the eastern United States. This included visits to the Moss Mine

---

BOX 1-3
Statement of Task

An ad hoc committee of the National Academies of Sciences, Engineering, and Medicine will evaluate the impacts of gold mining in Virginia, with an emphasis on potential impacts of gold mining on public health, safety, and welfare. The committee's final report will include conclusions and recommendations based on the study. The study will:

1. Briefly describe the geologic and mineralogical characteristics of the main gold deposits in Virginia, and the types of modern gold mining operations used with comparable deposits in other domestic or international locations.
2. Summarize the Commonwealth of Virginia's existing regulatory framework for gold mining and processing sites (for example, bonding, reclamation, closure, and long-term monitoring) and compare to other states with current or recently closed gold mining operations. This summary will include a discussion of relevant air and water quality regulations, as well as Chesapeake Bay watershed protections.
3. Evaluate the impacts of potential gold mining and processing operations on public health, safety, and welfare in the Commonwealth of Virginia. This evaluation will include:
   a. Discussion of current gold mining operations at sites with comparable geologic, mineralogical, hydrologic, and climatic characteristics to those found in the Commonwealth,
   b. Potential impacts of different leaching and tailings management techniques on downstream communities in the Commonwealth,
   c. Whether existing air and water quality regulations in the Commonwealth are sufficient to protect air and water quality, and
   d. Whether existing bonding, reclamation, closure, and long-term monitoring of sites for potential gold mining are sufficient to protect air and water quality.

in Virginia and the Haile Gold Mine and the Brewer Gold Mine Superfund Site in South Carolina, as well as examination of rock core from the Vaucluse mine in Orange County and samples obtained from exploration in Buckingham County. These activities helped the committee to better understand the geologic setting of gold deposits in Virginia (Moss Mine, Vaucluse Mine, and the exploration site in Buckingham County), presently operating gold mine operations (Haile Gold Mine), and historical operations that were unsuccessfully reclaimed (Brewer Gold Mine Superfund Site). The committee also benefited from public input obtained through two town halls, a tour of Buckingham County, and written comments. These written comments included a report from the Southern Environmental Law Center, the Chesapeake Bay Foundation, and Dr. Ann Maest, Vice President of Buka Environmental. The committee carefully considered all information sources—including scientific studies, peer-reviewed publications, regulatory documentation, case studies, site visits, and public comments—in determining the findings outlined in this report.

A few clarifications are noteworthy regarding the committee's interpretation of the Statement of Task. Although the impetus for the study had its roots in events related to exploration activities in Buckingham County, the Statement of Task indicates the study should have a statewide focus. The study's statewide focus limits the ability to draw site-specific conclusions. This report outlines general themes and procedures that are important to assessing potential gold mining projects in Virginia, but comprehensive and robust analysis of site-specific environmental baseline characteristics and evaluation of population and ecosystem health and social and environmental justice considerations are necessary in order to adequately assess the potential impacts of particular projects.

As described in Box 1-2, the committee interpreted "gold mining" to refer to the extraction of gold from a mineral reserve where gold is a primary metal of material economic interest. This includes historical mining sites in which waste rock[3] and tailings from earlier activities could be mined again in the future as "gold deposits" (e.g., Moss Mine, Box 1-1). The committee recognizes that there may be some gold deposits in Virginia that have not yet been discovered or well defined by exploration. Additionally, existing data on the geology and impacts of historic gold mines in Virginia is limited. Generally, modern geologists would typically visit an active (or recently active) local mine site to observe the mining operation, collect samples for laboratory analysis, and interview researchers and workers with intimate knowledge of the deposit geology. Unfortunately, such an approach is largely unfeasible for the gold occurrences in Virginia because much of the active mining occurred before the Civil War, and the most recent large-scale mining activities ceased before World War II. In addition, many former mine sites have been reclaimed back to a natural environment (see Figure 1-4), flooded, or covered by urban development—making them largely inaccessible for study. As such, available information about the geology of the state's gold mining sites is limited. Therefore, the committee acknowledges uncertainties in its findings, where relevant.

The committee interpreted the words "air and water quality," which appear several times in the Statement of Task, to confer an emphasis on human health and ecological concerns rather than other societal and economic impacts associated with potential mining. Neither the potentially negative nor the potentially positive societal and economic impacts associated with gold mining are considered in this report, even though these impacts are extremely important when doing a site-specific environmental assessment prior to permitting. For more information on the societal impacts and the economic benefits and costs of mining see Cust and Poelhekke (2015), Ivanova and Rolfe (2011), Ivanova et al. (2007), Lockie et al. (2009), Petkova et al. (2009), Que et al. (2015), and Sincovich et al. (2018).

Finally, the committee determined that the Statement of Task emphasized local and regional human health and environmental impacts, instead of global impacts like greenhouse gases. Additionally, based on guidance from the sponsor, the committee interpreted its Statement of Task to emphasize potential impacts of gold mining activities on the people and ecosystems living near mining operations, rather than on mining industry employees and professionals. The committee recognizes the importance of clear and consistent safety training and certification programs, review and updates to safety regulations, compliance inspections, coordination with federal partners, and promotion of a safety-driven culture in the mining industry, but the committee's evaluation did not focus on the adequacy of occupational safety and health aspects of the relevant codes and regulations.

---

[3] Waste rock is material which contains little or no gold and must be removed to access the ore from which gold will be extracted.

REPORT ROADMAP

The chapters that follow address the Statement of Task and present the committee's findings and recommendations. Chapter 2 summarizes the geologic setting of gold deposits in Virginia. Chapter 3 describes how those gold deposits might be mined and processed. Chapter 4 presents potential impacts associated with the mining of gold deposits in Virginia. Chapter 5 presents the committee's assessment of the regulatory structure of Virginia compared to other states and its assessment of whether the regulatory framework is adequate to ensure the protection of air and water quality. Recommendations and conclusions are provided at the ends of Chapters 4 and 5. Given that no new gold mines are currently proposed in Virginia and that several more years of exploration and development would be necessary before a mine could be proposed, there is ample time for the Virginia General Assembly and state agencies to consider the conclusions and recommendations reached by both the National Academies' committee and the state agency committee before the state would need to evaluate permit applications.

2

Geology and Geochemistry of Gold Occurrences in Virginia

Gold is the principal commodity or a major by-product in a greater variety of ore deposit types than any other metal (Sillitoe, 2020). Some gold deposits form in a shallow subaerial or submarine setting ("volcanogenic massive sulfide," "low-sulfide epithermal," or "high-sulfide epithermal" deposits), whereas others form as a result of remobilization of gold scavenged during metamorphism and redeposited elsewhere ("orogenic" deposits). Other gold deposits form from the movement of large amounts of hot water associated with magma chambers ("skarn" and "porphyry" deposits) and some gold deposits are the weathered products of any of these types of deposits ("placer" deposits). A robust understanding of the geology and geochemical characteristics of gold deposits in Virginia is required in order to assess the types of modern gold mining operations that might be used in such deposits and the potential environmental impact of those operations on the Commonwealth should gold mining occur in the future.

Several publications have summarized the available geologic and geographic data for gold mines and prospects in Virginia. Lonsdale (1927) described gold prospects and mines in a 600-square-mile area that included Culpeper, Fairfax, Fauquier, Orange, Prince William, Stafford, and Spotsylvania Counties. Sweet (1980) provided locations, current conditions at visited sites, and primary literature references for 245 gold mines and prospects in Virginia. Spears and Upchurch (1997) provided an updated listing of 492 mines, prospects, and occurrences and included 95 sites that had not been documented previously. Laney (1917) described ore deposits in the Virgilina district that spans the Virginia–North Carolina border. This chapter is necessarily a condensed and simplified summary of Virginia geology and its known gold occurrences. For the interested reader wanting to learn more, an extensive list of references on Virginia geology, particularly regarding the Piedmont province and its gold deposits, is included at the end of the report.

GEOLOGY OF VIRGINIA

Virginia is divided into five physiographic provinces (see Figure 2-1). From east to west, the provinces are the Coastal Plain, Piedmont, Blue Ridge, Valley and Ridge, and Appalachian Plateau provinces. The Coastal Plain is the youngest of the provinces. It is an eastward thickening wedge of unconsolidated gravel, sand, silt, and clay deposits eroded from the mountainous regions to the west that overlies the eastward extension of Piedmont rocks.

The Piedmont is the largest physiographic province in Virginia and consists primarily of deformed metamorphic and igneous rocks. This region also contains fault-bounded Triassic-age (252–201 million years [Ma]) sedimentary basins that formed during the breakup of supercontinent Pangea, which ultimately resulted in the current day configuration of continents. The Blue Ridge province, to the west of the Piedmont, consists of Precambrian-age

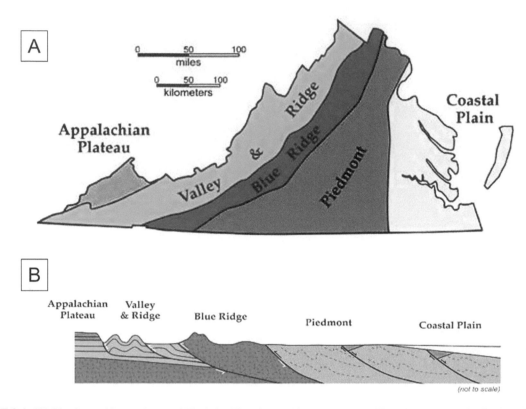

FIGURE 2-1 (A) Physiographic provinces of Virginia. Historical gold occurrences in Virginia, and deposits likely to be discovered in the future, are located primarily in the Piedmont. A smaller number of deposits occur in the Blue Ridge province. Gold is less likely to be found in the Coastal Plain, although some deposits may occur in Piedmont-related rocks that extend beneath the Coastal Plain sediments, and as surface or buried placer deposits, especially near the border between the eastern Piedmont and the Coastal Plain. The Valley and Ridge and Appalachian Plateau provinces are unlikely to host gold deposits. (B) Cross-section of Virginia physiographic provinces, with orange sections representing Triassic rift basins composed of sedimentary rocks. SOURCE: Images courtesy of Christopher Bailey.

(>541 Ma) metamorphosed and highly deformed igneous rocks and lava flows. The Valley and Ridge region consists of folded Paleozoic-age (542–252 Ma) sedimentary rocks, including limestone, sandstone, and shale, and the Appalachian Plateau province, in the southwest corner of Virginia, is composed of Paleozoic- and Mesozoic-age (252–66 Ma) sedimentary rocks that are less deformed than those in the Valley and Ridge province.

Most of the gold deposits in Virginia occur in the Piedmont province (see Figure 2-1), particularly in the gold-pyrite belt in the north-central part of the state and the Virgilina district in south-central Virginia and north-central North Carolina (see Figure S-1). The geology of this region is described in greater detail below. Some prospects and small deposits occur in the Blue Ridge province, but no lode deposits are known to occur in the Valley and Ridge or Appalachian Plateau provinces. Gold deposits may exist in buried Piedmont rocks that extend beneath the Coastal Plain sediments, but this region has not been explored.

Virginia Piedmont Province

The Piedmont province is a geologically complex region consisting of approximately eight geologic terranes[1] that became part of the North American craton (the continental core of North America) at different times. The accretion of the individual terranes was accompanied by intense deformation and varying degrees of metamorphism

---

[1] Terranes are coherent units of rock that have a distinct geologic history.

(Horton et al., 2016). Boundaries between the individual terranes are defined by shear zones where the rocks are highly deformed. Characterizing the complex geology of this area is complicated by a nearly omnipresent layer of deeply weathered rock (saprolite) up to 20 meters thick that blankets much of the region and obscures the underlying bedrock geology.

Virginia's Piedmont is divided into two regions by the central Piedmont shear zone (Hibbard et al., 2016), a major crustal break that runs north-south through much of the state. The Eastern Piedmont consists of multiple terranes, with the Carolina Terrane that includes a northern extension of the Carolina Slate Belt (Hackley et al., 2007) and hosts the Virgilina district gold deposits, being most relevant to this report. The Western Piedmont, stretching from the Washington, DC, area into North Carolina, is a tectonic transition zone that separates rocks of North American affinity (to the west) from rocks that originated offshore of present-day North America and were accreted to the eastern margin by plate tectonic processes (represented by the Eastern Piedmont; see Hughes et al., 2014).

Western Piedmont rocks consist of late Proterozoic- (2,500–542 Ma) and Paleozoic-age (542–251 Ma) metaclastics,[2] metavolanic,[3] and plutonic[4] rocks (Hibbard et al., 2016). Hibbard et al. (2016) separate the Western Piedmont rocks into the metaclastic tract and the magmatic tract. The metaclastic tract to the west is composed mostly of metaclastics of late Proterozoic and early Paleozoic age. These include the Potomac Terrane to the north and the Smith River regional thrust sheet to the south. The Potomac Terrane is the host for gold deposits in this region. The magmatic tract to the east has a higher proportion of Paleozoic metamorphosed volcanic rocks and includes the Chopawamsic Formation and the Ta River Metamorphic Suite to the north and the Milton Terrane to the south. Most gold deposits in this region are located near the top of the Chopawamsic Formation, near its contact with the Quantico Formation rocks, which are metasedimentary[5] rocks that overlie the magmatic track. This is particularly true where the Quantico units are silica rich and composed of rocks like quartzites and ferruginous quartzites (Hibbard et al., 2016). The contact between the metaclastic tract and the magmatic tract is represented by the Brookneal Shear Zone in the south and the Chopawamsic Fault in the north, with the southernmost boundary covered by Mesozoic sediments of the Danville Basin. These shear zones and faults likely played a major role in focusing fluid flow to produce gold mineralization, and many, if not most, of the gold occurrences in the gold-pyrite belt occur near these shear zones.

WHERE DOES GOLD OCCUR IN VIRGINIA?

The large majority of gold occurrences in Virginia are located in the Piedmont (see Figures S-1 and 2-2), specifically in the Western Piedmont. Virginia State Geologist David Spears notes that of 362 gold mines, prospects, and occurrences that are located in the Piedmont, 338 are located in the Western Piedmont and only 24 are located in the Eastern Piedmont (David Spears, personal communication, 2022). Gold occurrences are often spatially associated with metamorphosed volcanic rocks. The vast majority of Western Piedmont occurrences are hosted in the Ordovician-age (485–444 Ma) Chopawamsic Formation (magmatic tract) and only about 20, including Aston Bay's exploration property in Buckingham County (largest green dot in Figure 2-2), are hosted in Potomac Terrane rocks (metaclastic tract) (David Spears, personal communication, 2022).

Gold-Pyrite Belt

The Virginia gold-pyrite belt, which hosts the bulk of known gold occurrences in Virginia, is located in the Western Piedmont geologic province. It extends about 175 miles from just south of Washington, DC, to midway between the James and Roanoke Rivers, although the exact location of its southwest terminus is unknown because it is covered by Mesozoic-age (252–66 Ma) sedimentary rocks (Pavlides, 1981). This region hosts various types of ore deposits, including (1) low-sulfide, gold-quartz vein deposits with approximately 1–5 percent pyrite (see Box 2-1);

---

[2] Metaclastic rocks are clastic sedimentary rocks that have been metamorphosed.
[3] Metavolcanic rocks are igneous rocks that have been metamorphosed.
[4] Plutonic rocks are magmatic rocks that formed at great depth.
[5] Metasedimentary rocks are sedimentary rocks that have been metamorphosed.

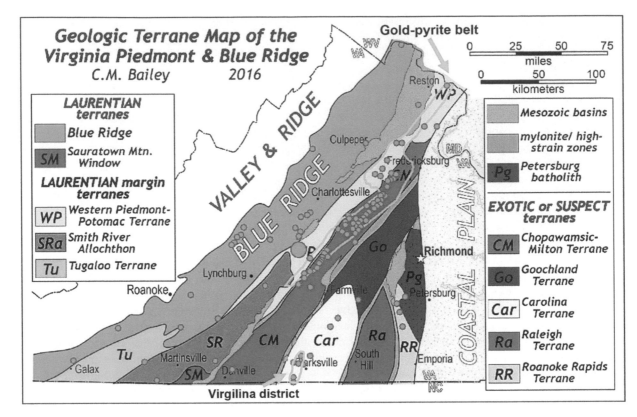

FIGURE 2-2 Simplified geologic terrane map of Virginia with locations of known gold occurrences (green dots) plotted. The largest green dot denotes Aston Bay's exploration property in western Buckingham County. Note that most of the gold deposits in the gold-pyrite belt are located in the Chopawamsic-Milton (southern part) and Potomac Terranes (northern part) of the Western Piedmont, while the deposits in the Virgilina district are located in the Carolina Terrane. Other gold occurrences are located in the Blue Ridge Terrane.
SOURCE: Modified map courtesy of Christopher Bailey.

BOX 2-1
Sulfide Minerals in Virginia

Metals of economic importance in ore deposits are often hosted in sulfide minerals. Sulfides that have been identified in both massive sulfide deposits and in quartz-gold veins in Virginia and which are most relevant to this report include

- Iron sulfides: pyrite (FeS_2); pyrrhotite ($Fe_{(1-x)}S$)
- Copper sulfides: chalcocite (Cu_2S), covellite (CuS), anilite (Cu_7S_4), digenite (Cu_9S_5)
- Copper and iron sulfides: chalcopyrite ($CuFeS_2$), bornite (Cu_5FeS_4)
- Zinc sulfide: sphalerite (ZnS)
- Lead sulfide: galena (PbS)
- Iron and arsenic sulfide: arsenopyrite ($FeAsS$)
- Copper, iron, and arsenic sulfide: tennantite-tetrahedrite ($(Cu,Fe)_{12}(As,Sb)_4S_{13}$)

(2) volcanic massive sulfide deposits with up to 90 percent pyrite (that produce little to significant amounts of base metals[6] and occasionally gold as a by-product) and (3) gold placer deposits generated by weathering and erosion of the in situ deposits (Park, 1936; Taber, 1913). The rocks that host the ores (called the "host rock") are dominantly interlayered metamorphosed volcanic and metasedimentary rocks of the Chopawamsic Formation and associated small intrusive bodies interpreted to represent an island arc[7] of late Proterozoic to early Cambrian age (~1,000–488 Ma) (Pavlides et al., 1982). Some deposits in the northern part of the gold-pyrite belt are hosted in Potomac Terrane rocks (see Figure 2-2).

The spatial association of the low-sulfide, gold-quartz veins and the massive sulfide deposits varies across the gold-pyrite belt. Gold-quartz veins generally occur to the west of massive sulfide deposits in the far north, spatially intermingled with massive sulfide deposits in the central part, and to the east of massive sulfides in the south (Pavlides et al., 1982; see Figures 2-3 and 2-4). Nevertheless, mining operations on gold veins and on the near-surface gossan (weathered and oxidized) iron ores usually reported bodies of pyrite at slight depths (Lonsdale, 1927). Perhaps the best example of the close relationships between gold mines and massive sulfide mines in the gold-pyrite belt is illustrated by an approximately 10-mile-long stretch in Buckingham County that includes the London and Virginia, Buckingham, and Williams Mines. The London and Virginia Mine near Dillwyn was first operated as a gold mine starting in 1853, producing gold from oxidized near-surface deposits using open cuts that were 20–40 feet deep and extending a distance of about 450 feet along strike (Taber, 1913). Discontinuous massive pyrite bodies, less than 1 meter thick and containing minor sphalerite, galena, chalcopyrite, tennantite, tetrahedrite, and native gold, occurred within the gold ore zone (Mangan et al., 1984). Brown (1969) reports that deep drilling at the London and Virginia and nearby Buckingham Mine properties in 1953 and 1955 identified 723,000 tons of ore containing 3.2 percent zinc, 20 percent pyrite, and fractional percentages of gold, silver, copper, and lead. The Williams Mine, which is located three-quarters of a mile along strike from the Buckingham Mine was explored for development of a pyrite mine. Material collected from the dump contains up to 80 to 85 percent pyrite, whereas the gold ores in these deposits only contain up to 4 to 5 percent pyrite (Taber, 1913). These examples show the wide variation in amounts of sulfides of closely spaced deposits that lie on an approximately 10-mile strike length and highlight the close spatial relationships of gold deposits and massive sulfide deposits in some regions of the gold-pyrite belt.

Given the close association between the two deposit types, it has been proposed that the low-sulfide gold-quartz veins and the massive sulfide deposits have a similar origin. Lonsdale (1927) outlined evidence that these two deposits may be derived from the same source, including observations that the gold veins and pyrite deposits are mineralogically similar and that they differ primarily in the relative proportions of quartz, feldspar, tourmaline, and sulfide minerals. In addition, Lonsdale (1927) highlighted observations from mine workers who describe quartz-gold veins associated with pyrite bodies, both of which are closely associated with granitic intrusions, and suggested that material in both quartz-gold veins and pyrite bodies may have been sourced from the igneous intrusions. Current theories, however, suggest that the igneous intrusions are not related to the formation of the ores, other than perhaps as providing a heat source or fluids that facilitated remobilization of metals from nearby volcanogenic massive sulfide deposits. Nevertheless, this remobilization of metals from the massive sulfide deposits, via the heat provided by the igneous intrusions or some other source, is the likely source of the gold and other metals associated with the gold-quartz veins (Good et al., 1977).

The Virgilina District

The Virgilina district, home to the second-largest concentration of historic gold mining sites in Virginia, is located in south-central Virginia in Halifax, Charlotte, and Mecklenburg Counties (see Figure S-1) and extends into Granville and Person Counties in North Carolina. The region is commonly known as the Virgilina Copper District because copper has been the main metal of economic importance; however, gold was also produced from some deposits in this area. The district lies within Precambrian- to Cambrian-age (>488 Ma) volcanic and metasedimentary

---

[6] Mostly copper, zinc, and lead.

[7] An island arc is a chain of volcanic islands that are found along tectonic plate margins.

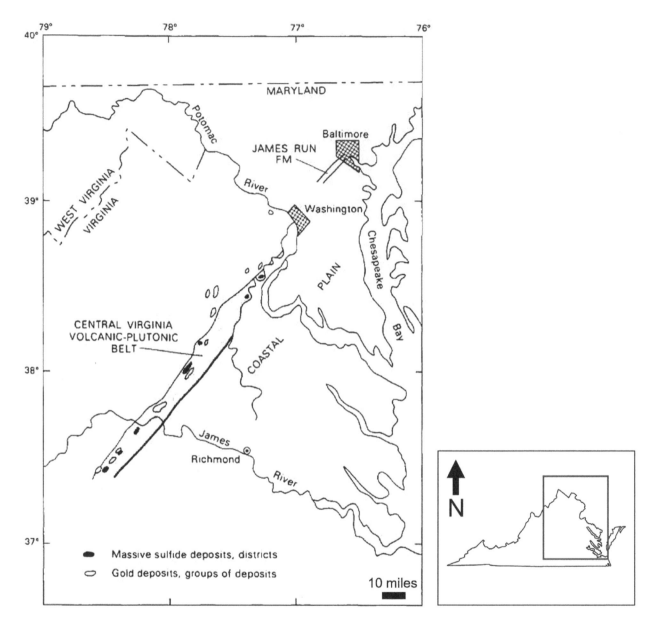

FIGURE 2-3 Locations of massive sulfide deposits and gold deposits in the central Virginia volcanic plutonic belt (which corresponds to the gold-pyrite belt). In the northern part, the gold deposits occur to the west of the massive sulfide deposits and are hosted in Potomac Terrane rocks, whereas in the central and southern part of the trend the gold deposits are intermingled or to the east of massive sulfide deposits and are hosted in Chopawamsic-Quantico rocks.
SOURCE: Image from Pavlides et al. (1982).

rocks of the Carolina Terrane, which originated offshore of present-day North America and later accreted to the North American craton, and includes a northward extension of the Carolina slate belt (Hackley et al., 2007).

Copper and gold are the main deposits that have been identified in the Virgilina district. Both the copper and gold deposits in the Virgilina district are hosted in low-sulfide quartz veins, but low-sulfide quartz veins that bear gold are primarily hosted within a single metamorphosed basaltic member of the Aaron Formation. Minor copper is present in the gold-bearing quartz veins, and gold is present in trace amounts in the copper ores, suggesting the formation of the two ore types are related, similar to the relationship between gold and base metal ores in

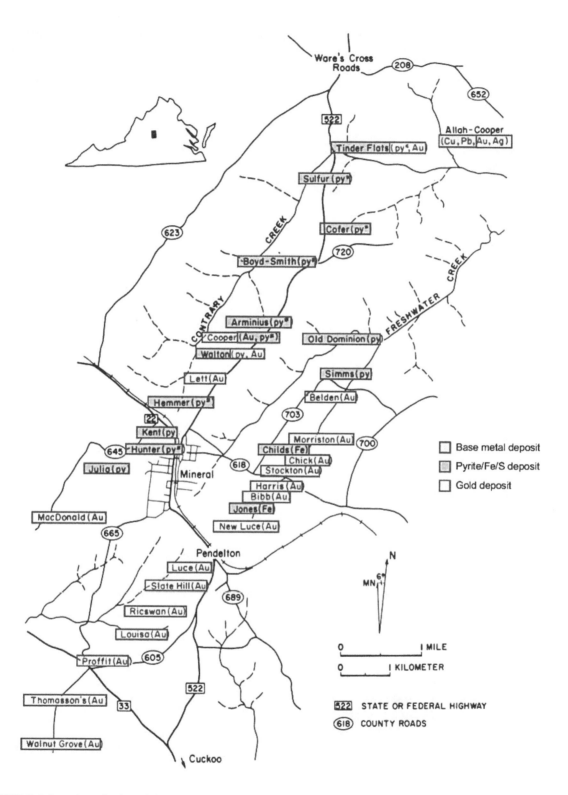

FIGURE 2-4 Location of selected deposits in the Mineral district, color-coded according to deposit type. Gold, base metal, and pyrite deposits are intermingled in this section of Virginia.
SOURCE: Image modified from Sandhaus and Craig (1986).

the gold-pyrite belt (Laney, 1917; Linden et al., 1985). Also, similar to the gold-pyrite belt, a consistent spatial distribution of base metal deposits and gold deposits is observed in the Virgilina district, where copper mines are located along a linear trend to the west of the trend containing gold deposits and prospects (see Figure 2-5). Kish and Stein (1989) report that no massive sulfide ore bodies (such as those that are common in the gold-pyrite belt) are present in the Virgilina district. Thus, unlike the gold-pyrite belt, metals were likely leached and remobilized

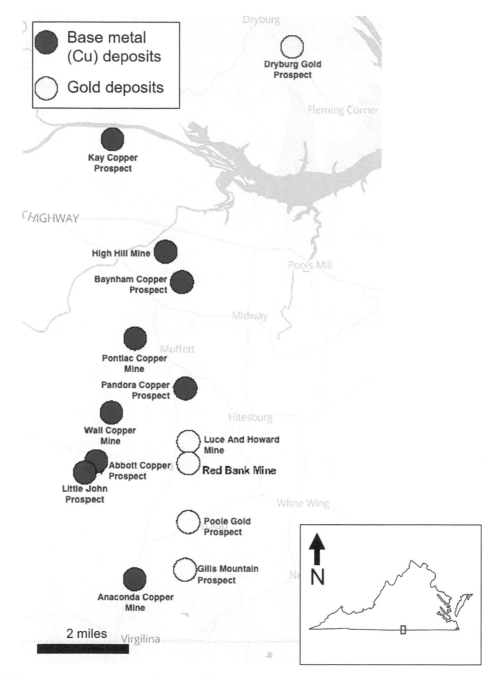

FIGURE 2-5 Locations of selected deposits in the Virgilina district of Virginia, color-coded based on the type of deposit. Gold-vein deposits are typically found to the east of the copper (Cu)-rich base metal sulfide deposits.
SOURCE: Map from The Diggings (2022).

from the surrounding magnesium- and iron-rich volcanic rock during alteration and metamorphism, instead of from nearby massive sulfide deposits as occurs in the gold-pyrite belt.

The Red Bank Mine, active in the early 1900s, is one of the few historic gold mines of significance in the Virgilina district. It consisted of a gold-bearing quartz vein that varied from a few inches to 6 feet in width, averaging 3.5 feet. Sweet et al. (2016) report that 1,064 ounces of gold were produced from the Red Bank Mine between 1903 and 1912.

Miscellaneous Occurrences

Sweet (1980) lists four locations in Floyd County (Blue Ridge Terrane) as gold occurrences, but little information is available on the sites. Two of the locations, at Brush Creek and Laurel Creek, are listed as placer operations. Sweet (1980) also reports extensive placer workings along Brush Creek in Montgomery County, a location that is probably the same deposit as the Brush Creek site listed for Floyd County because Brush Creek is located along the Floyd–Montgomery County line. These placer gold deposits likely represent gold that has been eroded from small gold-quartz veins in the Blue Ridge. Sweet and Trimble (1982) report an unusual occurrence of gold at the Walt Williams prospect in Grayson County in which gold is hosted by a quartz pebble metaconglomerate (metamorphosed coarse-grained sedimentary rock) in the early Cambrian-age (542–488 Ma) Unicoi Formation.

Other occurrences that are outside of the gold-pyrite belt and the Virgilina district are likely associated with Blue Ridge Terrane rocks and include prospects in Rockingham, Nelson, Rockbridge, Botetourt, Bedford, Warren, Carroll, and Franklin Counties (Sweet and Lovett, 1985; Sweet and Trimble, 1982; see Figure 2-2). This includes small occurrences in Nelson and Bedford Counties that are likely gold-bearing quartz veins in Precambrian-age (>542 Ma) Blue Ridge Terrane rocks, the Baker Branch Prospect in Grayson County with iron-stained quartz veins in the Precambrian-age Mount Rogers metasediments, and the Gold Hill mine in Grayson County where quartz veins are hosted by the Precambrian-age Elk Park Plutonic Group. Young (1956) describes an unnamed deposit in Blue Ridge Terrane rocks in northern Floyd County as primarily consisting of arsenopyrite that was mined for arsenic, not gold. Arsenopyrite is a characteristic mineral in orogenic gold deposits that formed at greater depths, such as those that might be expected to occur in the Blue Ridge Terrane.

CHARACTERIZATION OF GOLD OCCURENCES IN VIRGINIA

Sillitoe (2020) notes that most gold deposits worldwide show distinctive and defining combinations of geologic features that make the type of deposit readily recognizable. However, in some places, these characteristic features have been modified via ductile deformation[8] and metamorphism. Many sites in Virginia's gold-pyrite belt and, to a lesser extent, the Virgilina district fall into this latter category. Thus, in order to characterize Virginia's gold deposits and compare them with those found elsewhere, it is useful to consider the geologic history of Virginia's gold deposits, including both the original environment in which the gold was formed and how it has been modified over time.

Evidence suggests that most of the gold in Virginia's gold-pyrite belt and Virgilina district was originally deposited in submarine volcanic massive sulfide deposits or submarine equivalents of subaerial epithermal deposits, and subsequently remobilized during later metamorphism and deformation to form low-sulfide, gold-quartz vein deposits. This remobilization appears to have occurred after the host rocks experienced at least one episode of deformation and metamorphism (LeHuray, 1982). Gold deposits that are modified by, or form during, metamorphic and deformation processes associated with convergence of tectonic plates and formation of mountain belts are referred to as "orogenic" gold deposits. Orogenic gold deposits can form over a wide range of pressure/depth and temperature conditions (Goldfarb et al., 2005). While the Virginia gold deposits show many characteristics similar to greenschist-facies[9] orogenic deposits, they may fall into the category of deposits that Groves et al. (2003) refer to as "enigmatic metamorphic gold deposits" involving overprinting of more than one style of mineralization and alteration.

---

[8] Ductile deformation is when rocks bend and deform during intense pressure and temperature, instead of fracturing.

[9] Greenschist-facies refers to low to medium metamorphism corresponding to temperatures of about 300–500°C and pressures of 3–20 kbar, which is typical of continental collision tectonics (Arndt, 2011).

Structural Setting of Gold Occurrences

The structures associated with gold deposits in Virginia likely formed in two distinct stages. During the first stage, as various terranes were accreted to North America, the area of interest was far below the Earth's surface. At these depths, the rocks were hot and ductile. As the various terranes were assembled and slid past one another, the hot, ductile rocks at the interface underwent ductile deformation to form shear zones. Later, as the rocks were exhumed and the temperature and pressure decreased, the rocks became more brittle, resulting in the formation of cross faults that cut the shear zones at high angles. During this time, gold was remobilized and deposited with quartz to produce veins and brecciated[10] filled fractures that cut the shear zone.

This complex geologic and tectonic history means that at the regional scale Virginia gold deposits occur along shear zones that extend for tens of kilometers, whereas at the local or mine scale mineralization is most closely associated with later faults, especially in the more broken areas where the shear zones are crossed by tension cracks. Park (1936) states that "work at the Melville Mine and at other mines in the southern Appalachian region seems to indicate that a relation of importance exists between the northwestward-striking tension cracks (or cross faults) and the best ores." Similarly, in discussing protocols for exploration for orogenic gold deposits, Groves et al. (2020) report that "subparallel arrays of obliquely cross-cutting faults that develop where there are flexures or jogs on the first-order faults [i.e., in the case of the gold-pyrite belt the major NE-trending shear zone], in many instances provide the most important structural geometries in terms of predictive exploration." While this statement refers to exploration at the district scale, it also applies at the deposit scale.

Size and Geometry of Deposits

The size, shape, depth beneath the surface, and other properties of an occurrence of gold are critical for determining the economic feasibility of extracting gold and influence the type of mining and processing that must be undertaken to extract the ore. The size and shape of the ore body also determine how much surface area must be disturbed to mine the deposit and provide storage facilities for waste and tailings.

Little information is available concerning the surface footprint of historical gold mining operations in Virginia. Park (1936) reported that the Melville Mine tract in Orange County was developed on 844 acres leased by the Rapidan Gold Corporation, but did not state what portion of this area was actually occupied by mine facilities (shafts, trenches, waste dumps, processing facilities, etc.). The Vaucluse Mine tract in Orange County—noted to have the most extensive surface workings of all mines in the gold-pyrite belt—was described as occupying 200 acres (Park, 1936) but, again, the portion of this area occupied by mining-related infrastructure is unknown. A map of shafts, workings, and dumps at the Moss Mine in Goochland County circa 1935 (Pardee and Park, 1948) shows a disturbed area of about 500 × 700 square feet (8 acres), though the underground area may be smaller or larger than the surface area, depending on the extent of underground workings. Park (1936) reports that the two mined veins at Moss extended 1,500 to 2,500 feet—assuming that the veins were exploited along this entire length, the size (area) of the underground footprint could be much larger than the surface footprint.

In terms of shape, Lonsdale (1927) describes the gold veins in the gold-pyrite belt as being lens shaped, with long dimensions that range from a few inches to several hundred yards (see Figure 2-6). Mining showed that the lenses or veins are not continuous. Often, a lens (vein) will pinch out and another vein will begin at some distance away, or at some distance above or beneath the mined-out lens. The pinching and swelling of quartz veins is present at scales ranging from a few inches to thousands of feet. The veins did not fill open spaces or fissures, and instead replaced in situ material at considerable depth during metamorphism when open space (fissures) did not exist. Park (1936) also describes single quartz veins that branch to form numerous continuous smaller quartz veins that decrease in size along strike[11] (see Figure 2-7). Pyrite ore bodies have the same shape as the gold ore bodies in the same mine or district, but typically the pyrite bodies are larger and can reach up to 1,000 feet long (Lonsdale, 1927).

[10] Breccia is sharp-angled fragmented rock.
[11] Strike is the orientation of an imaginary horizontal line across the plan of a geologic feature.

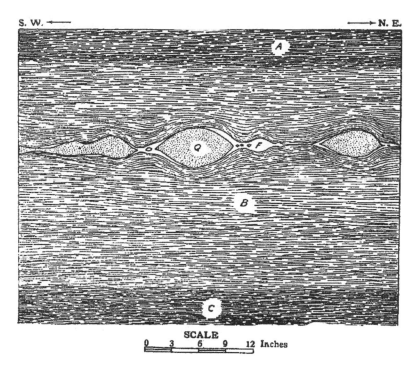

FIGURE 2-6 Sketch showing symmetrical quartz lenses in the middle vein at the Tellurium Mine, Fluvanna County. Note how the lenses pinch and swell, which is typical for gold-bearing quartz veins in the gold-pyrite belt.
NOTE: A = hanging wall schist; B = bed of lighter colored schist; C = footwall schist; F = feldspar; Q = quartz.
SOURCE: Image from Taber (1913).

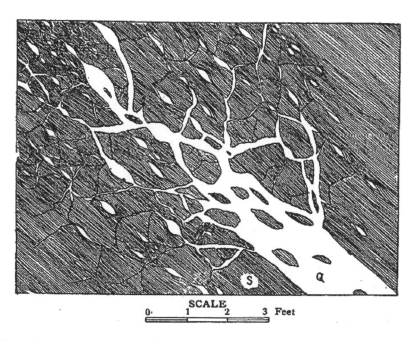

FIGURE 2-7 Sketch showing a larger quartz vein (lower right) that fragments to produce numerous smaller quartz lenses and veins that pinch and swell (upper left) along the strike of foliation (metamorphic layering) of the rock at the Morrow Mine in Buckingham County.
SOURCE: Image from Taber (1913).

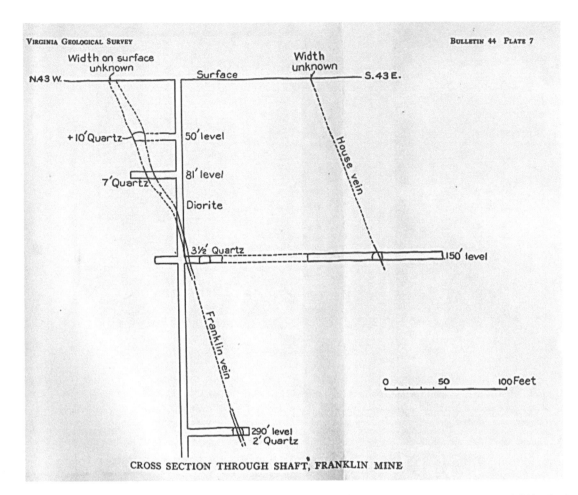

FIGURE 2-8 Cross-section through the Franklin Mine, Fauquier County, circa 1935 showing the shaft and drifts (horizontal tunnels) at 50-, 81-, 150-, and 290-feet depths and intersections with the Franklin and House veins. Note how the thickness of the Franklin vein varies with depth.
SOURCE: Image from Park (1936).

Descriptions of the Franklin Mine in Fauquier County illustrate the pinch-and-swell nature of the gold-quartz veins. This mine worked two steeply dipping[12] parallel veins, referred to as the Franklin vein and the House vein (Park, 1936; see Figure 2-8). These veins were initially accessed through shallow surface trenches, but a shaft to a depth of ~300 feet below the surface was developed to access deeper ore, with crosscuts excavated to intersect and mine the ore at depths of 50, 81, 150, and 290 feet. The Franklin vein was found to narrow from 16 feet on the 50-foot level to essentially zero at the 130-foot depth, then expand to 7 feet wide on the 150-foot level. On the 290-foot level, the quartz vein was very discontinuous and the bodies became more lens-like, varying from over 5 feet wide to a few inches over a distance of a few feet.

Another recent detailed description of a gold mine in the gold-pyrite belt is available for the Vaucluse Mine (sometimes referred to as the Grimes or Grymes Mine) in Orange County (Bass, 1940). The Vaucluse Mine was operated intermittently starting in 1832, producing gold from placers and near-surface oxidized ore. Starting in 1844, gold was mined from two open cuts, each about 60 feet deep, 75 feet wide, and 120 feet long. By 1854, six shafts had been sunk to access deeper ore. The mine closed during the Civil War and was not reopened until

---

[12] The dip is the angle of inclination measured from horizontal of a planar geologic feature.

FIGURE 2-9 Photos of the Vaucluse core shown to the committee in Charlottesville on March 30, 2022. The core was collected from beneath the deepest parts of the deposit that were mined. Note the pyrite-rich areas at the quartz vein–wallrock contact. Most of the gold in these samples is contained in the pyrite.
SOURCE: Photos by Robert J. Bodnar.

the 1930s. Between December 1935 and December 1938 the mine produced a total of 25,452 tons of ore containing 4,305.3 ounces of gold. Ore was contained in steeply dipping veins averaging 4 feet wide by 50 feet long, and varying in width from a few inches to 30 feet, and in length from a few inches to more than 200 feet. Exploratory drilling beneath the 300-foot level indicated that the ore continued to a depth of at least 600 feet. More recently, in the 1980s, four exploration drill holes were drilled to depths ranging from 601 to 754 feet and intersected the gold-bearing veins well below the deepest levels that had been previously mined (see Figure 2-9).

Gold Production in Virginia

Park (1936) reports total gold production from the gold-pyrite belt during the period 1829–1934 as 91,208 total ounces, with the maximum in any one year being 6,259 ounces in 1849. After 1860, total production in a single year never exceeded 1,000 ounces. Production in the Virgilina district (mostly from the Red Bank and Luce-Howard Mines) through 1912 amounted to a little over 1,000 ounces of gold in total. Park (1936) also reports the gold production for the entire state from 1829 to 1934 was valued at $3,318,388. The fixed price of gold from 1834 to 1933 was $20.67 per ounce; therefore, the reported gold production in dollars corresponds to about 160,000 ounces of gold produced in Virginia over a 105-year period. For comparison, the Turquoise Ridge Gold Mine in Nevada produced 287,144 ounces of gold in 2020, the Carlin trend gold mines in Nevada (which include the Arturo JV, Betze Post, Carlin Trend Operations, and Meikle mines) produced more than 1.6 million ounces of gold in 2020 (Nevada Division of Minerals, 2020), and the Haile Mine in South Carolina produced 137,413 ounces of gold in

TABLE 2-1 Calculated Average Grade Based on Reported Average Ore Value per Ton and Average Gold Price from 1834 to 1933 of $20.67 per Ounce

	Average Ore (dollars/ton)	Average Calculated Grade (oz/ton)	Average Calculated Grade (g/t)
Culpeper Mine vein (Culpeper County)	$6	0.29	9.94
Franklin Mine (Fauquier County)	$12	0.58	19.89
Grasty Tract (Orange County)	$6–$32	0.29–1.55	9.94–53.14
Vaucluse Mine (Orange County)	$8	0.39	13.37
Red Bank Mine (Virgilina district)	$8	0.39	13.37

SOURCES: Linden et al. (1985); Lonsdale (1927).

2020 (OceanaGold, 2022a) and 189,975 ounces in 2021 (Junior Mining Network, 2022). Thus, the Haile Mine produces as much or more gold in one year as the total production during the 105-year history of gold mining in Virginia. As such, individual gold deposits in Virginia are very small compared to gold deposits that are currently being mined in the western United States and in South Carolina, and the scale of mining operations in Virginia would be commensurately smaller compared to mines operating in the western United States and the Haile Mine in South Carolina.

Grades for most gold mines that operated in Virginia during the 1800s through the mid-1900s are reported in dollars, rather than in the more conventional units of weight (grams/ounces) used today. While it is not stated directly, the values are also likely reported in short tons, rather than in metric tons. Thus, estimated grades of ore mined in the gold-pyrite belt and in the Virgilina district ranged from a few tenths to over an ounce per ton, or roughly 10–50 g/t (see Table 2-1). Recent exploration drilling in Virginia has intercepted similar or greater grades in quartz veins over short distances (see Box 2-2). These grades are comparable to some of the highest-grade underground gold mines operating in the United States today, including the Fire Creek Mine (44 g/t), Turquoise Ridge Mine (16.9 g/t), Pinson Mine (13.8 g/t), and Midas Mine (11.1 g/t), all of which are located in Nevada (Basoy, 2015). In contrast, the Haile Mine in South Carolina reports an average grade of 1.37 g/t for the open pit, with a cutoff grade of 0.45 g/t for December 2019 through June 2020 (Cision PR Newswire, 2020).

Although data are limited, the total cumulative amount of ore extracted at Virginia gold mines is estimated to range from one hundred to several hundred thousands of tons of ore. For comparison, in 2021 the Haile Mine in South Carolina produced 3,214,000 tons of ore, an amount estimated to exceed the total amount of ore produced in the more than 100-year history of gold mining in Virginia. Thus, while gold grades being mined in the 1800s and early 1900s in Virginia were comparable to grades being mined elsewhere in the United States today, the total ore tonnage mined at the Virginia gold mines was small compared to most mines operating in the United States today.

Mineralogy of the Gold Deposits

Some amount of host rock must necessarily be disturbed and extracted along with the ore during mining, and this material eventually ends up in waste rock piles or tailings. As such, the lithology and mineralogy of the host rock can affect the local environment, including surface water and groundwater. The host rocks for Virginia's gold deposits are mostly metamorphic rocks containing (in decreasing order of abundance) quartz, sericite, potassium- and sodium-rich feldspars, chlorite, hornblende, biotite, garnet, tourmaline, and kyanite. Many of these same minerals are also included in the alteration assemblage associated with gold ore as described below, and all of the minerals listed above are common rock-forming minerals that are relatively stable at surface conditions and contain few elements that might contribute to contamination of the local soils and waters during mining. Host rocks for gold mineralization in the gold-pyrite belt are dominantly highly deformed metamorphic rocks, and less often igneous plutonic rocks that have been intruded into the metamorphic rocks. All of the host rocks show distinct, foliated fabrics characterized by parallel bands of minerals, especially quartz and fine-grained mica (muscovite, illite, or sericite) that are indicative of their strongly deformed nature. The dominant host rocks for gold deposits

BOX 2-2
Gold Grades Reported for Aston Bay's Western Buckingham County Project

In recent years, Aston Bay Holdings, Ltd., has initiated a gold exploration program in Virginia. The property is located in Cambrian-age metasediments in western Buckingham County, outside of the area that would normally be considered a part of the gold-pyrite belt. The project involves both surface sampling and shallow drilling to depths of less than about 410 feet.

The Buckingham site includes gold in quartz veins and disseminated gold mineralization associated with sericite-quartz-pyrite alteration. The northwest-southeast trending vein is perpendicular to the strike of the main shear zone hosting deposits in the gold-pyrite belt, and comprises a series of gold-bearing quartz vein outcrops (see Figure 2-10) in which visible gold can be observed in hand samples. These veins extend over 492 feet and have yielded values up to 701 g/t gold in surface samples. Exploration drilling by Aston Bay and a previous operator has intersected significant gold mineralization in the quartz vein at depth, but often only over short distances (e.g., 24.73 g/t over an estimated width of 9.35 feet, including an estimated width of 3.64 feet that has a grade of 62.51 g/t). Drilling that has targeted the broader zones of sericite-quartz-pyrite mineralization around the vein has intersected lower grade, but significant, gold mineralization (e.g., 2.2 g/t over a core length of 59.06 feet and 0.37 g/t over an estimated width of 138.39 feet; Aston Bay Holdings, Ltd., 2022a,b).

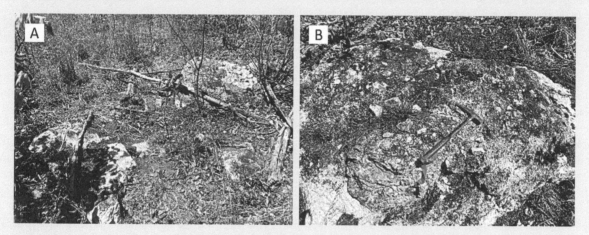

FIGURE 2-10 Photographs of surface outcrops of the gold-bearing quartz vein being explored by Aston Bay Holdings in western Buckingham County. (A) Quartz vein outcrops. (B) Quartz vein outcrop with hammer for scale.
SOURCE: Photos by Robert J. Bodnar during committee visit on April 28, 2022.

in the gold-pyrite belt (Chopawamsic Formation and Potomac Formation) show a higher degree of metamorphism and deformation compared to Carolina Terrane host rocks in the Virgilina district. As a result, the Virgilina deposits retain more of the original characteristics of formation compared to deposits in the gold-pyrite belt.

The potential contamination of local soils and waters during mining is not only dependent on the nature of the rocks that host the gold deposit, but also the mineralogy of the gold ore and associated alteration minerals. Gold is the only ore mineral that is sought in the gold deposits of the gold-pyrite belt. In the near-surface weathered and oxidized zone of the gold-pyrite belt, gold occurs as free native gold, and below the water table, where the rock has not been oxidized, gold mostly occurs within pyrite. Taber (1913) reports that pyrite is the main sulfide in Virginia gold-quartz veins and is always present in small quantities in gold ores, often averaging ~1 percent. This observation agrees with the committee's study of drill cores from the Vaucluse Mine to Virginia Energy in Charlottesville on March 30, 2022, as well as the committee trip to the exploration property of Aston Bay Holdings, Ltd., on April 28, 2022. Nevertheless, as discussed above, the gold-quartz veins can be in close proximity to massive sulfide deposits, which may have >90 percent pyrite. As noted by Park (1936), the mineralogy of gold-containing ores in Virginia is relatively simple. The most abundant gangue (non-commercially valuable) mineral is quartz, which varies from white to light gray to bluish gray in the nonoxidized zone and in the oxidized zone takes on a reddish color from iron oxides along cracks. Sericite (fine-grained white mica) is the second-most abundant gangue mineral, and is most abundant near the vein walls and in the sericitized host rock that surrounds the quartz lens. Carbonates, including ankerite (Ca-Mg-Mn-Fe carbonate) and calcite, are the next most common phase, and in some deposits represent the main gangue mineral. Ankerite often occurs between the quartz lens and the silicified country rock. Chlorite is observed in all deposits, and often is partially replaced by sericite. In some deposits, biotite, garnet, tourmaline, and hornblende have been identified in the host rock adjacent to veins and lenses. Hydrous iron and manganese oxides are common in veins and lenses that are near the surface (above the water table). Chalcopyrite is often present in small amounts, and minor ilmenite, magnetite, tetradymite (Bi_2Te_2S), and arsenopyrite have been reported. Pardee and Park (1948) suggest that the sericite, quartz, and chlorite immediately adjacent to the quartz veins represent alteration of the original host rock.

At the Red Bank and Luce-Howard Mines in the Virgilina district, very fine-grained native gold occurs with hematite (iron oxide) near the surface. With increasing depth, pyrite becomes more common, and the gold is contained in pyrite, similar to gold occurrences in the gold-pyrite belt. In contrast to the gold-pyrite belt, quartz veins in the Virgilina district contain minor bornite and other copper-bearing minerals that represent the major ore minerals in the copper deposits in the district (Johnson, 1983).

Geochemistry of the Gold Deposits

Minor and trace elements that may occur in gold ores and adjacent host rocks associated with the low-sulfide, gold-quartz deposits in the gold-pyrite belt and Virgilina district include antimony, arsenic, cadmium, chromium, cobalt, copper, bismuth, boron, lead, molybdenum, nickel, platinum-group elements, scandium, selenium, tellurium, thallium, tungsten, vanadium, and zinc (Ashley, 2002). Of these elements, those that are most likely to be present in the ores at concentrations exceeding 100 milligrams per kilogram (mg/kg), and perhaps reaching a few thousand mg/kg, include arsenic, copper, zinc, and lead. The absolute abundances of these elements vary depending on the specific geologic environment and historical use of the site. For example, in some low-sulfide, gold-quartz vein deposits, the arsenic content of the ore can be as high as 1 percent, whereas arsenic-bearing minerals (e.g., arsenopyrite) are rare in the Virginia deposits. Additionally, even though no natural mercury-rich minerals have been reported in gold deposits in Virginia, the historic use of mercury in the amalgamation process has contaminated soils, water, and waste materials in historic gold mining regions (Hammarstrom et al., 2006; Seal et al., 1998; VDH, 2022b; Virginia Energy, 2022e). For example, elevated levels of mercury have been identified in the pond sediment at the site of the Mitchell Gold Mine (Hammarstrom et al., 2006), soil near the Greenwood Gold Mine (Seal et al., 1998), and in stream sediments near the Vaucluse Mine (Virginia Energy, 2022e).

The base metal content of deposits is generally correlated with the amount of pyrite in the rock (Plumlee et al., 1999). Ores in low-sulfide gold deposits consist mostly of quartz containing a few percent pyrite, often with only trace or minor amounts of copper, zinc, and lead sulfide minerals, and often with significant amounts of carbonate minerals. Thus, mining these deposits generally results in insignificant acid rock drainage with low to undetectable metal contents in the waters (see Box 2-3). Conversely, volcanogenic massive sulfide deposits can contain more than 90 percent total sulfides, with most being pyrite but sometimes containing up to a few tens of weight percent of copper, zinc, and lead sulfides, with very little carbonate present. As such, mining of these volcanogenic massive sulfide deposits can lead to significant acid rock drainage and elevated metal contents in waters that traverse the mine site.

Data reported from the Greenwood Mine low-sulfide, gold-quartz vein deposit (Prince William County), and the massive sulfides at the Cabin Branch Pyrite Mine (Prince William County) and Valzinco Lead-Zinc Mine (Spotsylvania County) provide some useful geochemical information to assess the potential water quality impacts of gold mining in Virginia. Many of the elements that are of interest to human health and the environment (e.g., arsenic, cadmium, copper) are found at low levels in the host rock and quartz veins of low-sulfide gold deposits at the Greenwood Mine in Prince William County (Seal and Hammarstrom, 2002; see Table 2-2). Conversely, the nearby massive sulfide deposits—the Cabin Branch Pyrite Mine and the Valzinco Lead-Zinc Mine—have elevated

---

BOX 2-3
Chemistry of Acid Rock Drainage

Acid rock drainage (ARD), also known as acid mine drainage (AMD), refers to acidic (low-pH) waters produced when iron-bearing sulfide minerals such as pyrite (FeS_2) and pyrrhotite ($Fe_{(1-x)}S$) are oxidized in the presence of oxygen and water (see Equations 2-1, 2-2, and 2-3). The end result (see Equation 2-4) is the generation of the characteristic acidic water (H^+) and the red-yellow-brown iron(III) oxide, hydroxide, and oxyhydroxide minerals (e.g., $Fe(OH)_3$) that coat the surface in areas of ARD (Stumm and Morgan, 1996).

$14\ Fe^{3+} + FeS_2 + 8\ H_2O = 15\ Fe^{2+} + 2\ SO_4^{2-} + 16\ H^+$	**Equation 2-1**
$15\ Fe^{2+} + 15\ H^+ + 15/4\ O_2 = 15\ Fe^{3+} + 15/2\ H_2O$ (usually mediated by microorganisms)	**Equation 2-2**
$Fe^{3+} + 3\ H_2O + Fe(OH)_3 + 3\ H^+$	**Equation 2-3**
Overall: $FeS_2 + 7/2\ H_2O + 15/4\ O_2 = Fe(OH)_3 + 4H^+ + 2SO_4^{2-}$	**Equation 2-4**

The distinction between ARD and AMD lies in the source of the acidic water. While AMD refers specifically to acidic waters produced from a mine site, ARD is a broader term, referring to acidic waters produced after the exposure of iron sulfide minerals to air and water at any site, such as at an outcropping mineral deposit, road-cuts, or where iron sulfide–containing material is used as aggregate. This report uses the more general term, ARD.

---

metal contents in the ore, especially arsenic, cadmium, cobalt, copper, nickel, lead, and zinc (Hammarstrom et al., 2006; Seal and Hammarstrom, 2002; see Table 2-2). While neither the Cabin Branch nor the Valzinco massive sulfide deposits produced gold, the much higher concentrations of some elements of interest in massive sulfide deposits compared to gold-quartz veins illustrates the potential impact that massive sulfides could have on water quality should one of these bodies be intentionally or unintentionally disturbed during mining to access the gold-quartz veins.

In addition to considering the minor and trace element content in gold deposits, it is also important to consider the stability of the minerals that host those elements. If an element of interest occurs in a stable mineral phase that will not be altered during mining, processing, and later long-term storage of waste rock and tailings, it will have less of an impact on water quality compared to a mineral that is easily altered during these events. For example, pyrite and arsenopyrite, which can host many metals of concern (e.g., arsenic, cadmium, selenium, thallium), are unstable in a humid, oxidizing near-surface environment. These minerals will quickly break down to produce various iron oxyhydroxide mineral phases, with the concomitant release of the metals and other trace elements to the local environment. Other sulfides (e.g., galena, pyrrhotite, sphalerite, chalcopyrite) are also susceptible to oxidization and can release metals (e.g., cadmium, copper, lead, selenium, thallium, zinc) when exposed to air, albeit at a slower rate than pyrite (Koski et al., 2008). Conversely, minerals such as native copper, chalcocite, cuprite, malachite, and azurite are more stable at near-surface conditions and may release insignificant amounts of elements of interest into the environment (see Appendix C). Another major factor that determines abundances of dissolved metals in mine drainage waters and natural waters draining unmined mineralized sites is the formation of acid rock drainage (see Box 2-3), which results in low-pH water. Concentrations of iron, aluminum, manganese, zinc, and copper have been shown to increase by up to six orders of magnitude as the pH decreases from slightly alkaline to near-neutral values to highly acidic pH values (Plumlee et al., 1999). Nevertheless, although acidity tends to increase the concentrations of metals solubilized and transported in drainage, metals can be mobilized and released in the absence of ARD (Ashley, 2002; Ashley and Savage, 2001).

While few data are available for mine drainage waters associated with gold-quartz vein deposits in Virginia, one study examined this issue for the Greenwood Mine in Prince William County (see Table 2-3), as part of a plan to incorporate the mine area into the Prince William Forest Park (Seal et al., 1998). While the focus of the study was on mercury contamination, various water samples were collected and analyzed for an extensive suite of elements of environmental concern. Water collected from two shafts at the Greenwood Mine had pH values of 5.9 and 6.1 (similar to the pH of rainwater). Metal concentrations were below current EPA Maximum Contaminant

TABLE 2-2 Metal Concentrations in Solid Material Collected at Low-Sulfide Quartz Vein Gold Deposits and at Massive Sulfide Deposits

Deposit Type	Mine (sample type and name)	Al (wt.%)	As (mg/kg)	Cd (mg/kg)	Cu (mg/kg)	Pb (mg/kg)	Sb (mg/kg)	Se (mg/kg)	Tl (mg/kg)	Zn (mg/kg)
Low-sulfide, gold-quartz vein deposit	Greenwood granite wallrock (sample PW-GM1)	6.3	<10	<2	8	27	—	—	—	—
	Greenwood vein quartz (sample PW-GM2A)	1	<10	<2	3	6	—	—	—	—
	Greenwood quartz from waste pile (sample PW-GM3)	0.03	<10	<2	<2	<4	—	—	—	—
Massive sulfide deposit	Valzinco (sample 99VLZN9)	0.13	62	160	15,000	260	4.2	33	<2	69,000
	Valzinco tailings	1.6–8.6	19–46	0.4–110	280–2,000	2,400–16,000	5.5–21	—	0.75–1.9	230–21,000
	Cabin Branch (sample CB-ORE-1)	0.19	262	173	3,840	2,560	—	—	—	—
	Cabin Branch (sample CB-ORE-2)	0.22	149	19	11,000	2,010	—	—	—	—

NOTES: Where applicable, data are reported as range. Dashes indicate not analyzed.
SOURCE: Data from Seal and Hammarstrom (2002).

TABLE 2-3 Metal Concentrations and pH in Water Samples Collected at and Near the Greenwood Low-Sulfide Quartz-Vein Gold Mine in Prince William County

Sample Name	pH	Al (mg/L)	As (mg/L)	Cd (mg/L)	Cu (mg/L)	Pb (mg/L)	Sb (mg/L)	Se (mg/L)	Tl (mg/L)	Zn (mg/L)
VA Surface Water Quality Criteria (9 VAC 25-260-140)	—	—	0.01	0.005	1.3	0.015	0.0056	0.17	0.00024	7.4
MCL (40 CFR § 141.62)	—	—	0.01	0.005	1.3	0.015	0.006	0.05	0.002	—
Collected from main shaft (PWGM-1-1)	6.1	0.290	0.0004	0.0001	0.008	0.0083	0.0002	0.0003	<0.00005	0.027
Collected from small shaft north of main shaft (PWGM-4-1)	5.9	0.220	0.001	<0.00002	0.0007	0.0004	0.00009	0.0006	<0.00005	0.003
Collected from Quantico Creek, downstream of mine (PWGM-2-1)	6.2	0.045	<0.0002	0.0002	0.003	0.0002	0.0001	0.0005	<0.00005	0.041
Background value: Collected upstream from mine (PWGM-3-1)	6.4	0.180	<0.0002	0.00003	0.002	0.0003	0.0002	0.0004	<0.00005	0.005

NOTES: The water data are compared to the Virginia Criteria for Surface Water for public water supply and the EPA Maximum Contaminant Level (MCL) shaded in grey. None of the water samples from Greenwood exceeds either standards. Dashes indicate not analyzed or not applicable.
SOURCE: Data are from Seal et al. (1998).

Levels (MCLs; 40 CFR § 141.62) and Virginia Criteria for Surface Water (9 VAC 25-260-140) that are set to protect human health. The total base metal content of the mine shaft waters at the Greenwood Mine range from 0.014 to 0.046 mg/L. Seal et al. (1998) concluded that the mine drainage waters posed no significant environmental threat. The committee views the concentrations measured in water samples at the Greenwood Mine as reflective of a scenario in which only the low-sulfide, gold-quartz vein and adjacent wallrock are disturbed during gold mining in Virginia.

Owing to the close spatial relationship of the gold-quartz vein deposits with massive sulfide deposits in the gold-pyrite belt, it is possible that some sulfide-rich material could be intersected and disturbed during mining of the low-sulfide, gold-quartz vein deposits. This could cause greater potential for acid generation and other environmental impacts (see Box 2-3). Should this occur, significant ARD may result and cause mobilization of metals into local waters. This may be the cause of the "extremely acidic drainage" noted in 2019 at the historical Vaucluse Gold Mine site (Virginia Energy, 2022e) despite the low concentration of pyrite observed in the ore. The committee considered water samples taken downstream of the Cabin Branch Mine and Valzinco Mine massive sulfide deposits (see Table 2-4) as representative of the ARD that could occur if these massive sulfide deposits are disturbed. These samples show water pH as low as 2.4 and concentrations of cadmium, copper, lead, thallium, and zinc that exceed the U.S. Environmental Protection Agency MCLs (40 CFR § 141.62) or Virginia Criteria for Surface Water (9 VAC 25-260-140). The total base metal content of groundwater samples from the Cabin Branch massive sulfide mine ranges from 0.058 to 14.434 mg/L, much greater than that measured at the low-sulfide Greenwood Mine. The potential ecological and human health impacts of the mobilization of these metals is considered further in Chapter 4.

The committee notes that two samples from the Valzinco Mine (VLZN-10-2RA and VLZN-10-2FA) were not considered relevant to our analysis (see Appendix D). These samples were collected from stagnant puddles immediately on top of mine tailings. The committee determined that these two samples represented an anomalous geological setting that should not be directly compared to surface water or groundwater standards, given that these puddles likely experienced some degree of concentration due to repeated evaporation and little dilution.

COMPARABLE DEPOSITS AROUND THE WORLD

As directed by the Statement of Task (see Box 1-3), the committee sought to identify gold deposits that displayed comparable geologic, mineralogical, hydrologic, and climatic characteristics to those in Virginia. When examined in detail, subtle to significant differences are observed in every deposit (see detailed descriptions of deposits in Appendix E), such that no single gold deposit is fully comparable with the known deposits in Virginia. For example, while South Carolina gold mines are the closest modern commercial gold mines and occur under a similar climate and hydrology, they are not a good analogue for the geologic characteristics of known Virginia deposits (see Box 2-4). Nevertheless, South Carolina gold deposits are discussed here and in subsequent chapters of the report because of the similar hydrology and climate of South Carolina, and because various stakeholders—including Aston Bay Holdings, Ltd.—have referenced South Carolina mines when discussing the potential for gold mining in Virginia (Vogelsong, 2021).

As approximately 75 percent of all gold mined in the United States is mined in Nevada, it is worth noting why the many dozens of active gold mines in Nevada are not considered to be comparable deposits in those in Virginia. As noted above, many different types of gold deposits exist, with each type characterized by specific geological, geochemical, mineralogical, and other features. The largest and most-well-known gold deposits in Nevada are classified as Carlin-type deposits. Carlin deposits, also sometimes referred to as "invisible gold" deposits, are hosted in carbonaceous sedimentary rocks that have undergone little to no metamorphism and deformation. The geological and geochemical characteristics of this type of gold deposit are very different from gold deposits in Virginia and will not be considered further.

As discussed in this chapter, gold occurrences in the gold-pyrite belt and the Virgilina district most closely resemble the low-sulfide, gold-quartz vein deposit type (i.e., orogenic gold deposits). We note here that a common characteristic of orogenic gold deposits is that the gold in the deposit has been remobilized from some other source—this source might be gold-bearing country rocks or a previously formed gold deposit that has

TABLE 2-4 Metal Concentrations and pH in Water Samples Immediately Downstream from the Valzinco Mine (Spotsylvania County) and the Cabin Branch Mine (Prince William County) Massive Sulfide Deposits

Sample	Comments (sample name)	pH	Al (mg/L)	As (mg/L)	Cd (mg/L)	Cu (mg/L)	Pb (mg/L)	Sb (mg/L)	Se (mg/L)	Tl (mg/L)	Zn (mg/L)
VA Surface Water Quality Criteria (9 VAC 25-260-140)	—	—	—	0.01	0.005	1.3	0.015	0.0056	0.17	0.00024	7.4
MCL (40 CFR § 141.62)	—	—	—	0.01	0.005	1.3	0.015	0.006	0.05	0.002	—
Valzinco downstream water	Collected in Knights Branch immediately downstream of the tailings (sample VLZN-3)	2.6–3.9	0.70–19.47	<0.0002–0.001	0.0032–0.088	0.049–2.2	0.17–1.3	<0.00003–0.00069	<0.0002–0.0007	<0.00005–0.0003	1.9–7.0
Valzinco water downstream	Collected in Knights Branch ~1 km downstream of the tailings (sample VLZN-11)	2.4–3.6	0.52–31.15	<0.0002–0.0009	0.0017–0.099	0.038–2.8	0.13–1.6	<0.00002–0.00031	<0.0002–0.0004	<0.00005–0.00009	0.99–27.0
Cabin Branch water	Groundwater	4.1–7.0	<0.00001–5.50	—	0.2–20.0	0.003–3.3	<0.00005–12.0	—	—	—	0.036–11.0
Cabin Branch water	Seep	5.7–6.9	<0.00001–1.10	—	0.2–35.0	0.001–1.3	<0.00005–0.0028	—	—	—	0.39–13.0

NOTES: The water data are compared to the Virginia Criteria for Surface Water for public water supply and the EPA Maximum Contaminant Level (MCL) shaded in grey. Water values are highlighted in red if they exceed the EPA MCL or Virginia's surface water quality criteria for human health. Where applicable, data are reported as a range. Dashes indicate not analyzed or not applicable.
SOURCES: Valzinco data are from Seal and Hammarstrom (2002) and Seal et al. (2002). Cabin Branch data are from Seal et al. (1998).

BOX 2-4
Differences Between South Carolina and Virginia Gold Deposits

Although the climate and regional geology of Virginia and South Carolina are similar, South Carolina deposits are not a good geologic analogue for those in Virginia. The known ore bodies in Virginia occur along narrow, linear structures with little disseminated mineralization away from quartz veins. Several historic mines in the gold-pyrite belt, each owned and operated by a different company or individual, were developed along a single semi-continuous quartz vein that may have extended for a few kilometers along strike. Each mine may have involved a few open cuts at the surface, each measuring perhaps a few tens of feet wide and a few hundred feet in length, and to depths of several tens of feet. After the near-surface ores were depleted, underground mines were developed that followed the quartz veins, but these usually did not exceed more than several hundred to a thousand feet in total length, and rarely exceeded more than about 300–400 feet in depth. Examples of multiple mines that were extracting gold from the same quartz vein at different locations along strike included the Melville and Vaucluse Mines in Orange County and the London and Virginia and Buckingham mines in Buckingham County. In contrast, deposits in South Carolina consist of multiple parallel (i.e., en echelon) veins with disseminated mineralization between individual veins. The Haile Gold Mine, for example, consists of numerous quartz veins or lenses with disseminated gold, producing ore bodies that are wider and larger than those in Virginia (OceanaGold, 2022b; see Figure 2-11). The Vaucluse Mine, which was noted to have the most extensive surface workings of all historical mines in the gold-pyrite belt, occupied 200 acres (Park, 1936). This area is far smaller than the 4,552 acres comprising the Haile operation (OceanaGold, 2022b). Although the surface workings of both mines would constitute a fraction of the total acreage, the difference suggests that the footprint of the Haile Mine is very roughly 20 to 25 times larger than the footprint of the historical Vaucluse Mine.

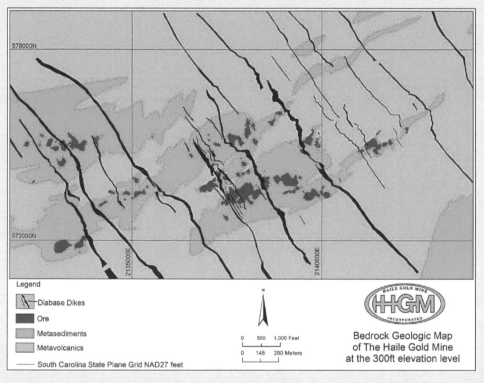

FIGURE 2-11 Map of Haile Gold Mine ore bodies. The open pits at Haile span across multiple en echelon veins that lie within a ~3.5 km by ~1 km area.
SOURCE: Image from Mobley et al. (2014).

been metamorphosed. As such, the Virginia gold deposits are thought to have originally formed as submarine volcanic massive sulfide deposits or submarine equivalents of subaerial epithermal deposits. Similarly, the Haile and other deposits in South Carolina are thought to have originally formed as epithermal deposits. However, both have subsequently been metamorphosed and the gold remobilized to form the orogenic-style type of mineralization observed today. These orogenic types of deposits are commonly associated with regional shear zones. For example, gold deposits in the gold-pyrite belt are associated with shear zones and the gold occurs in brecciated lodes, in veins, and disseminated in wallrock. This makes the known Virginia deposits comparable to a large number of intermediate-depth orogenic gold deposits (Goldfarb et al., 2005). As noted above, Virginia low-sulfide, gold-quartz vein deposits are found in close association with massive sulfide deposits (gold-pyrite belt) or with copper-rich base metal deposits (Virgilina district) and the gold in the gold-quartz veins may have been remobilized from those deposits. A similar source for gold has been proposed for the Mic Mac and Mooshla A and B deposits in the Doyon-Bousquet-LaRonde gold camp of Quebec, where Tourigny et al. (1989) suggest that the gold was remobilized from gold-bearing sulfides in the volcanic massive sulfide deposit and deposited in quartz veins to produce low-sulfide, gold-quartz veins. A similar occurrence is reported for deposits in the Nubian Shield region of northeast Africa, where volcanic massive sulfide deposits are comingled with and overprinted by orogenic gold veins, similar to what is observed in the gold-pyrite belt (see Figures 2-3, 2-4, and 2-5).

The mineralogy and geochemistry of the gold deposits in the gold-pyrite belt are characterized by post-metamorphic mineralization consisting of quartz-carbonate-sericite wallrock alteration, with common opaque minerals that include pyrite and iron and titanium oxide minerals. These are features that characterize orogenic deposits formed at shallow to intermediate depths of 5–12 km (Goldfarb et al., 2005). Minerals that are common to essentially all gold deposits in the gold-pyrite belt of Virginia include quartz, sericite (fine-grained white mica), pyrite, calcite, ankerite (Ca-Mg-Mn-Fe carbonate), and chlorite (often gives the rock a greenish color). In some deposits, biotite, garnet, tourmaline, and hornblende have been identified in the host rock adjacent to veins and lenses. Hydrous iron and manganese oxides are common in veins and lenses that are near the surface (above the water table). Chalcopyrite, sphalerite, and galena are often present in small amounts, and minor ilmenite, magnetite, tetradymite (Bi_2Te_2S), and arsenopyrite have been reported (Bass, 1940; Lonsdale, 1927; Pardee and Park, 1948; Park, 1936; Taber, 1913). As such, the mineralogy and geology of gold deposits in the gold-pyrite belt are comparable to that of a large number of orogenic deposits, including the Racetrack, Granny Smith, Mt. Charlotte, Golden Mile, Lancefield, Porphyry, Sons of Gwalia, Great Eastern, and Norseman deposits in the Yilgarn block of Australia (Goldfarb et al., 2005). Nevertheless, gold deposits in the gold-pyrite belt do not conform to all common characteristics of orogenic gold deposits. For example, many orogenic deposits show significant enrichment in arsenic, with arsenopyrite as a major sulfide phase, whereas arsenopyrite is rare in the Virginia deposits.

Finally, in terms of the size and scale of the deposits, the Kensington Mine in southeast Alaska is a currently active mine that is a good analogue. The Kensington Mine consists of several spatially associated ore bodies. The vein system in these ore bodies is a steeply dipping network of quartz extension veins and shear veins. Measured and indicated resources[13] at the Kensington Mine as of December 31, 2021, include 660,000 ounces measured and 323,000 ounces indicated, for a total of 983,000 ounces (Pascoe et al., 2022), which is approximately one to two orders of magnitude larger than those of the known deposits in Virginia.

SUMMARY OF FINDINGS

Several hundred gold mines and prospects have been documented in Virginia, with the large majority located in the gold-pyrite belt. The amount of information available about these sites is highly variable. For some, the documentation merely provides evidence that a small gold mining operation existed at some time in the past, often in the early to mid-1800s. Other sites have been documented in much greater detail. For example, the Vaucluse Mine in Orange County is described in several publications (Bass, 1940; Lonsdale, 1927; Pardee and Park, 1948; Taber, 1913), and drilling and exploration activities at the site in the 1980s provided additional data and samples. The lack of comprehensive information for most mines and prospects makes it challenging to understand some of

---

[13] An "indicated mineral resource" has a lower level of confidence than a "measured mineral resource."

the more detailed characteristics of Virginia's potential gold mining sites, especially given that most gold mining activity in the state ceased in the early to mid-1900s, before modern analytical techniques for obtaining detailed mineralogical and chemical data on host rocks and ores were developed. However, based on available information, the general characteristics of gold occurrences in Virginia are summarized below.

Most known gold occurrences in Virginia are associated with metamorphic and igneous rocks in the Piedmont physiographic province, except for a few small occurrences in the Blue Ridge province (see Figure S-1). New gold deposits are unlikely to be found outside of the Piedmont and Blue Ridge physiographic regions and are unlikely to be hosted in sedimentary rocks such as those that occur in the Triassic sedimentary basins within the Piedmont.

Gold deposits in Virginia consist of one or more lens-shaped quartz veins that dip steeply in the subsurface and that vary in width and grade. The large-scale plate tectonic processes that shaped the geology and geography of Virginia are responsible for the northeast-southwest orientation of Virginia's two gold districts—the gold-pyrite belt and Virgilina district. All known gold deposits in these districts are associated with shear zones in highly deformed and metamorphosed rocks. The gold-quartz veins characteristically "pinch and swell," and a given vein may decrease in width to only a few inches before widening to several tens of feet. In addition, the amount of gold per ton of rock, known as the grade, varies significantly along a given quartz vein, which introduces a large uncertainty into estimating the economic viability of the undeveloped portions of veins.

In many gold-quartz vein deposits in Virginia, total pyrite represents an average of 1 percent of the ore, and rarely exceeds about 4–5 percent. Most of the gold is contained within pyrite, which is captured during processing. This, combined with the presence of carbonate minerals in the veins, suggests that gold-quartz veins themselves are unlikely to cause substantial acid rock drainage (ARD). In contrast, Virginia massive sulfide bodies may contain more than 90 percent total sulfides. As demonstrated by the London and Virginia, Buckingham, and Williams Mines, massive sulfide bodies can occur in close proximity to gold-quartz vein deposits in Virginia, and could release acid rock drainage and metals if disturbed during mining.

The rocks surrounding the known gold deposits are composed of common rock-forming minerals, which are unlikely to release significant amounts of harmful metals or compounds to the environment, even in the presence of ARD. Minor phases that have been reported in gold ores themselves that could release harmful elements or compounds to the environment, especially during ARD, include galena (PbS) chalcopyrite ($CuFeS_2$), and arsenopyrite ($FeAsS$). No mercury-rich minerals are reported in any known Virginia gold deposits, but most historical mining sites show significant mercury contamination from historical gold processing methods.

3

Modern Gold Mining Operations

As the supply of gold from existing mines wanes, it is largely replaced by new geologic sources. Although recycling is a way of helping to meet demand for many materials like paper, aluminum, or iron, the amount of gold recovered from recycling does not meet current market demand. As described in Chapter 2, mineral resources do not occur everywhere. Unlike a factory or a manufacturing plant that has multiple options for building sites, mines can only be developed where the mineral resources are located. Because high-grade ore deposits exposed at the surface have largely been discovered and depleted by mining, mining companies are increasingly looking deeper in the Earth or at lower-grade deposits to maintain supply. Recent improvements in technology have also led to profitable production from previously uneconomic, mineralized deposits, or from remining of material that was historically considered "waste." For example, remining of waste rock is currently being considered at the Brewer Gold Mine Superfund Site in South Carolina.

The mere presence of gold does not mean that a mine will be developed. Mining companies must first consider the technical and economic suitability of a deposit for extraction and processing. Developing a new mine can be part of a long, complicated, and expensive endeavor that involves multiple stages, and mining may never prove feasible. Best practices for mine evaluation, planning, and operations use a life-cycle approach, where the entire life of a mining project (see Figure 3-1) is considered at the earliest stages. With these practices, a mine is designed for closure and reclamation from the beginning, and quantitative analyses of environmental and social impacts of all activities during the entire life cycle are considered (Farjana et al., 2021). Mines developed with a life-cycle approach—including meaningful engagement with stakeholders (see Chapter 5), rigorous consideration of local conditions and community concerns, and transparent and ongoing communication—have resulted in mutually beneficial projects (Schoenberger, 2016). In doing its work, the committee learned of projects where failure to consult communities resulted in costly delays, or even abandonment of the project after significant financial investment (Davis and Franks, 2011). The various stages in the life of a mine, from initial exploration to closure and long-term stewardship, are shown in Figure 3-1.

EXPLORATION, EVALUATION, AND DEVELOPMENT

Before a mine can begin operating, a mineral deposit must first be identified, evaluated, and then developed. The identification and development of a gold mine is challenging and complex and requires significant time and financial resources. Only 1 in 1,000 to 1 in 10,000 of prospected sites will become a productive mine (USFS, 1995; World Gold Council, 2022). Before a mine is operational, there are often years to decades of exploration, evaluation, and development (see Figure 3-2). In this section, we will describe these initial stages in the life cycle

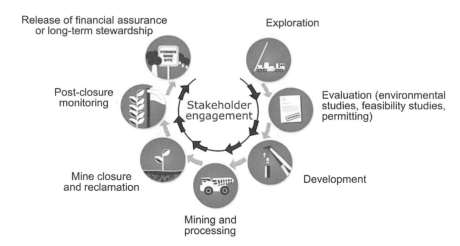

FIGURE 3-1 The life cycle of a gold mine. Stakeholder engagement is a necessary component during all stages of the life cycle and is shown in red, whereas technical, regulatory, and economic considerations by the industry are shown in black/gray. SOURCE: Image modified from Minerals Council of Australia (2014).

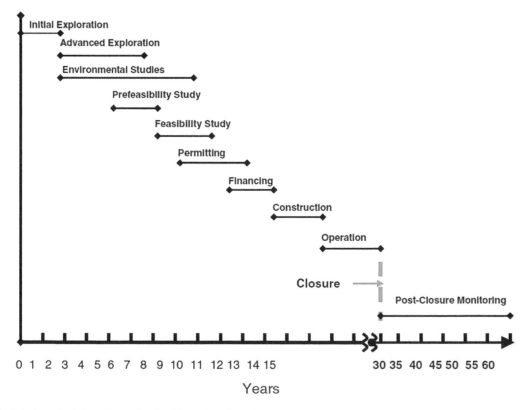

FIGURE 3-2 A typical time frame for the life cycle of a mine.
SOURCE: Image from Ramani (2012). Originally created by Tom Crafford, Alaska Forum on the Environment, Anchorage, Alaska (2008).

of a mine, assess the impact they have on the surrounding communities and environment, and consider any engineering controls that can mitigate the impact.

Exploration

Mineral exploration is a set of activities used to discover economically viable deposits. The methods employed for gold exploration are very similar to exploration for other metallic and nonmetallic mineral resources. Typically, exploration efforts for new mineable prospects are extensions of currently operating mines or nearby regions with similar geologic conditions. Historical mining records, as in the case of recent gold exploration activities in Virginia, are also an indication that viable mineral resources may exist. In the initial stages of prospecting, geologists will analyze regional maps, aerial photography, aerial geophysics, remote sensing imagery, and other available geologic information such as that presented previously in this report. Valuable insight into probable ore formation processes will drive the exploration plan for subsequent exploration data collection needs. A project proponent can begin acquiring environmental baseline data during exploration for permitting decisions and to establish the baseline against which to measure future environmental impacts.

After existing information has been analyzed, and land acquisition and/or permitted access has been established for the area of interest, geologists will examine rock outcrops, map major structures such as faults and folds, and collect detailed soil, stream sediment, and rock samples and other surface information that involves little or no surface disturbance. Exploration geochemistry—the measurement of chemical or chemically influenced properties of a potential ore deposit—can help outline major rock types and the extent of alteration patterns that typically accompany mineralizing systems (Jaacks et al., 2011). Geophysical techniques for estimating physical properties from surface measurements are another means of acquiring exploration information. Some of the techniques used to gather geochemical and geophysical data may involve little or no surface disturbance, and might include collecting small samples of rocks, soil, or sediment that are exposed on the ground surface. Other techniques may involve measuring the seismic, electrical, magnetic, or gravitational properties of the subsurface material with devices that are in contact with the ground surface or aboard vehicles or aircraft. Detailed information on geophysical prospecting techniques can be found in Dentith and Mudge (2014).

While major advances have occurred in remote sensing and geophysical techniques, drilling is the most reliable technique used to confirm or deny the presence of an ore deposit. Some information can be obtained from reverse circulation drilling, which returns only rock fragments, but the best data for assessing a mineral deposit are obtained from core drilling. Core drilling uses a special drill bit that excavates and preserves a tubular section of rock typically 2–3 inches in diameter (see Figure 3-3). The details of each drill hole are extensively logged for

FIGURE 3-3 Initial characterization of an ore deposit often involves core drilling. (A) A portable, truck-mounted core drilling rig and (B) core samples that have been cataloged and placed in storage boxes.
SOURCE: Images in public domain from U.S. Geological Survey (USGS, 2017b).

the rock type, structure, alteration, and ore minerals, and portions of the core are chemically analyzed to determine the location and grade of gold and to identify other associated minerals. Initial exploration drilling (typically core drilling) is focused on discovery and, if the results are favorable, is almost invariably succeeded by "definition drilling" that involves more closely spaced drill holes, to better define the resource. Larger-diameter drill holes may be needed for hydrologic testing and/or to obtain samples for engineering measurements and metallurgical testing. Depending on the deposit and material properties, trenching and excavation may be needed to obtain samples from relatively shallow depths (typically 3–12 feet) for geochemical and metallurgical testing. Discovering and defining a subsurface ore deposit, particularly complex ones that may have been faulted, folded, and deformed, based on a relatively small number of drill holes from the surface can be a very difficult task. It is common for an individual deposit to be explored by multiple companies with different exploration concepts before a potentially economic discovery is made.

Drill pads are sometimes constructed to make room for drilling equipment and, if appropriate access is not available, roadways for moving equipment may also have to be constructed. Depending on the methods and extent of drilling, shallow pits or sumps may be constructed in the drill pad to manage cuttings (fragments of rock) and water associated with drilling. The sumps may include a liner to contain the cuttings or water to control infiltration or sediment transport off-site.

After drilling has ended, best practices for the reclamation of the drill pads, sumps, and access roads include plugging drill holes with concrete or other approved materials, backfilling any excavations or sumps, water management measures, covering any drill cuttings, recontouring the areas so that soil and suitable vegetation can be reestablished, and postreclamation monitoring to ensure successful reclamation.

Evaluation and Development

While it may be technically feasible to mine a mineral deposit and process the mined rock to extract gold, it may not be economically feasible to do so. Therefore, before a mine site is developed, companies complete comprehensive mine project evaluations that include resource evaluation studies, feasibility studies, due diligence reviews, economic evaluations, and risk assessments. Once mineral resources have been estimated, feasibility studies assess processing, metallurgical, economic, marketing, legal, environmental, infrastructure, social, and governmental factors. Engineers determine the mining methods based on the geology, economics, regulatory restrictions, and community concerns. Whether mining engineers select surface or underground mining methods depends mainly on the depth, geometry, and grade of the deposit. Near-surface and lower-grade deposits are mined with surface mining methods while deeper and/or higher-grade deposits are generally mined with underground mining methods. The quantity and quality of analytical data (i.e., the level of confidence in the size and grade of the deposits), the mineralogy, the accessibility of the site, the climate, local resources and availability of supplies, infrastructure such as power and water, property access, permitting costs, and costs associated with environmental compliance can also have a profound impact on the cost of mining and processing and, hence, the economic feasibility. In cases where there has been previous mining or other disturbance, the evaluation includes characterizing past use, evaluating its effect on cumulative environmental impacts, and estimating the cost of mitigating any such impacts during mining and on reclamation. For example, if the mine is in an area with previous gold mining that resulted in mercury contamination, the evaluation should include soil and stream sediment surveys to characterize the extent of contamination and develop approaches to mitigate impact during mining and reclamation (Wang et al., 2012).

Mine development, or the process of constructing site facilities and the infrastructure to support the operation of the mine, typically follows or overlaps with the exploration and permitting phases. This infrastructure generally includes roads, offices, utilities, drainage control structures, processing plants, water treatment facilities, and waste disposal facilities.

Environmental Risks Associated with Exploration and Development

Impacts on the environment during initial exploration are generally minor, localized, and easily reclaimed, but advanced exploration methods and mine site development will be associated with potentially greater impacts. For example, closely spaced "definition drilling" requires more roads and drill pads than may be required during initial exploration, and the development of a mine site (e.g., roads, buildings) has an even larger disturbance footprint. These larger-scale activities must have expanded plans for mitigating impacts.

Potential impacts of exploration and development can include traffic, lights, and noise. These impacts can be variable, based on the hours of operation and the length and intensity of activity at the site. Although some of these impacts may be unavoidable, the activity schedule and anticipated conditions can be clearly detailed in permits that incorporate the reults of consultation with government agencies and affected communities. The equipment and materials used for exploration are typically mobile and easily removed from the site, leaving no permanent facilities or waste management.

While active exploration and development is occurring, the surface disturbances (e.g., roads, drill pads, trenches, material stockpiles) may be a source of air pollution including fugitive dust, which is comprised of particulate matter. The primary dust control method used by modern mines is water sprays in different forms to suppress dust emissions. Often, mines add surfactants to the water to improve wetting, which increases the capacity to control dust emissions.

Exploration and mine development can create disturbed areas that may increase sediment transport and stormwater runoff. Best management practices can successfully control runoff and reduce erosion during these activities, thereby mitigating potential impacts to surface water. These best management practices may include sediment traps or barriers (e.g., fences, liners, filters, straw or fiber products), diversion channels, landscaping or vegetative controls, and detention ponds or basins. It is best practice to identify and control all major sources, to make it a priority to control "at source," and to have proper runoff control (EPA, 2021c; Virginia Water Resources Research Center, 2022). For stormwater control, mines often build a series of drainage ditches to intercept runoff (particularly from haul roads) into sediment ponds that allow sediments to settle before the water is discharged. The development and implementation of runoff controls can also include response plans for potential spills or releases of process solutions, fuel, or other substances that might be used at the exploration site. These potential impacts are often short term, as exploration disturbances are easily reclaimed through backfill and recontouring, followed by the reestablishment of soil and suitable vegetation. Although there are limited peer-reviewed studies regarding impacts at mineral exploration drilling sites, some longer-term impacts to local vegetation have been observed (Chambers and Zamzow, 2019).

Surface water or groundwater systems might be locally affected by exploration drilling if the drill holes are not plugged or abandoned appropriately, or if any reactive minerals within the drill cuttings from the hole are not covered and reclaimed appropriately. Groundwater within open drill holes may be affected by lower-quality water or other substances entering the hole at the surface, although the extent of effects would likely be limited and controlled by the quantity of contamination, dilution, and properties of the aquifer. Conversely, artesian groundwater flowing out the open hole or runoff that encounters reactive geologic materials could affect any nearby surface water (e.g., pH, metals, other solutes), depending on the flow rates and chemistry of the mixing waters. Although the potential risks associated with these scenarios are likely minor and limited in extent, the risks can be reduced by using best practices for drill hole plugging and reclamation methods.

Exploration drilling does not require any pumping of groundwater, and drill holes are typically plugged after completion, so groundwater levels should not be substantially affected. Interaquifer mixing can affect water quality in certain hydrologic settings if improperly plugged drill holes intersect multiple aquifers. However, this is not a major concern in the majority of the Blue Ridge and Piedmont regions of Virginia, because most of the region is composed of crystalline-rock and undifferentiated sedimentary-rock aquifers where interconnected fractures host groundwater and mixing naturally occurs. Additionally, groundwater quality in the Piedmont and Blue Ridge regions is generally similar, so extreme changes in aquifer chemistry after mixing would not be expected (Trapp and Horn, 1997).

MINING METHODS

Engineers use many different methods to mine geologic deposits. Mining engineers select and design the mining methods based on the geology of the deposit and surrounding rock (i.e., "host rock"), the configuration of the ore deposit, regulatory restrictions, economic and technological feasibility, and social and community perspectives. The two main categories of mining methods are surface and underground mining, with different mining methods within each category. Mines use surface mining methods for near-surface deposits and underground mining methods for deposits that are too deep to access with surface mining methods. For the gold deposits of

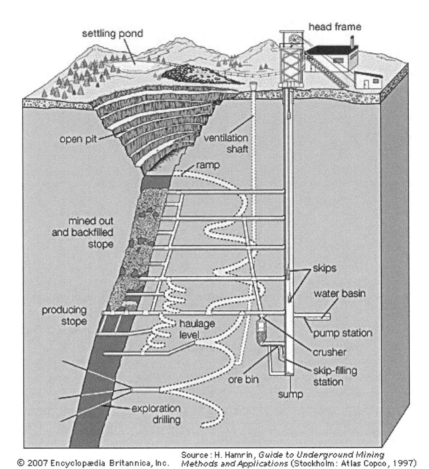

© 2007 Encyclopædia Britannica, Inc.   Source: H. Hamrin, *Guide to Underground Mining Methods and Applications* (Stockholm: Atlas Copco, 1997)

FIGURE 3-4 Different types of mining used to extract ore, including open pit and underground mining methods. Access to the underground may be via a vertical shaft (elevator) that allows personnel and equipment to be transported between the surface and underground, and allows ore to be brought to the surface. Access to underground workings may also be via a ramp, whereby personnel and equipment are moved between the surface and underground using wheeled vehicles (trucks, tractors, etc.) and ore is brought to the surface in haul trucks.
SOURCES: Image from Atlas Copco (2007) and Encyclopedia Britannica (2022a).

Virginia and the geology described in Chapter 2, the likely surface mining method is open pit mining while the likely underground mining methods are those suitable for vein-type gold deposits (see Figure 3-4), such as sub-level stoping, cut and fill, and shrinkage stoping (see the "Underground Mining Methods" section for definitions).

It is important to note that, in some instances, gold may be recovered as a by-product during mining of some other primary metal, and the mining methods used to extract the primary metal might differ from those discussed here. However, the methods discussed in this section represent the methods mining companies are most likely to use when extracting the gold deposits likely to be discovered based on the geology and deposit characteristics described in Chapter 2. This section of the report describes modern large-scale gold mining methods (and not small-scale[1] or recreational[2] gold mining, which may occur in Virginia) and their possible environmental impacts. Some descriptions of current and historical small-scale/recreational gold mining are discussed in Chapter 1.

---

[1] Small-scale mining can be defined as "low-tech, labor-intensive mineral extraction and processing carried out mostly by local people" (Hilson and Maconachie, 2020). We use this definition of small-scale mining to differentiate it from industry-scale, or commercial mining.

[2] Mining, often by a few individuals, primarily for recreation. This is often limited to panning for alluvial gold in streams.

Open Pit Mining

Open pit mining is a surface mining method in which the excavation can expand laterally and vertically. The method is characterized by mining a series of terraces, called benches (see Figure 3-5), with the ore transported by mining haul trucks or conveyor belts from the mine to the process plant or leach dump piles. Waste rock or any non-ore material that is encountered through mining is also transported out of the pit to waste dumps. The typical processes (the so-called unit processes) of open pit mining are ground fragmentation (to loosen the material and make it possible to excavate), materials handling, and auxiliary processes such as ground control and mine drainage.

In instances where the gold deposit is in weathered rock, ground fragmentation may not be necessary. In other instances where the weathered rock is slightly more difficult to dig, mechanical tools such as bulldozers may be adequate to "rip" the rock prior to excavating the material. However, in most cases, ground fragmentation is achieved by using explosives to break up the rock by blasting. To blast in an open pit, miners drill holes into the rock on a bench (flat surface) and place explosives in the drill holes. Modern blasting agents have almost entirely replaced outdated explosives such as nitroglycerine dynamite which was used from the mid-1800s to the mid-1900s. Blasting agents are classified as (1) dry blasting agents, (2) emulsions, (3) water gel, or (4) slurry blasting agents, which offer varying degrees of water resistance. The most common blasting agents contain from 70–94 percent ammonium nitrate fuel oil (ANFO) (Revey, 1996), which is set off with detonators and boosters. The use of explosives for blasting means mining companies may store and/or transport explosives to the mine sites. The manufacturing, sales, use, possession, storage, and transportation of explosives are heavily regulated by the Bureau of Alcohol, Tobacco, Firearms and Explosives and federal regulations (CFR Title 27) and are also subject to additional State of Virginia Regulations (4VACS25-40 Part VI) and the Virginia State Fire Marshal (§ 27–97 of the Code of Virginia). There are several best practices governing explosives handling and design criteria that guide engineers to ensure on-site worker safety and to minimize the possible impacts of blasting, which include vibration impacts on nearby structures, flyrock from blasts, dust emission, and noise (DOL, 2022; 30 CFR Part 56 Subpart E).

The most common way to load and transport ore and waste rock in open pit mines is via the use of excavators and mining haul trucks (see Figure 3-6). In some instances, the broken rock is crushed (see the "Gold Processing

FIGURE 3-5 Open-pit mine at the Touquoy Gold Mine in Moose River, Nova Scotia. The size of this open pit (approximately about 2,000 feet in the longest diameter) is the rough approximate size of open pit that would be used to mine gold from known deposits in Virginia, with deeper ore accessed by underground methods using a shaft or ramp (see Figure 3-4).
SOURCE: Photo by Simon Ryder-Burbidge.

FIGURE 3-6 A 64.6 metric ton haul truck. The truck contains large blocks of rock collected from the open pit after blasting and is preparing to dump the large rocks into a crusher that will break the rocks into smaller pieces for further processing. SOURCE: Photo in public domain from Brodowsky (2019).

Methods" section), if necessary, to place the mined material on conveyor belts for transportation. In some instances, mine haul roads cross public roads and require traffic control to ensure the safety of all road users.

Besides the main processes of ground fragmentation and materials handling, other activities are required to ensure the smooth running of the mining operation. For open pit mines, these include ground control (the process of ensuring that all excavations are stable, using techniques such as slope stability analysis, slope monitoring and maintenance, meshing, and dewatering) and hydrogeologic assessments. Depending on the hydrogeology, mine drainage systems are designed, built, and operated to collect water that flows into a pit, pump it out, treat where necessary, and discharge the water into the environment (some of the water is used in the processing plant). Mines operate under water discharge permits that stipulate compliance requirements for any water discharged by mines.

Underground Mining Methods

Deeper deposits in Virginia would most likely be mined via selective underground mining methods that would minimize the amounts of material that would need to be handled. There are many underground mining methods, but given the geometry and size of the vein-type gold deposits (i.e., the deeper deposits) likely to be discovered in Virginia (see Chapter 2), sublevel stoping, cut and fill, and shrinkage stoping are the most likely methods (Haptonstall, 2011; Pakalnis and Hughes, 2011; Stephan, 2011). See Figure 3-7 for illustrations of these methods. Sublevel stoping is a large-scale open stoping method sometimes referred to as long-hole or blasthole stoping. This method usually is applied to regular shaped and well-defined ore bodies that require minimal support as they are surrounded by strong country rock. Cut-and-fill mining is a highly selective method fit for steeply dipping high-grade deposits in weak host rock. Many variations of the general cut-and-fill technique exist and mining may occur in horizontal or vertical horizons. Cut-and-fill mining is more expensive than other mining methods because mined-out areas are completely backfilled. The backfill may simply be broken rock, but more likely it is a mix of cement and waste rock, a hydraulic sandfill, or a cemented paste fill (possibly made with spent mine tailings). Shrinkage stoping is most suitable for steeply dipping orebodies (70° to 90°) when the ore and host rock are reasonably competent. In shrinkage stoping, mining proceeds from the bottom upward, in horizontal slices with the broken ore being left in place as a work platform for the next level. Once the ore is removed, the stope may be backfilled or left empty, depending on the rock conditions. While the methods differ in some key respects, they have similar impacts on the environment.

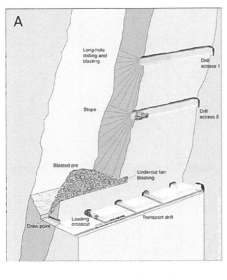

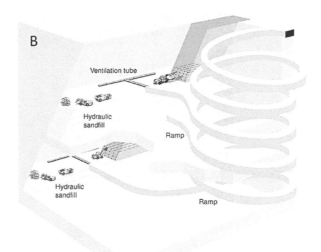

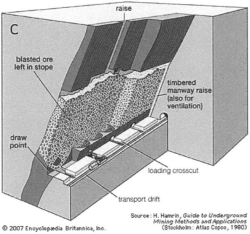

Source: H. Hamrin, *Guide to Underground Mining Methods and Applications* (Stockholm: Atlas Copco, 1980)

© 2007 Encyclopædia Britannica, Inc.

FIGURE 3-7 Different underground mining methods including (A) sublevel stoping, where broken rock produced by blasting material falls to a lower level where it is collected and removed from the bottom of the mine, (B) cut-and-fill operation, where an area that had previously been mined out (opened stope) is backfilled with waste materials from another area of the mine as ore is removed, and (C) shrinkage stopage, where broken ore is removed from the draw points at a deeper level in the mine. SOURCES: Images modified from Atlas Copco (2007) and Encyclopedia Britannica (2022a).

Similar to surface mining methods, underground metal mining methods involve ground fragmentation, material handling, and auxiliary processes. In underground mining methods, the rock is fragmented during drilling and blasting. Ammonium nitrate-based blasting agents, similar to open pit blasting, are used underground. However, because underground mining uses much smaller drill hole diameters, the use of pre-packaged explosives (in small diameter, flexible, water-resistant packaging) or a pumped emulsion or slurry are more common underground than the dry bulk ammonium nitrate commonly used in surface mining. Packaged explosives typically contain 45–75 percent ammonium nitrate, 8–28 percent sodium nitrate, and small amounts of aluminum (0–10 percent) (Dyno Nobel, 2022). Loading and hauling ore is generally completed using the "load-haul-dump" machine (see Figure 3-8) small haul trucks, or conveyor belts and skips (large "buckets" that move vertically through a mine shaft).

The main auxiliary processes for underground mining are ground control (the process of ensuring the stability of underground mine openings), mine drainage, mine atmosphere control (mainly ensuring good mine ventilation), and personnel transport. Ground control processes vary for the mining methods discussed here. There are engineering guidelines for the design and operation of these mining methods to ensure successful ground control. Owing to the

FIGURE 3-8 Load-haul-dump machine used in an underground environment.
SOURCE: Image from PJSC Gaysky GOK (2017).

thick regolith (weathered rock) in the Virginia Piedmont that can extend to depths of more than 150 feet in some areas (Pavich et al., 1989), it was historically difficult to maintain good ground control in shallow underground workings in the gold-pyrite belt. At greater depths where the rock is more competent, it is easier to maintain good ground control in underground workings. Mines also build and operate mine drainage systems to collect water that flows into an underground mine, pump it to the surface, treat where necessary, and discharge the water into the environment. Successful mine dewatering ensures the health and safety of mine workers as well as meeting water effluent requirements stipulated by water discharge permits for the protection of the environment. Mines use various techniques to control the mine atmosphere to allow miners and machines to work safely and efficiently. This mainly involves the use of fans and ventilation controls to provide fresh air to working areas and to vent exhaled air and exhaust fumes out of the mine.

Environmental Risks Associated with Mining

Various air pollutants can be generated from mining activities. Since mine roads are generally not paved, truck traffic can lead to significant dust generation on the haul roads. Dust emissions also come from the in-pit crushers that are used to reduce the sizes of the rock, conveyor belts as they transport the ore or waste, and transfer points (where one conveyor transfers material to another or to a receiving hopper or crusher). There are also dust emissions from wind activity over the large areas of exposed land and rock piles (waste dumps and stockpiles of ore). As described in the exploration section, the primary dust control method used by modern mines is water sprays in different forms to suppress dust emissions, but mines can also limit dust emissions by covering conveyor belts and promptly revegetating disturbed areas. Given the smaller surface footprint of underground mines and the wet nature of underground mines, the dust emissions from underground mines are much lower than those from surface mines.

In addition to dust, pollutants such as carbon monoxide, oxides of nitrogen, and volatile organic compounds will be emitted from combustion in fuel-burning vehicles and machines. They can be emitted directly from machinery in surface mines, or from ventilation systems that remove exhaust fumes out of underground mines. Current diesel and gasoline fuels contain low levels of sulfur and emissions of carbon monoxide, and volatile organic compounds and oxides of nitrogen can be reduced through the use of diesel and gasoline engines that meet the U.S. Environmental Protection Agency's (EPA's) Tier 4 emission standards (EPA, 2022l).

Individuals living near open pit mines are likely to experience some impacts from blasting, including vibrations and noise. Mines may monitor the noise from blasting and the schedule of blasting may be arranged to minimize impacts to the nearby residents. Mines will also often install seismographs between the mine and nearby structures to monitor for vibration and regulators use this information to monitor the mine's compliance. Mines use engineering design criteria (e.g., ISEE, 2011) to minimize vibration and other impacts of blasting such as flyrock. Because of variations in geology, these impacts can still occur in some circumstances.

Because mining requires the clearing of vegetation and the potential rerouting and changes to nearby waterways, it can result in significant habitat loss, which can lead to adverse impacts to biodiversity. Additionally, because the exposed land has no or limited vegetative cover, precipitation can lead to sediment loading of nearby streams. Similar to exploration, modern mines use various techniques including vegetative buffers, sediment traps, and sediment ponds to control sediment loads into nearby streams. For stormwater control, mines often build a series of drainage ditches to intercept (particularly from haul roads) and divert runoff into sediment ponds that allow sediments to settle. Water is tested before being discharged to ensure water quality requirements are met. The impacts of underground mines on biodiversity and stream sediment loading are similar to those of surface mines, although the scale tends to be smaller because the amount of disturbed surface land is less.

Surface and underground mining activities can also affect water quantity and quality. Large quantities of water may need to be pumped out of the ground if the open pit and underground mining horizon is lower than the groundwater table. This can affect the water table and the other users who rely on the same groundwater. The effects of dewatering depend mostly on the local hydrogeology and scale of the operation, and less on surface or underground methods employed. Mining can also lead to the creation of mine-influenced water, which is defined as any surface water or groundwater whose chemistry has been affected by mining or mineral processing (e.g., Durand, 2012; Mashishi et al., 2022). Mines use engineering controls, monitoring, water treatment, and contingency plans to ensure they do not violate their permit conditions that stipulate discharge requirements. Nevertheless, examples of alterations to water chemistry associated with mining include acid rock drainage (ARD), elevated concentrations of dissolved metals, and elevated nitrates associated with blasting activities.

As described in Chapter 2, the mineralogy of ores in the gold-quartz veins in Virginia is characterized by low total sulfide content and ubiquitous carbonate minerals (ankerite and calcite). These veins, by themselves, are unlikely to be significantly acid generating (Seal et al., 1998). However, if massive sulfide bodies are located nearby and are disturbed during mining of the gold-quartz veins, then acidic drainage could be generated. Accurate prediction of acid-producing potential through chemical and mineralogical characterization and geochemical modeling is critical to avoid potential groundwater contamination both during the active mining phases and after the mining and reclamation have been completed. Documents that describe best practices related to the prediction and treatment of acid rock drainage and metal leaching are numerous and include the Global Acid Rock Drainage Guide (INAP, 2014); documents from SME (Gusek and Figueroa, 2009; McLemore, 2008); the Mine Environmental Neutral Drainage Program (Price, 2009); as well as Maest et al. (2005), Paktunc (1999), and White and Jeffers (1994). Modern mines use several techniques to minimize the likelihood of ARD from sulfide minerals in the ore or waste rock, including alkaline addition to ore or waste rock before disposal, lining facilities containing potentially acid-generating material, or controlled placement within buffered material.

Another potential impact that mining may have on surface water or groundwater is the increase of nitrates from explosives used in blasting. Identification of blasting-related nitrates can be complicated as other sources, including agriculture and wastewater disposal, are also significant sources of nitrate (Degnan et al., 2016). However, poor blasting practices and/or poor hydrologic containment around a mine site can also contribute to nitrate loading via surface runoff and to groundwater. ANFO is not water resistant, so wet blastholes must be dewatered, and a blasthole liner or water-resistant product should be used to ensure proper detonation occurs. Proper detonation will ensure the blasting product is wholly consumed, thereby minimizing contaminant effects. Other operational best practices to avoid nitrate contamination from explosive use include loading the blastholes in a manner that avoids excess spillage; following loading practices and procedures for cleanup, disposal, or use of spilled product; documenting any voids, cavities, or fault zones encountered; detonating explosives as soon as is practical; and preventing unnecessary drainage off-site through properly designed runoff and stormwater systems.

GOLD PROCESSING METHODS

Gold has historically been classified as a "noble" metal, namely one that is relatively inert and unreactive. Geologically, it usually occurs in the "native" state—that is as elemental metal, rather than combined with other elements to form a gold-bearing mineral. Many modern mines around the world extract gold from ores in which native gold is finely dispersed, in close association with other minerals (such as sulfides). The gold in some of these ores is sufficiently fine-grained that it is invisible to the naked eye. In many such ores, the gold is encapsulated by pyrite. It is not practical to extract fine or encapsulated gold directly from the surrounding rock using traditional "gravity" methods (such as panning or sluicing) that rely on the fact that gold is much denser than the minerals with which it is associated. Instead, the ore, or a "concentrate" that is enriched in the minerals with which the gold is associated (such as pyrite), are treated hydrometallurgically. This involves leaching the gold, typically using dilute, alkaline aqueous cyanide solutions, followed by processes to recover the gold in the metallic, elemental state. The leaching agents capable of solubilizing gold cannot attack sulfides. Hence, if the gold in the ore or concentrates is encapsulated by sulfides, it must be oxidized before gold can be leached, unless it is feasible to grind it finely enough to liberate the gold.

As discussed in Chapter 2, the principal known gold deposits in Virginia are low-sulfide, gold-quartz vein deposits with about 1–5 percent pyrite. Volcanic massive sulfide deposits may also contain sufficient gold to warrant extraction, but base metals, not gold, would be the primary metal of material economic interest in Virginia massive sulfide deposits. Low-sulfide, gold-quartz veins would probably be relatively straightforward to treat and if the gold is coarse enough the veins may be amenable to concentration by gravity methods. In contrast, any gold in volcanic massive sulfide deposits is more likely to be encapsulated, or occur with various base metals that might dominate the processing route adopted. Because, as discussed in Box 1-2, the committee considered "gold deposits" to be those deposits where gold is a primary metal of material economic interest, the following discussion of gold processing does not delve into the metallurgical approaches that may be adopted to treat volcanic massive sulfide deposits.

Figures 3-9 and 3-10 are schematic, simplified flowsheets depicting two different processing routes deemed most likely for treating gold ores in Virginia. Figure 3-9 shows the route whereby ore is leached, in large,

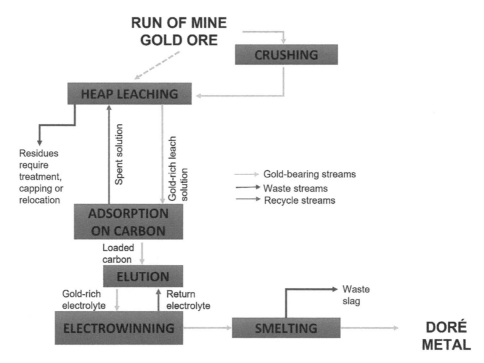

FIGURE 3-9 Schematic simplified flowsheet for processing of gold ore by heap leaching. Most run of mine gold ore—ore that has been mined but had no further processing—is first crushed and after agglomeration is stacked on heaps for leaching. In some mines, run of mine gold ore reports directly to heap leaching (dotted line). Following heap leaching, the leach residue is treated, capped, or relocated, whereas the "rich" gold-bearing leach solution moves to adsorption on carbon. The "spent," or gold-free, solution from adsorption on carbon is recycled to heap leaching, while the gold-bearing stream moves to elution and electrowinning. Following electrowinning, the electrolyte is returned to elution, while the gold-bearing stream moves to smelting to produce gold-rich doré metal.

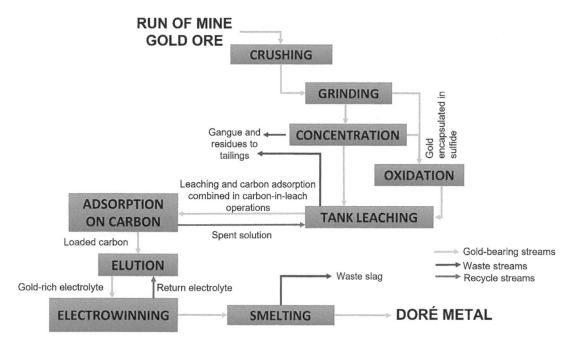

FIGURE 3-10 Schematic simplified flowsheet for processing of run of mine gold ore by tank leaching, with or without pre-oxidation of sulfide minerals or flotation. Run of mine gold ore is first crushed and ground, then may undergo gravity separation or flotation to yield a concentrate. The ore or concentrate is then leached in tanks, with or without oxidation. The gangue and residues from these processes report to tailings, whereas the gold-bearing streams move to carbon adsorption, which is combined with leaching in carbon-in-leach operations. The "spent," or gold-free, leach solution is recycled to leaching. The gold-rich carbon moves to elution and electrowinning. The electrolyte is recycled to elution and the electrodeposited gold alloy is smelted to remove impurities and produce a gold-rich doré metal.

open heaps. Figure 3-10 shows the route whereby ore or concentrate is leached in tanks, with or without preoxidation or flotation. The choice between these alternatives depends on the exact mineralogy of the ore, the size of the ore body, the availability of land near the mine, and the climate, among other considerations that would be explored rigorously during the planning stage. Although heap leaching requires careful water management in regions with high rainfall (Manning and Kappes, 2016), it was used at both the Brewer Mine and Barite Mine in South Carolina (see Box 3-1). Because of this relatively local adoption of heap leaching, it is discussed in this report. It is also possible, given the relatively small scale of many of the gold ore deposits in Virginia and the capital costs associated with building a processing plant, that a mining company would only produce a concentrate on site, using either gravity separation methods, flotation, or both, and then ship it elsewhere, including to another state or country, for leaching and gold recovery. This approach was adopted at the Jamestown Mine in northern California, with concentrates being shipped to Nevada for processing, as well as in the Melville gold mine in Orange County, Virginia, in the 1930s, with concentrates being shipped to the American Metal Company smelter in Carteret, New Jersey, for processing. Because site-specific factors determine the optimal processing route, the committee is not able to state at this time what might be more likely. The individual stages in Figures 3-9 and 3-10 are discussed in more detail below.

Crushing

Mined ore typically comprises a wide range of fragment sizes, from millimeters to meters. Regardless of the subsequent processing route, the ore must first be reduced in size. Crushing, which may be done in covered buildings, underground, or in an open pit to limit emissions of fine particles, is the first step in size reduction, called "comminution." Crushing involves the application of compressive forces to induce fracture. Jaw crushers,

BOX 3-1
Operations at South Carolina Mines

Although the climate and regional geology of Virginia and South Carolina are similar, South Carolina gold deposits are not a good analogue for those in Virginia (see Box 2-4). Importantly, the known ore bodies in Virginia occur along narrow, linear structures with little disseminated mineralization away from the quartz veins, whereas deposits in South Carolina consist of multiple parallel (en echelon) veins with disseminated mineralization between individual veins. This difference affects the mining method that would be used to extract the ore. The narrow, steeply dipping veins in Virginia deposits are more likely to be mined using shallow trenches or pits, followed by underground mining to access deeper ore. In contrast, the much greater thickness and areal extent of the ore zones at South Carolina deposits require large open pit mining methods. Several modern mines in South Carolina are described below, but because of the differences in shape and size, these mining methods used are unlikely to be adopted at new gold mines in Virginia.

Haile Gold Mine
- Mining methods: Gold ore is currently mined in open pits. Haile has proposed an expansion of its operations, including underground mining under its Horseshoe deposit, using sublevel stoping (USACE, 2022).
- Processing methods: Ore is crushed and ground, then undergoes flotation to yield a concentrate. After thickening, the concentrate is leached using cyanide in tanks, with gold undergoing carbon adsorption and elution, followed by recovery by electrowinning. Leach residue is treated to destroy cyanide before deposition in the tailings storage facility (USACE, 2022).

Brewer Gold Mine
- Mining methods: The modern mine used open pit methods to remove more than 12 million tons of ore, although the pit intersected underground workings from historical operations (EPA, 1994b).
- Processing methods: Ore was crushed, agglomerated with cement, and underwent heap leaching using cyanide. The leachate underwent carbon adsorption and elution, followed by electrowinning prior to smelting to form doré bars (EPA, 1994b).
- Closure: The operator backfilled pits with waste, and a limestone-filled drainage conduit was installed to drain the water expected to collect in the pit and increase the pH as a mitigation measure. In fact, water short-circuited this engineered structure; acidic drainage seeped through unmapped fractures in rock. Reclamation/closure planning did not begin until shortly before mining ceased (Candice Teichert, personal communication, 2022; EPA, 2022a). Thus, the Brewer Gold Mine stands today as an example of the importance of early mine closure planning—the mine is now a Superfund site, with average ongoing annual costs of $1.18 million to treat acidic drainage (EPA, 2021d).

Barite Hill Gold Mine
- Mining methods: Open pit operation from 1991 to 1994 extracted 64,700 oz of gold and 119,500 oz of silver (EPA, 2020a).
- Processing methods: Crushed ore underwent heap leaching with cyanide solution.
- Closure: From 1994 to 1999, reclamation activities sought to isolate acidic drainage from acid-generating waste rock into the pit lake. Surface water migration from overflows affected nearby streams. The site was abandoned in 1999. Treatment of the pit lake water and waste rock piles is anticipated in cleanup activities (EPA, 2020a).

Ridgeway Gold Mine
- Mining methods: Open pit operation from 1988 to 1999 extracted 60 million tons of ore (averaging 1 g/t gold and silver for a total of 1.47 million oz of gold and 900,000 oz of silver) and 40 million tons of waste rock (Duckett et al., 2012).
- Processing methods: Crushed and ground ore was leached by cyanide leaching in tanks. The leachate underwent carbon adsorption and elution, followed by electrowinning prior to smelting to form doré bars.
- Closure: Lime was added to high-sulfur waste and encapsulated in a clay-covered tailings management facility embankment. Waste rock was also mixed with lime, deposited in pits, and allowed to fill with water (Duckett et al., 2012).

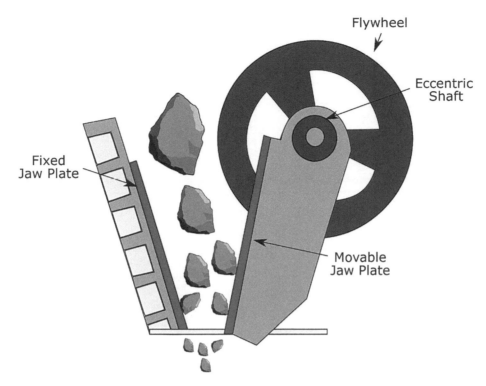

FIGURE 3-11 An image of a jaw crusher showing the movable and fixed jaw. The movement reduces the size of the material or "feed" that is introduced.

in which ore is fed into a tapering gap between two plates or jaws (see Figure 3-11), and gyratory crushers are most common for primary crushing, the first crushing step. Alternatively, primary crushing may use roll crushers, particularly high-pressure roll mills, where ore is fed continuously between two long rolls moving in opposing directions. Secondary crushers, usually cone crushers, produce particles that have approximately equal dimensions in all directions, while also reducing the size further to below about 15 mm.

Referring to Figures 3-9 and 3-10, if the ore is to be treated by heap leaching, comminution would generally end with secondary crushing. This achieves particles small enough that the leach solution and air can access all parts of the ore particle, and transport gold away, over the duration of treatment in the heap. If large amounts of fines are produced during crushing, they would be reformed into a larger size (agglomerated) prior to being placed on heaps, to ensure that they do not block leach solution flow paths through the heap. If the ore is to undergo concentration and/or oxidation ahead of leaching in tanks, it will proceed to grinding, the final stage of comminution, to achieve the desired particle size.

Grinding

Grinding is generally done in tumbling mills—drums that rotate continuously as a charge of ore, along with a grinding medium if used, passes through the mill (see Figure 3-12). The ore is generally supplied as a wet slurry, which helps control dust. Steel balls or rods are the most common grinding media; these themselves undergo grinding, particularly if the ore is hard, and must be replenished regularly. Some ores are hard enough that milling can be "autogenous," with no grinding medium. Most gold mines use semi-autogenous grinding (SAG milling), in which falling grinding media, in conjunction with ore particles, are responsible for attrition. SAG mills typically have a high aspect ratio (diameter/length), whereas ball and rod mills have a low diameter-to-length ratio.

Because of the uncontrolled nature of particle breakage in tumbling mills, there can be a large range of particle sizes in the discharge from a mill. The slurry would then be screened or passed through a hydrocyclone to separate

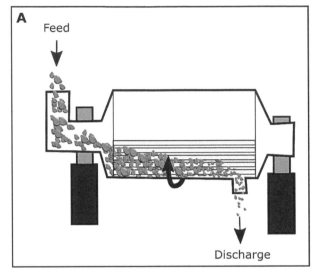
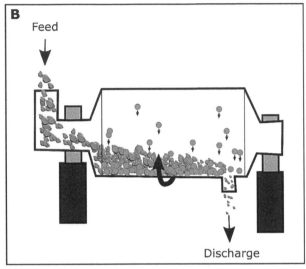

FIGURE 3-12 Cross-section of (A) rod and (B) ball mills.

material that is fine enough for the next stage from material that requires further size reduction. The latter is either retained or fed back through the mill for further grinding.

Concentration

Depending on the mineralogy and several site-specific factors, some gold ores would proceed directly to leaching after grinding. Others would be amenable to concentration, or could be leached more effectively as a concentrate. Gravity concentration routes, such as panning (historical), tabling, and centrifuging (Fullam et al., 2016) may be appropriate for ores with relatively large gold grains, and a mineralogy that allows effective liberation of gold from other minerals in the ore. This approach would be used directly after grinding. Overall, however, flotation is likely to be used for upgrading any ores in Virginia because of its efficacy, particularly for ores containing sulfides, such as the pyritic ores in Virginia.

Froth flotation is a physical separation method whereby particles of different minerals in an ore/water slurry are separated on the basis of differing abilities to adhere to the surface of rising air bubbles. Adherence is determined by the relative affinity of a particle for air or water. A hydrophilic (water-loving) surface will tend to remain immersed in water, while a hydrophobic (water-hating) surface has a stronger affinity for air, and hence will adsorb onto the surface of any air bubble that it encounters. A few minerals are intrinsically hydrophobic. More often, it is necessary to use various reagents to induce hydrophobicity in the mineral of interest, and hydrophilicity in the gangue minerals. With an appropriate surfactant that stabilizes the bubbles as a froth (a "frother"), the adhering particles will be retained in a froth phase above the original pulp and can be physically separated from the pulp. For flotation to be effective, the ore should have been ground sufficiently finely that mineral particles rich in gold (the "values") are physically "liberated" from mineral particles that have little to no gold (the "gangue").

Table 3-1 presents some of the reagents likely to be used in flotation of gold ores. In addition to frothers, hydrophobicity is induced by adsorption of reagents called "collectors." These are organic reagents containing a

TABLE 3-1 Reagents Used in Flotation of Minerals, with Examples from Gold Processing

Reagent Category	Function	Example Likely to Be Used for Gold	Typical Dosage
Frothers	Mild surfactants that stabilize froth above the mineral/water pulp, into which mineral particles being floated report	Polypropylene glycol (PPG); methyl isobutyl carbinol (MIBC)	60 g/t
Collectors	Organic reagents that selectively adsorb on the surface of the mineral to be floated, rendering it hydrophobic	Xanthates, especially potassium amyl xanthate (PAX), along with dithiophosphates and xanthate/dithiophosphate mixtures for free gold and sulfide ores; oily collectors such as thiocarbamates used for native gold	25 g/t
Activators	Reagents that facilitate adsorption of collectors on the mineral to be floated	Copper sulfate—very common; lead nitrate	100 g/t
Depressants	Reagents that adsorb selectively on the minerals that are not to be floated, to promote hydrophilicity	Uncommon when gold is being floated, but guar gum occasionally used; other reagents used when gold or gold-bearing sulfides are being depressed, but this is unlikely for Virginia deposits	
pH regulators	Reagents that maintain an optimal pH in the mineral pulp, to control the surface potentials of the different minerals and the dissociation/hydrolysis of other flotation reagents	Mildly alkaline conditions usually needed, achieved with lime additives	As needed to achieve pH of 8–10, usually 8–9

SOURCES: Dunne (2016), Kappes et al. (2013), and Teague et al. (1998).

functional group that specifically interacts with the mineral surface, along with at least one hydrophobic hydrocarbon segment that extends out from the surface onto which the collector has adsorbed. With a sufficiently high adsorption density of these molecules or ions, the mineral effectively becomes hydrophobic. Some minerals require "activators," usually salts whose cations adsorb onto the desired minerals and induce adsorption of the collector. In addition, it is sometimes necessary to use "depressants," which selectively adsorb onto gangue minerals, inducing hydrophilicity. Finally, the sorption of reagents often depends on the pH of the medium, so pH regulators may be needed to optimize flotation. It should be noted that when processing some complex ores, gold may be depressed while other sulfides are floated (Dunne, 2016). This approach might conceivably be used for processing complex ores containing copper and other base metals, along with gold, should the local mineralogy and processes to be used make it feasible and advantageous to produce both a copper and gold concentrate. In general, the quantity of flotation reagents is very low, especially compared with that of reagents used for leaching. The reason for this is that flotation reagents are used to modify only the *surfaces of minerals* in a flotation pulp, whereas leaching reagents must chemically dissolve the valuable material *throughout* the mineral particles. In addition, the process water with which minerals are mixed to create a pulp is usually recycled within the flotation circuit, retaining unused reagents. This minimizes both the cost of water and reagents and the amounts of process water that must be treated and discharged.

Figure 3-13 depicts a continuously operating flotation cell. Feed comprising ground ore mixed with recycled process water is admitted into the cell, where it is stirred to keep the ore particles suspended. Air is supplied to the bottom of the cell, where the stirrer creates small air bubbles. As these rise through the pulp, hydrophobic mineral particles adhere to the bubbles and report to the froth layer at the top of the cell. This layer is removed continuously, yielding a concentrate containing the desired minerals. The gangue minerals report to the tailings, which flow out from the base of the flotation cell. These tailings go to solid-liquid separation in "thickeners" where some process water is recycled and the remaining solid-water mixture goes to an impoundment. As noted above, it is possible that the processing at small mines ends with the production of a gold concentrate, which is transported elsewhere for leaching and gold recovery.

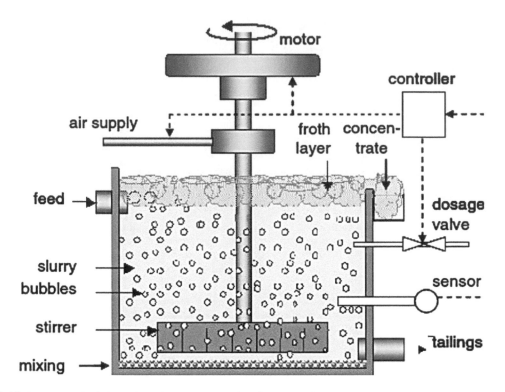

FIGURE 3-13 Schematic depiction of continuously operating flotation cell.
SOURCE: Image from Kramer et al. (2012).

Oxidation

Referring back to Figure 3-10, if the gold concentrate contains sulfide minerals that encapsulate gold particles, the concentrate is either ground finely enough to expose the gold, or is oxidized to destroy the sulfide minerals surrounding the gold. Either process ensures that the lixiviant for gold can physically access the gold particles that must be dissolved during leaching. Oxidation is typically completed using either roasting or autoclaving, both of which are described below. In addition, biological oxidation of sulfides has been used to treat gold ores at a limited number of sites globally. A fourth type of oxidation, using chlorine or hypochlorite, was used at a few mines in Nevada to treat the Carlin-type deposits in Nevada. These ores contain carbonaceous material that readsorbs gold after leaching, thereby significantly reducing gold recoveries, and oxidation of these ores is aimed at destroying the carbonaceous material. Such carbonaceous-rich gold ores are not known in Virginia, and chlorine/hypochlorite leaching is no longer used, hence this approach is not discussed further here.

Roasting

Roasting involves heating ore or concentrates in the presence of air to oxidize pyrite and other sulfides, creating iron (and other) oxides and sulfur dioxide, for example:

$$2FeS_2 + 5.5O_2 = Fe_2O_3 + 4SO_2$$ **Equation 3-1**

No other reagents are needed. Oxidation is typically done in fluidized bed roasters, at temperatures in the range 550–720°C. The reaction in Equation 3-1 is exothermic (meaning that the reaction generates heat), and it is common to blend ore or concentrate with additional pyrite if this is necessary to maintain the roasting temperature

(Hammerschmidt et al., 2016). Fluidized bed reactors create significant dust, which is captured using cyclones, then electrostatic precipitators. Because gold may be present in the dust, it then proceeds to gold leaching, along with the rest of the roasted material. It is evident from Equation 3-1 that large amounts of sulfur dioxide are generated during roasting—this must be captured, along with any other volatiles such as gaseous metallic mercury that may be released during roasting, and is usually used to make by-product sulfuric acid, or absorbed by lime or dolomite (Hammerschmidt et al., 2016).

Pressure Oxidation

Pyrite and other sulfides can also be oxidized by atmospheric oxygen (or oxygen-enriched air) in aqueous solution at temperatures in the range 180–225°C, which is much lower than the temperatures needed for roasting. As for roasting, the exothermic oxidation of pyrite provides the heat needed to maintain optimal operating temperatures, provided there is enough pyrite in the feed. Sufficient oxygen is admitted to ensure that sulfide is fully oxidized to sulfate. No other reagents are needed. Elemental sulfur, which forms under moderately oxidizing conditions and at lower temperatures, must be avoided because it encapsulates gold and reacts with cyanide during leaching to form thiocyanate, thereby increasing cyanide consumption and costs (Thomas and Pearson, 2016). The sulfate appears as sulfuric acid, not as a sulfate salt, because of the hydrogen ions generated during oxidation, as seen in Equation 3-2. If the ore or concentrate contains enough carbonate minerals, the acid is neutralized. Otherwise, the oxidized feed must be neutralized before proceeding to leaching, which usually requires alkaline conditions. The overall reaction occurring during acidic pressure oxidation is

$$2FeS_2 + 7.5\ O_2 + 4\ H_2O = Fe_2O_3 + 4\ H_2SO_4 \qquad \textbf{Equation 3-2}$$

Autoclaves (gas-tight, sealable pressure vessels) are needed to withstand the pressure generated by aqueous solutions at temperatures above the boiling point. Figure 3-14 shows a plan and elevation view of a horizontal, multicompartment autoclave typical of those used in gold processing. Each compartment is separated from adjacent compartments by a weir. The height of the weirs decreases in the direction of the slurry flow, thereby affording control of the overall residence time. Conditions within an autoclave are extremely corrosive. As a result, the pressure vessels are usually lined with a protective layer of lead, along with acid-resistant brick that contacts the slurry.

When pressure oxidation is done on concentrates, rather than whole ore, the amount of material passing through the autoclaves can be much lower, which may save capital and operating costs, because much of the gangue from the ore has been eliminated, provided concentration is efficient with minimal loss of gold to the tailings. However, whole ore has sometimes been treated by pressure oxidation. At the Homestake McLaughlin Mine in Northern California, and the Barrick Goldstrike Mine in Nevada, whole ore high in pyrite was, or continues to be, oxidized prior to cyanidation.

Biooxidation

An alternative approach to oxidation of sulfides that encapsulate gold is to use biooxidation by sulfide-oxidizing microorganisms. These microorganisms are present in the natural environment, where they are responsible for the weathering of sulfide minerals. Sulfidic gold ores may be biooxidized in heaps (the engineering and operation of which is discussed below), in which case the ore is agglomerated with an acidic solution containing appropriate microorganisms selected to withstand the operating temperature, which in turn would depend on the sulfide content of the ore (Brierley, 2016). Newmont Mining Company used this process at Gold Quarry, Nevada, from 1999 to 2006. Newmont discontinued biooxidation because the pyrite was needed to maintain the heat balance in their autoclave (Corale Brierley, personal communication, 2022). Alternatively, instead of oxidizing on heaps, oxidation can be done in stirred tanks. This process of bioxidizing concentrates in tanks was first used commercially in 1986 in South Africa and the largest plant currently utilizing this process is in Kokpatas, Uzbekistan (Metso:Outotec, 2022).

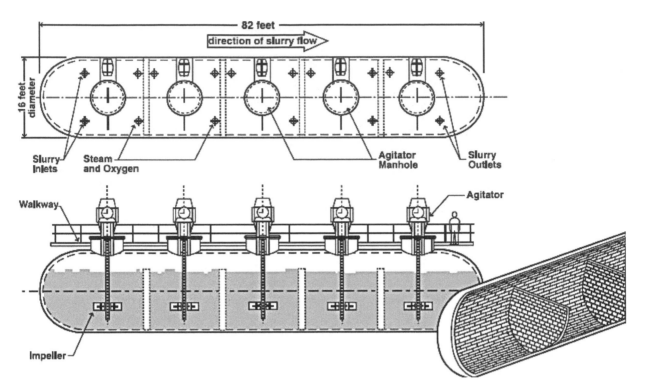

FIGURE 3-14 Horizontal autoclave used by Barrick Goldstrike, USA. Top: plan view. Bottom: side view. Inset shows brick-lined compartments separated by weirs. Preheated slurry is admitted to the autoclave at the left, and passes through a series of compartments, separated by weirs. Each compartment is stirred by an agitator. Oxygen and cooling water (or steam, if required) can be injected into each compartment to ensure appropriate thermal control and oxygen partial pressure. Slurry passes out of the autoclave, via flash vessels to reduce the pressure. Flash steam is used for heating.
SOURCE: Image from Thomas and Pearson (2016).

Leaching Agents (Lixiviants)

Any gold ores mined in Virginia would almost certainly be treated by leaching, although, as mentioned above, this need not be done at the mine site. While mercury was previously used in historic mines for gold recovery, it is no longer employed in modern commercial gold mining operations because of its toxicity, propensity to bioaccumulate in food webs, and persistence in the environment (see Box 3-2).

Gold is, of course, a noble metal. It is valued because it is extremely unreactive, and many uses of gold depend on this lack of reactivity. As such, it is difficult to dissolve gold in neutral-pH water or aqueous solutions at ambient temperatures—this requires an agent that can oxidize the gold, and another that can stabilize gold in aqueous solutions by complexation. Aqua regia, a mixture of concentrated hydrochloric and nitric acids, has traditionally been used to dissolve gold—in this scenario the nitric acid serves as a powerful oxidizer, while chloride complexes gold. However, this reagent would be impractical to use on the scale needed for processing gold ore or concentrates. Alkaline cyanide solution (pH 9.5 to 11) (Manning and Kappes, 2016) is by far the most common leaching agent (i.e., lixiviant) used for gold. Cyanide forms extremely stable gold complexes, allowing gold to be oxidized by atmospheric oxygen in relatively dilute cyanide solutions, typically 100 to 600 ppm NaCN (0.01 to 0.06 percent) for heap leaching (Bleiwas, 2012). Unfortunately, cyanide itself is acutely toxic to humans and non-human biota. In addition, it forms stable complexes with other metals such as copper, which significantly increases reagent costs should these metal ions be present in the ore (e.g., as sulfides such as chalcopyrite or sphalerite).

Because of the toxicity of cyanide, significant effort has been made to identify alternative lixiviants. The most widely researched of these are thiourea, thiocyanate, thiosulfate, and halides (iodine or bromine). None of these

BOX 3-2
Gold Recovery with Mercury

Although mercury is no longer used in modern gold mining operations in the United States, it is discussed here because earlier use in Virginia has left residual mercury at many historic mine sites that may be under consideration for remining. In addition, because mercury is used to recover gold elsewhere in the world, particularly at unregulated, artisanal mining sites, its devastating health impacts are widely known.

Amalgamation describes the process whereby gold (and many other metals) forms an alloy called amalgam when mixed with mercury. In the case of historic gold mining in Virginia, crushed ore containing fine-grained gold was often passed over a table in which mercury had been placed in the grooves; as the gold contacted the mercury, it was removed as amalgam that was collected for further processing. Alternatively, a metal plate coated with mercury was placed at the base of stamp mills where the crushed ore was released, and the gold particles were removed as the ore passed over the mercury-coated plate. To extract the gold from the amalgam, the amalgam was heated in retorts to vaporize the mercury, which was recovered by condensation. The process was inefficient, resulting in mercury contamination of tailings, along with air emissions of mercury which then contaminated soils and streams. Even today, one can sometimes find elemental mercury pooled at the bottom of Virginia streams and rivers in areas where amalgamation was used to extract gold in the past. The amalgamation process was used (following roasting) to extract gold at the Culpeper, Embry, and Morgana Mines (Culpeper County), Kelly Mine (later redesigned to use cyanide treatment), Fauquier County, the Melville and Vaucluse Mines in Orange County, and the United States Mine in Spotsylvania County, among others.

Many of the waste materials at historic mines contain relatively high concentrations of gold by modern standards. Before processing such materials, it is important that such materials are analyzed carefully to determine whether they contain mercury and, if so, its concentration. This would determine how best to safely process the materials and capture the mercury. For example, if a mine site were to produce a gold concentrate, it would be important that the mercury content of this be carefully characterized and disclosed to the entity processing the concentrate. Alternatively, if cyanidation were part of processing, mercury would be dissolved as a cyanide complex, and follow gold through the processing circuit. Care would be needed at all stages where mercury cyanide complexes could decompose and release mercury, and mercury would have to be removed ahead of smelting using modern retorts with scrubbers.

is as effective as cyanide; the complexes that they form with gold ions are orders of magnitude less stable, necessitating higher concentration of the lixiviant than used for cyanide (Aylmore, 2016). For example, Aylmore (2016) cites typical concentrations of 0.13M for thiourea, 0.01M to 0.05M for thiocyanate, and 0.05M for thiosulfate. More strongly oxidizing conditions are also needed, as seen in Figure 3-15, which shows the typical operating regions for different lixiviants on an Eh-pH diagram (also known as a Pourbaix diagram). The operating windows for Eh (a measure of oxidizing potential) for the noncyanide lixiviants are narrower than that for cyanide, which dictates closer engineering control during leaching. Thiosulfate and thiourea themselves also oxidize at the potentials needed to oxidize gold, which increases reagent consumption (Aylmore, 2016) and generation of decomposition products, all of which are skin irritants. Thiourea and thiocyanate are also toxic, and thiourea is a suspected carcinogen. Thiosulfate has been used commercially at Nevada Gold's Goldstrike Mine operation, using copper ions to catalyze oxidation (Aylmore, 2016). Despite extensive research, neither thiourea nor thiocyanate has proven viable enough for commercial adoption. Given the dominance of cyanide as a lixiviant for gold, the remaining discussion on gold leaching and subsequent processing focuses on cyanide solutions.

Heap Leaching

As discussed earlier, heap leaching is most commonly used in arid regions, because high rainfall levels necessitate careful water management. In the United States, and globally, stirred tank leaching produces significantly more gold than heap leaching. Nevertheless, because two gold mines in South Carolina (Brewer and Barite; see Box 3-1) used heap leaching, it is discussed first here. As shown in Figure 3-9, heap leaching would typically occur after crushing of gold ore. If the crushed ore contains significant quantities of fine particles, it is advantageous to agglomerate, or reform the fine particles into larger particles, ahead of leaching. This prevents localized

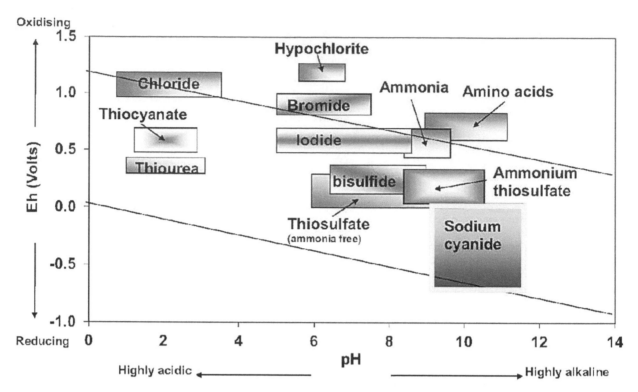

FIGURE 3-15 Eh-pH diagram showing typical operating regions for gold lixiviants. Eh is the oxidation-reduction potential. Note that sodium cyanide (highlighted in yellow) requires relatively alkaline (high-pH) operating conditions. The diagonal lines indicate the stability of water.
SOURCE: Image from Aylmore (2016).

accumulation of fines that impair flow of the leach solution through the stack. Prepared ore is stacked into layers called "lifts" on pads lined with impermeable material to prevent leakage, with leachate collection systems (see the section on tailings management, below, for more detail on liners and leachate collection systems). "Spent" leach solution, or solution that does not contain gold, is applied using drippers or sprinklers. The leach solution percolates through the heap, dissolving gold, and when it reaches the base of the lined pad it is collected and sent into a pond. The gold-bearing solution, called a "rich solution," is treated to recover gold (see below), after which the solution returns to the spent pond for reuse, after adjustment of its solution chemistry. Heap leaching has low capital and operating costs, but it is slow, with leach cycles typically spanning 45 to more than 100 days (Bleiwas, 2012). After leaching, additional ore may be added to create a new lift (see Figure 3-16A, which shows multiple lifts on a single pad). Occasionally, the ore is removed to a spent ore pile, as depicted in Figure 3-16C. Valley-fill heap leaching (see Figures 3-16A and 3-17) is used when there is little flat ground available for construction of flat leach pads. Instead, leach piles are constructed within valleys, with an engineered dam to retain ore during leaching along with the liners needed to retain gold-laden leach solutions. Given the typical landscape in the Virginia Piedmont, any heap leaching that may be used would most likely be on flat leach pads or valley fill, depending on the local topography at the site.

When ore is removed from reusable heaps, or heaps are decommissioned, the ore is rinsed thoroughly to remove residual cyanide. The rinse solution is reused as make-up water for leach solutions, so that the residual dissolved gold is recovered. During mine decommissioning, the rinse solution can be treated with carbon to remove residual gold, and the remaining cyanide is destroyed by oxidation, which can be accelerated using either chemical or microbial processes. Chemical treatment options include the use of hydrogen peroxide or alkaline chlorination with bleaching powder, sodium hypochlorite, or liquid chlorine (Dong et al., 2021).

Legend
→ Rich solution
→ Spent solution
→ Ore

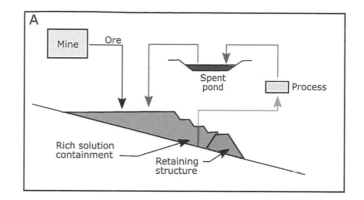

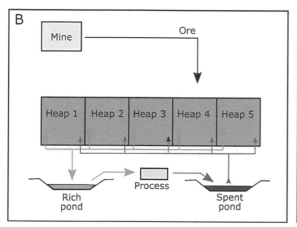

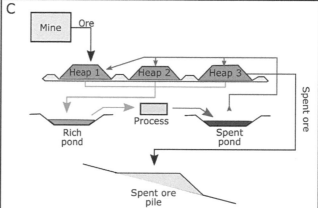

FIGURE 3-16 Schematic depictions of three major gold heap-leach methods: (A) valley-fill leach, (B) conventional leach pad, and (C) reusable leach pad.
SOURCE: Image from Bleiwas (2012).

FIGURE 3-17 (A) Expanding heap-leach operation with multiple lifts. (B) Valley-fill heap leach operation, with impoundment dam toward the bottom of the image.
SOURCE: Images from Bleiwas (2012).

FIGURE 3-18 Cyanide leaching in tank at gold processing plant.

Tank Leaching

As shown in Figure 3-10, tank or vat leaching would follow crushing, grinding, and flotation of ore, possibly with oxidation of the resulting concentrate to expose gold grains. Leaching is usually done at ambient temperature, and requires aeration to oxidize the gold, which is dramatically facilitated by complexing by the lixiviant. Generally, a series of tanks is used (see Figure 3-18) and a gravity feed controls the overall residence time during leaching. Agitation is usually provided with impellers, although Pachuca tanks, one particular design of tank, are engineered to provide agitation through the convective flow induced by forcing air into the base of the tank. In contrast with heap leaching, tank leaching has higher capital and operating costs, but is much faster, on the order of a day or so, and may yield higher recovery. As such, it may be economically comparable to heap leaching, depending on site-specific considerations.

Carbon (or Resin) Adsorption and Elution

Whether gold ore or concentrates have undergone heap or tank leaching, the resulting rich lixiviant solution is too dilute for efficient recovery of the gold by electrolysis, the most common method used now to recover gold (discussed below). Because of this, activated carbon is usually used to concentrate the gold, although synthetic ion-exchange resins have also been developed for this purpose. Carbon has a strong affinity for gold cyanide complexes and will adsorb these from relatively dilute rich leach solutions. Granular coconut-shell carbon or extruded carbon derived from peat are commonly used (Staunton, 2016); this carbon must have a high capacity for gold and be mechanically strong and abrasion resistant. When loaded, the carbon is then eluted in columns with an alkaline cyanide solution. Because the volume of eluent is much smaller than that of the rich leach solution, the resulting solution obtained by elution (eluate) is much more concentrated in gold, which can then be recovered efficiently, as described below.

In the case of tank leaching, carbon adsorption or ion exchange is usually integrated with leaching, in carbon-in-leach (CIL), carbon-in-pulp (CIP), or resin-in-pulp (RIP) configurations. Carbon in pulp is the most common of these, although CIL is increasing in popularity (Staunton, 2016). After some of the leaching has occurred, the pulp flows into a series of tanks, where it encounters carbon, which adsorbs gold from the leach solution. Countercurrent movement of the carbon and pulp streams ensures optimal utilization of the carbon, and optimal leaching. Resin-in-pulp operations are very similar, but use ion-exchange resin rather than carbon. In a carbon-in-leach process, carbon is present in all leach tanks.

After heap leaching, clarified rich leach solution flows through columns loaded with carbon, which adsorbs gold from the leach solution. After gold removal, the spent leach solution returns to leaching. When a column is fully loaded, it is eluted with eluent (a more concentrated cyanide solution that reverses the sorption reaction), thereby generating a (relatively) concentrated solution for gold recovery. The column can then return to adsorption mode after reactivation (Costello, 2016). Similarly, carbon (or resin) from CIP, CIL, or RIP processes would undergo elution before being returned to leaching. Various reagents have been used or studied for eluting resins. The selection of these depends on the nature of the resin, and the nature of any non-precious metals that may be present in the leach solution. The interested reader is referred to Kotze et al. (2016) and references therein for further details.

Gold Recovery

Returning to Figures 3-9 and 3-10, once an eluate concentrated in gold has been generated by elution of loaded carbon or resin, it proceeds to gold recovery. The crucial part of this stage is reduction of the dissolved gold to the elemental, metallic state, which results in its precipitation from solution.

Electrolysis

Electrowinning is the most commonly used method of recovering gold from eluate. The eluate is electrolyzed in cells in which gold is deposited on the cathode, while oxygen forms at the anode:

Cathode: $Au(CN)_2^- + e^- = Au + 2CN^-$ **Equation 3-3**
Anode: $4OH^- = O_2 + H_2O + 4e^-$ **Equation 3-4**

Good fluid flow and cathodes with high surface areas are needed to prevent concentration polarization (which would result in hydrogen forming at the cathode and low current efficiencies). Woven stainless steel cathodes are now common, with periodic pressure jetting to remove deposited bullion-sludge (Costello, 2016).

Merrill-Crowe Process

The oldest commercial method of recovering gold from solution, the Merrill-Crowe process, uses powdered metallic zinc as a chemical reducing agent. Being more reactive than gold, the zinc is oxidized, and dissolves as zinc cyanide complexes:

$2Au(CN)_2^- + Zn = 2Au + Zn(CN)_4^{2-}$ **Equation 3-5**

This process yields a powdery gold deposit, contaminated by undissolved zinc along with other contaminants. The deposit is filtered, dried, and usually proceeds to smelting. The Merrill-Crowe process does not require preconcentration of leach solutions from heap or tank leaching, but does require good clarification and filtration, which are costly. Although largely superseded by electrolysis (and not shown in Figures 3-9 and 3-10), Merrill-Crowe recovery remains useful for treating gold ores with relatively high silver or mercury content, or eluates that contain high levels of dissolved organic compounds (Walton, 2016).

FIGURE 3-19 Doré (gold-rich) metal being poured from a refining furnace.
SOURCE: Image by Schoemaker (2022).

Smelting

After metal recovery, the impure gold is smelted, typically in a crucible furnace, at high temperature with fluxes under oxidizing conditions to remove the nonprecious impurities, such as steel that may have come from cathodes in the case of electrolysis, or undissolved zinc after Merrill-Crowe recovery. These oxidized impurities report to a slag phase, which is immiscible with gold, and is tapped off during casting. If the gold-bearing sludge from electrolysis or Merrill-Crowe treatment contains mercury, this must be removed ahead of smelting in a retort furnace operating at moderate temperature, which vaporizes mercury. The mercury is collected in a condenser for sale or disposal, and the gases leaving the condenser are scrubbed using either a wet scrubber or sulfur-impregnated carbon (KCA, 2020). The residual liquid metallic phase is an alloy of gold with any silver called doré metal. This is cast into bars (see Figure 3-19), which are then transported to a specialist gold refinery.

Environmental Risks Associated with Modern Processing

The potential impacts associated with gold processing include noise generated by machinery, the emission of air pollution, and the potential release of chemicals to surface or groundwater. A properly designed and operating plant will control such emissions. As discussed above, process solutions are recycled as much as possible. Tailings and leach residues are treated to reduce cyanide concentrations before emplacement in tailings storage facilities. Nonetheless, despite such best practices, human error and extreme climatic events may lead to rare failures.

Air quality could be affected by crushing, grinding, roasting, and smelting, and requires controls to prevent or reduce release of fugitive dust, sulfur dioxide, and other gaseous elements such as mercury. These emissions are generally controlled through the use of water sprays to depress fugitive dust production, sulfur dioxide capturing

---

BOX 3-3
Case Study of Cyanide Release at Brewer Gold Mine Site, South Carolina

Brewer Gold Mine in South Carolina was operated until 1995. On October 28, 1990, an earthen dam failed at the Heap Leach Pad 6 holding/overflow pond. The 17.5 million gallon pond was double-lined and had a leak detection sump (EPA, 1991). Approximately 11.5 inches of rain fell on the site in the 24 hours before the dam failed and was likely the cause of the dam's failure (DOI, 1991). The breach of this earthen dam caused 10 million gallons of cyanide solution to be released to Little Fork Creek and Lynches River (EPA, 1991). Fish were killed in the Little Fork Creek and Lynches River for 50 miles downstream (EPA, 2022p). The holding pond was about 2,000 feet from the Little Fork Creek and 2 miles from Lynches River (Associated Press, 1990). Taxa richness and abundance of aquatic invertebrates were reduced for months downstream of the point of cyanide release, but other signs of recovery were beginning to become evident months after the spill (Shealy Environmental Services Inc., 1991). Numerous following studies completed between 1991 and 2004 by Environmental and Chemical Sciences Inc., Shealy Environmental Services Inc., and the South Carolina Department of Health and Environmental Control reported gradual improvement to near full recovery of the macroinvertebrate community (EPA, 2005).

---

systems, and other air scrubbers, such as mercury abatement control devices. In contrast to roasting that releases sulfur dioxide, autoclaving converts sulfide (in pyrite) to sulfuric acid, which does not become airborne.

Surface water and groundwater quality could also be affected upon engineering failure and spills via the release of cyanide or flotation reagents. The release of significant quantities of flotation reagents is highly unlikely given the low concentrations in which they are used; the committee is unaware of any references to cases where accidental release of flotation solutions caused environmental damage, but did find studies that indicated that many flotation reagents were nonlethal to fish at concentrations typically used (Fuerstenau et al., 1974).

Because of both its acute toxicity and historical cyanide release events (see Box 3-3), extreme care is needed in managing all materials and solutions containing cyanide. Cyanide could potentially be released from failures of leaching tanks, heaps, ponds or from accidental spills from CIP, CIL, or RIP circuits. Although cyanide degrades rapidly in the surface environment, it can persist for longer periods of time in complexed forms and in groundwater, and some of its degradation products can be problematic. Cyanide also forms volatile hydrogen cyanide when the pH falls below 9, which necessitates careful pH control, including neutralization of any acid that may have been formed during oxidation of sulfidic ores or concentrates.

Beyond engineering failures, another potential impact of cyanide is the possible exposure of wildlife, particularly migratory birds, to leach solution at the mine site during regular operations. This exposure is most likely from either spent or rich ponds, particularly in arid regions where there are few other drinking water sources, along with tailings facilities. Numerous strategies have been devised to deter wildlife (e.g., netting and fencing, noise deterrents) and are generally effective. Virginia hosts diverse wildlife including many species that are dependent on surface water, such as pond breeding amphibians, that might require modified types of deterrents to ensure their efficacy. Tank leaching has less potential than heap leaching to accidentally harm wildlife, because the leach solutions are stored in tanks in operating plants that are less accessible/attractive to wildlife.

Many mining companies have voluntarily been certified as compliant with the International Cyanide Management Code (2022) to ensure that they are following best practices to minimize risks to human health and the environment when using cyanide. The Cyanide Code was one of the earliest standards and certification programs developed for the mining industry and it has been successfully adopted around the world to help companies improve their safe management of cyanide, in order to limit the risks to human health and the environment. Standard of Practice 4.4 of the International Cyanide Management Institute (International Cyanide Management Code, 2021) recommends weak acid dissociable (WAD) cyanide levels of 50 mg/l or lower for exposure of birds, other wildlife, and livestock. This limit, based on evidence that WAD cyanide up to 50 mg/l is typically non-lethal to wildlife, applies to water in tailings impoundments, heap leach facilities, other open ponds and impoundments accessible to wildlife, including process solutions, open solution channels at a heap leach pad, as well as leach solution ponded on the surface of heaps due to poor infiltration. Participants unable to comply with the 50 mg/l WAD cyanide requirement are required to use deterrent systems, such as netting or bird balls. Participants in the Cyanide Code have their operations audited by an independent third party for certification and the audit results are made public. The audit considers the production, transport, handling, operations, decommissioning, worker safety, emergency response, training, and dialogue with stakeholders as part of its certification. As of 2021, the Cyanide Code has 358 global participants in mining, cyanide production, and cyanide transport. Around 80 percent of these participants have been successfully certified. The International Cyanide Management Institute reports that no catastrophic events have occurred at certified mines since the inception of the code in 2005 (International Cyanide Management Institute, 2021).

WASTE MATERIAL

All mining and processing operations produce some waste materials, which might be waste rock, soil, tailings, spent ore/leach residue, or some combination. Most operations will place any topsoil and soil suitable for reclamation into stockpiles for later use. Waste rock is typically stacked in large piles proximal to the mine site. In some instances, the waste may be segregated into piles of weakly mineralized rock that might warrant further processing in the future, pending development of new technologies or increases in the price of gold. Below are described engineering practices related to waste rock and tailings, as well as the potential impacts, and engineering controls to mitigate impacts.

Waste Rock

In open pit mining, the volume of waste rock can be considerable, while in underground mining the volume is relatively limited. Waste rock typically is piled and stored on the mine property in nonmineralized rock facilities, also known as waste dumps or waste rock facilities. These facilities are engineered and constructed in lifts with design that can be sloped and seeded to blend in with topography after mining is complete. Many operations will slope and reclaim waste dumps concurrently with mining operations. If the waste rock contains sulfide-bearing minerals, the waste rock piles may produce ARD and must be stored on an engineered pad to control seepage water (leachate) drainage. Alternatively, if carbonate material is present on site, careful blending of sulfide-bearing rock with carbonate (e.g., as limestone) will minimize ARD generation. If the waste rock is chemically stable (i.e., non–acid generating), it can be placed directly on the ground surface, and it may be used for various mine operations. Waste rock can be further processed and returned to an underground mine as backfill, or placed in an open containment pit. Such a containment pit also may have an engineered liner to prevent inflow of subsurface water and an engineered cover to prevent inflow of precipitation and surface water. Preventing the influx of water and oxygen reduces the risk of creating ARD that could be released into the environment.

Tailings

Following ore processing to recover the desired resources (gold in this case), the remaining solid waste (processed rock) is termed tailings. Depending on the character of the ore and the mechanical and chemical processing methods employed at the mine, tailings particles may range in size from fine sands to clay-sized grains. However, the majority of the particles typically are microscopic silt-sized grains. Geochemically, tailings commonly retain the chemical signature of the parent ore body, but, depending on the processing technique, they may also include some amounts of reagents, flocculants, or other additives used during processing.

Following processing, tailings typically are thickened (i.e., partially dried) to reduce the water content and disposed of in a tailings storage facility (TSF). Alternatively, the thickened tailings can be mixed with cement (and/or other additives) and placed in exhausted portions of underground mines, open containment pits, or in impoundments. The addition of cement and/or other binders helps to prevent groundwater contamination as well as oxidation and acid generation in pyritic tailings for both underground and surface placement. In addition, in underground applications, the tailings-cement mixture provides support to the opening that improves stability and allows ore-rich pillars to be excavated. The degree of thickening utilized for both above-ground and below-ground storage impacts the processing costs, the disposal techniques, and the long-term behavior of the stored tailings. Furthermore, the volume of tailings produced at the mine strongly affects the thickening technologies that can be practically and economically employed. In general, more energy usage and higher processing cost are associated with producing drier processed tailings. However, the benefits of producing drier processed tailings include reduced storage space requirements and, in arid climates, improved water balance. Today, decisions related to the degree of thickening also depend on corporate risk tolerance, as drier tailings may be more geotechnical stability and have higher hydrological resilience (as related to water balance changes and climate change).

As illustrated in Figure 3-20, processed tailings generally can be described (in order of increasing percent solids) as conventional or slurry tailings, thickened tailings, paste, and filtered tailings. Conventional slurry tailings typically have solids contents of 15–35 percent and can be transported via pipeline using standard centrifugal pumps. Thickened tailings typically have solids contents of 40–60 percent and for the higher solids contents require positive displacement pumps to transport the tailings via pipelines. Paste typically has solids contents of 75–80 percent and requires positive displacement pumps for transport. Filtered tailings typically have solids contents of 80–90 percent and require either conveyors or conventional haul trucks for transport. Pipelines and conveyors are subject to occasional breaks and leaks, but generally this type of damage can be readily identified with regular inspection and repaired with minimal impacts to the environment.

The types and characteristics of tailings storage facilities have changed significantly in recent decades. Historically, slurry tailings have been pumped and deposited hydraulically behind above-ground perimeter dikes or in natural topographically low-lying areas with little or no treatment to either the tailings or the foundation materials. However, in recently constructed tailings storage facilities, tailings often are (1) placed in lined (or occasionally unlined), above-ground engineered storage facilities; (2) placed in lined (or occasionally unlined), engineered pits; or (3) mixed with cement (and/or other additives) and placed as backfill in exhausted portions of underground mines (e.g., Lucky Friday and Stillwater Mines; Williams et al., 2007). In the first two cases, the tailings can be mixed with waste rock in various proportions prior to disposal. When placed above ground, various methods can be employed to construct the perimeter compacted soil or tailings. In addition, the design and construction of above-ground storage facilities depend on the consistency of the processed tailings. For example, filtered tailings may require only small perimeter dike structures constructed prior to tailings placement (i.e., starter dikes). When stored above ground, conventional slurry tailings, thickened tailings, and pastes require full-height perimeter dikes to provide stability. Mine operators commonly select the storage method based on a variety of factors, including the type of mine (open pit or underground), the volume of tailings produced at the facility, available space, available borrow materials, water-balance requirements (and climate), potential environmental impacts, potential for future remining, and costs.

FIGURE 3-20 Different types of tailings include conventional or slurry tailings, thickened tailings, paste, and filtered tailings, which are defined by percent solids. Examples of conventional and filtered tailings are shown in the images.
SOURCE: Image on the right modified from Vargas and Campomanes (2022). CC BY 4.0.

As illustrated in Figure 3-21, perimeter dikes used for above-ground storage can be constructed using downstream, centerline, or upstream methods, although combinations of these methods may be employed. In general, all of these construction methods can produce safe storage facilities, provided that high-quality engineering and operations/maintenance are provided consistently during design, operation, and closure. However, upstream construction methods typically involve the highest level of risk. In some areas worldwide (e.g., Chile [high seismicity] and Brazil [as a result of recent failures]), upstream-constructed TSFs have been prohibited by law. While all types of structures can be constructed and operated safely, all types of tailings dams also can fail, often as the result of engineering or operations oversights (Morgenstern, 2018). For example, in 2019 a TSF that utilized upstream construction methods failed at the Corrego do Feijao mine in Brazil (Robertson et al., 2019), while the 2014 a TSF that utilized modified centerline construction methods failed at the Mount Polley Mine in Canada (Morgenstern et al., 2015). Such failures can result in extensive environmental damage and potential loss of life (as discussed in Chapter 4). In both cases, failure was attributed to inadequate engineering.

Today, best practices often require new above-ground or in-pit tailings storage facilities to be lined using a combination of leachate collection (drainage) materials, geosynthetic liner materials, and low-permeability soils. Depending on the local requirements and the tailings geochemistry (or specific processing application), bottom liners may be "single composite" or "double composite" systems. Figure 3-22 schematically illustrates typical bottom liner systems.

Other best practices are provided by Morgenstern et al. (2015). This report analyzed one particular TSF failure and recommends implementing best available technology for TSF operations and closure, appointing an independent technical review board, better defining the role of regulators, and making improvements to professional practice. The report indicates that the best available technology for existing TSFs are to (1) eliminate surface water from the impoundment, (2) promote unsaturated conditions with drainage provisions, and (3) achieve dilative conditions by compaction (i.e., the granular material volume increases with increasing shear deformation, generally yielding high shear strengths). As noted by the Canadian Dam Association, every TSF is unique and has site-specific conditions and requirements that must be considered for any technical analysis, particularly for a breach analysis. For example, "Guidelines for dam breach studies are available for water retaining dams, but none of these guidelines deal with the hydrodynamic, geotechnical and rheological considerations specific to tailings outflows" (Canadian Dam Association, 2021). The technologies and best practices for tailings dam breach analysis are still evolving, but technical bulletins provide a consistent basis for analysis from the current state of knowledge.

Some mine owners are adopting (in whole or in part) new industry guidelines for tailings management, termed the Global Industry Standards for Tailings Management (GISTM; Global Tailings Review, 2020). These standards were produced through a collaboration of the International Council on Mining and Metals, the United Nations Environment Programme, and the Principles for Responsible Investment following the 2019 TSF failure at Corrego do Feijao in Brazil. Published in 2020, the GISTM lays out 6 topics, 15 principles, and 77 auditable requirements to reach this goal. These standards (1) address project-affected communities; (2) require operators to consider the social, environmental, and local economic context of a proposed or existing tailings facility; (3) aim to improve the design, construction, operation, maintenance, monitoring, and closure of a TSF; (4) standardize ongoing management and governance of a TSF; (5) cover emergency preparedness; and (6) require public disclosure of information while also protecting operators from the need to disclose confidential information.

Environmental Risks Associated with Waste Management

The potential impacts associated with waste material include the production of fugitive dust from the transport of waste material, the potential failure of tailings dams releasing toxic material downstream, and the failure to capture acidic, nitrate-laden, and metal-laden drainage from waste material. Of these, the potential failure of tailings dams is likely the least common, but most serious.

Azam and Li (2010) compiled and reviewed global tailings dam failures over the past 100 years and estimated a failure rate of 1.2 percent, which is very high compared to the 0.01 percent failure rate of conventional water retention dams. They reported that failures are most common in small- to medium-sized dams (15–30 m height) with maximum tailings volumes of more than 10 million cubic meters (Azam and Li, 2010). Recent studies have

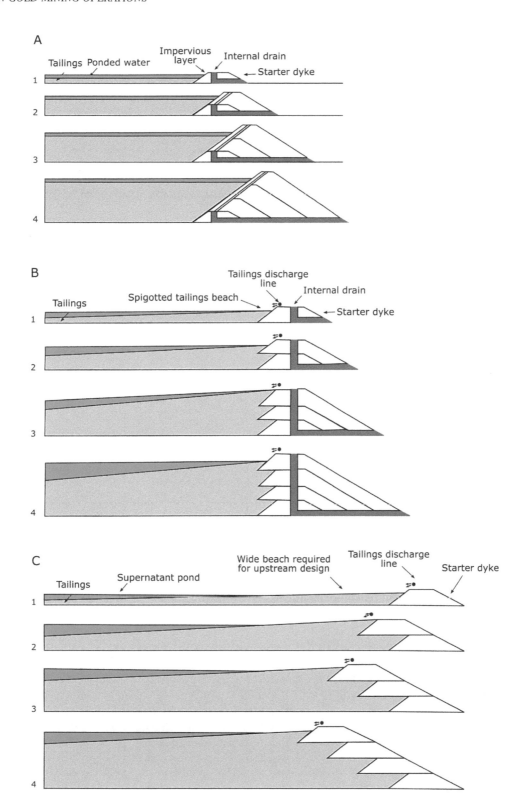

FIGURE 3-21 Above-ground tailings storage facility construction methods: (A) downstream, (B) centerline, and (C) upstream. Numbering of each schematic represents order of construction.
SOURCE: Adapted from Engels (2021).

Single composite liner system   **Double composite liner system**

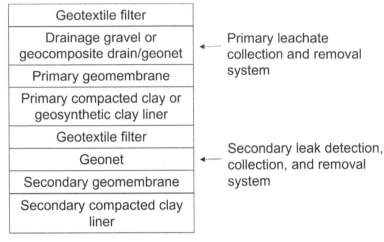

FIGURE 3-22 Single and double composite liner systems.

shown that failures generally are associated with inadequate engineering, operational issues, or a lack of communication between engineering and operations (Morgenstern, 2018). Some failures have occurred during more normal operations and have not been associated with extreme events, but others have been triggered by extreme events (earthquakes, heavy rainfall, etc.).

Earthquakes and seismic activity can damage mining infrastructure, including processing facilities and tailings storage facilities and waste dumps (Lenhardt, 2009). For example, two tailings dam failures at the Mochikoshi Gold Mine were associated with the 1978 Izu-Ohshima-Kinkai magnitude 7.0 earthquake in Japan (Ishihara, 1984) and tailings dam failures were also associated with earthquakes at the El Cobre Mine in Chile (1965), the Tapo Canyon Mine in the United States (1994), and the Kayakari Mine in Japan (2011) (Lyu et al., 2019). Probability theory can be used to quantify rigorously the magnitude of extreme events. The U.S. Geological Survey (USGS) uses historical and prehistoric earthquake events to project the seismic hazard associated with future events, typically in terms of the likelihood of a particular peak ground motion parameter associated with a particular return period, or the probability of exceeding a certain strength of shaking in a given period of time (Petersen et al., 2020; USGS, 2022a). The seismic hazard levels used for design commonly are related to the risk associated with a particular TSF or dam. These seismic hazard levels are an 80-year return period (50 percent probability of exceeding [PE] a particular level of shaking in 50 years), a 476-year return period (10 percent PE in 50 years), or a 2,475-year return period (2 percent PE in 50 years). As noted in Chapter 1, the largest magnitude earthquake in Virginia was the magnitude ~5.9 Giles County earthquake in 1897, although there is evidence for higher-magnitude earthquakes within about the past 3,000 years (Tuttle et al., 2015, 2021).

Extreme floods also impose a significant risk, not just for TSFs but also for holding ponds, heap leach pads, and waste dumps (see Box 3-3). Azam and Li (2010) identified that the most common cause of dam failure was "unusual weather," and it was stated that failures due to "unusual rain" increased from 25 percent pre-2000 to 40 percent in the time interval 2000–2009, leading the authors to suggest that dam design needed to better incorporate the effects of climate change. Another review of tailings dam failures, in this case restricted to Europe, found that the most common cause of failure was also "unusual rain" and found that 90 percent of failures occurred in active mines with only 10 percent in abandoned ponds (Rico et al., 2008). Furthermore, in some regions extreme weather-related events (precipitation, floods, etc.) may be exacerbated by climate change. Failure of some dams has been linked to the drying-wetting seasonal variation, but few account for the potential impacts of climate change, which will likely impose higher risks than anticipated. The National Weather Service of the National Oceanic and Atmospheric Administration (NOAA) uses historical meteorological and hydrologic data to project probable maximum precipitation (PMP) events (NWS, 2022). It has been recognized that many of the current PMP estimates require updating. Congressional bills providing the necessary funding have been proposed, but have not passed (e.g., House Report 1437 and Senate Bill 3053). Recently, a National Academies of Sciences, Engineering, and Medicine consensus study sponsored by the National Oceanic and Atmospheric Administration was announced that will "consider approaches for estimating PMP in a changing climate, with the goal of recommending an updated approach, appropriate for decision-maker needs."

WATER MANAGEMENT

Waters that require management during mine operation include naturally occurring water, precipitation, and waters that are used for processing, called "process water." Management techniques include the diversion, acquisition, recycling, treatment, and eventual discharge of water.

Naturally Occurring Water and Precipitation

The diversion of naturally occurring springs, seeps, and surface water features around mine disturbance areas is sometimes necessary during development and operation. Other management techniques for naturally occurring surface water, groundwater, and precipitation may involve the removal of water from within or near pits, as well as from tailings storage facilities or underground mines. Water management methods for impoundments or ponds could include the construction of spillways, decant structures that operate via gravity (see Figure 3-23), or barge systems that use power to pump out water. In some instances, forced evaporation is utilized to control water levels from precipitation. Methods for forced evaporation can include spraying water into the air or routing it to shallow

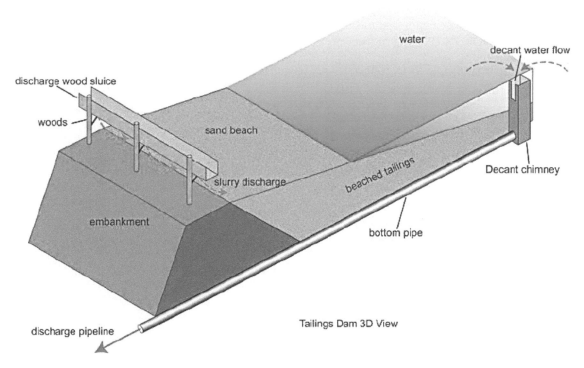

FIGURE 3-23 Internal structure of a typical tailings pond, which includes a decant structure or chimney, which allows water to be pumped out of the chimney via gravity via a buried discharge/button pipeline.
SOURCE: Image from Martinez-Pagan et al. (2009).

evaporation ponds. Spraying water into the air is currently a utilized method at Haile Gold Mine tailings pond in South Carolina, but generally this technique may be less effective on the humid East Coast than it is in arid settings.

Process Water

Depending on the operational needs for water use and the conditions allowed under different regulatory systems, some mines might use surface water or groundwater to support mining and processing activities. These waters then become process solutions, which are typically regulated and managed separately. These solutions are generally recycled for reuse during processing, to optimize utilization of reagents. Nevertheless, in humid climates such as Virginia where precipitation can dilute process water, the excess water may sometimes need to be eliminated to retain the necessary chemistry in process solutions. This may be achieved by forced evaporation, or by treating and discharging excess process water.

Treatment of Water

All waters must be treated prior to discharge from the mine when contaminant concentrations exceed permissible levels. These permissible levels often consider the naturally occurring or "background" concentrations of some elements, or parameters such as pH, salinity, or biological oxygen demand. The treatment methods that may be appropriate to comply with water quality requirements are site-specific factors, depending on the potential contaminants and concentrations that occur at the site. Contaminants might include metals and metalloids, acidity, high total dissolved solids (including dissolved sulfate), sediment loads or total suspended solids, nutrients (e.g., nitrate/nitrite, ammonia, or phosphorous, which can be derived from blasting agents or the degradation of reagents like cyanide), and limited amounts of organic contaminants (e.g., fuels, lubricants, and reagents). The appropriate treatment methods depend on the variety and concentrations of the contaminants of concern, for example, pH neutralization and precipitation to remove metals, biological or chemical treatment for nitrate or cyanide, ion exchange, filtration, reverse osmosis, or adsorption media for metalloids and other solutes (EPA, 2014c). Best practices related to the prediction and treatment of acid rock drainage are numerous and include the Global Acid Rock Drainage Guide (INAP, 2014), documents from SME (Gusek and Figueroa, 2009; McLemore, 2008) and the Mine Environmental Neutral Drainage Program (Price, 2009), and Maest et al. (2005). Passive treatment techniques like engineered wetlands (discussed below in the section "Reclamation, Closure, and Monitoring") may also be appropriate for treating modest flows of ARD.

Discharge of Water

Treated water that is discharged (sometimes called "effluent") must meet the applicable standards established by state and/or federal regulatory agencies for the approved method of discharge or disposal. Discharge or disposal methods could include "discharging the water into a stream channel, storing the water in a lined impoundment for future use, conveying the water to an unlined impoundment that serves as an infiltration gallery or ground water recharge facility; or, re-injecting the water through a well" (Virginia DMME, 2011). However, the last option would be administered by EPA through the federal Underground Injection Control program and is unlikely to be permitted in the fracture-controlled aquifers of the Virginia Piedmont. The methods of water discharge are typically coordinated with other state or federal permits to protect water quantity and quality (e.g., National Pollutant Discharge Elimination System/Virginia Pollutant Discharge Elimination System permit; see Chapter 5). Monitoring the resulting treated water quality and the discharge or disposal infrastructure are critical aspects of the water treatment plans.

Environmental Risks Associated with Water Management

The diversion or depletion of surface water sources or removal of groundwater from mine areas could have impacts on downstream or down-gradient areas (see Chapter 4) unless flows are returned to the systems from which the water was diverted or collected. Risks to water quality may result from inadequate facility designs, management, or treatment strategies, and/or implementation that can lead to release of water or process solutions. Water quality impacts can occur from seepage from impoundments or ponds, storm runoff or infiltration on disturbed surfaces or waste disposal areas, reactive mineral surfaces within underground mines, or open pits that affect adjacent groundwater. Additionally, water quality impact may occasionally occur from loss of containment from impoundments, ponds, tanks, or other retaining structures, which may be triggered by seismic or storm events. As discussed in more detail in Chapter 4, the extent, severity, and duration of water quality impacts are related to the type of contaminants and concentrations that may be released.

Water treatment facilities may produce waste by-products or residual solutions that must be disposed of or managed properly to prevent further environmental impacts. This may occur inside an on-site repository or at a licensed off-site facility. The waste by-products could include high- or low-density sludges and precipitates, brine or flushback solutions from filtration or reverse osmosis methods, and spent treatment media from filtration, adsorption, or ion-exchange methods.

RECLAMATION, CLOSURE, AND MONITORING

Decades ago, little thought was given to how a gold mine would be closed when it was being designed. Today, there is widespread recognition by regulators and operators alike that a gold mine must be designed for reclamation and closure from the beginning. Reclamation is defined as the "restoration or conversion of disturbed land to a physically and chemically stable condition that minimizes or prevents adverse disruption and the injurious effects of such disruption and presents an opportunity for further productive use if such use is reasonable" (§ 45.2–1200 of the Code of Virginia 2022). There are examples throughout the mining industry of innovative approaches to such planning, wherein the mine is designed and operated to enable specific postmining land uses, including commercial, recreational, residential, species habitat, and renewable energy facilities (Wheatley, 2020). Stakeholder involvement is a key aspect of such planning. When practicable, best practices should include reclamation that is concurrent with mining activities in order to reduce the length of time that disturbances are exposed, demonstrate the efficacy of reclamation techniques, and expedite the completion of reclamation activities at the end of mining. However, mine production may be intermittent during the life of an operation due to economic or other factors. In lieu of permanent closure and reclamation, such sites may be managed in a state termed "care and maintenance," where environmental and safety standards are met but the site is not yet fully reclaimed. For example, as gold prices fluctuate, a drop in the price may make the active operation of a mine uneconomic. If the price is likely to increase again in the future, the operators would want to be able to resume operations following a temporary cessation.

An essential component of project design and permit applications is a reclamation and closure plan that includes engineered drawings, designs, and plans detailing how reclamation and closure will be accomplished. The plan generally documents standards and a postclosure monitoring plan for measuring reclamation success. These plans are periodically updated to reflect the changes that inevitably occur as a mine operation changes over time. Fundamentally, an effective reclamation and closure plan is of sufficient detail to allow a third-party contractor to step in and complete reclamation and closure if the project operator is unable to do so. Because personnel and project ownership may change over the life of the project, the plan also serves as a record and guide for achieving reclamation and closure success for those successors. The potential scenarios for gold mining in Virginia in the previous sections suggest that the following aspects may be important in reclamation of such operations.

Recontouring

The goal of recontouring is to return the land surface to premining topography, or to create a new, stable topography that facilitates the planned postmine uses. Areas to be recontoured may include portions or the entire open pit, waste rock disposal areas, other fill features, ripped/reclaimed road surfaces, or disturbed slopes. The designs for recontouring and potential cover systems are based on the geotechnical and geochemical characteristics of the materials within the feature to be reclaimed. Long-term stability of waste piles is often ensured with the prohibition of steep slopes. Long uninterrupted slopes require drainage control structures to accommodate surface

FIGURE 3-24 Highwalls remaining adjacent to backfilled, contoured, and vegetated surfaces at the Zortman-Landusky Mine, Montana.
SOURCE: Photo courtesy of Wayne Jepson.

water and minimize surface erosion, and permanent vegetative or rock covering for stabilization and protection. Pit highwalls (see Figure 3-24) constructed in nonreactive bedrock are likely much more stable than unconsolidated waste rock or fill materials described above. If the approved postmining land use would allow highwalls to remain at closure, then the highwalls may remain uncontoured at a steeper angle than unconsolidated material.

Recontouring activities may also include grading tailings surfaces or embankment slopes during the reclamation of an impoundment, as long as the established slopes are consistent with acceptable designs and do not impact the stability of the facility. For example, an impoundment may have excess fill or soil material stockpiled along the embankment crest, so that the material can be pushed down the slope during reclamation in order to achieve the final angle. This is in contrast to cutting material out of the embankment itself to achieve a different slope at closure, which is a method to be avoided as it could potentially jeopardize the function and stability of the embankment.

Cover Systems

Following the recontouring steps, the reclamation of waste rock and/or tailings disposal areas typically includes dry or wet cover systems. To achieve long-term stability and limit water quality impacts, the best practices for covering waste materials should reflect any special handling or segregation methods that occurred during mining (described above). Cover system designs are based on the geochemical properties of the contained waste materials

Cover liner system

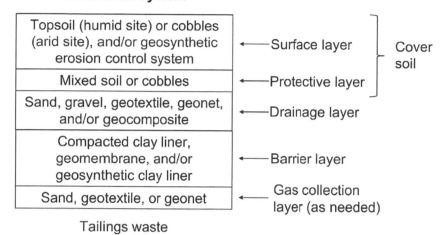

FIGURE 3-25 Cover liner system for tailings waste.
SOURCE: Adapted from Daniel and Wu (1994).

and the need to limit the flux of oxygen and water into the materials in order to reduce their potential reactivity and leachability (INAP, 2014). Mines often rely on "dry covers" of soil layered with low-permeability, natural (e.g., clay) or engineered materials (e.g., synthetic liners), while also utilizing evapotranspiration from established vegetation to consume water. The "dry cover" system concept could be applied to the reclamation of waste dumps, tailings disposal areas, features constructed with fill materials, and other disturbed areas around the mine through the use of covers made with nonreactive material (e.g., inert overburden, soil, or other approved growth media). The overall thickness of the cover system and the complexity of any internal liners or capillary barriers depend on the characteristics and reactivity of the underlying material being reclaimed; that is, a thicker cover and barriers would be needed for increasingly reactive material. Figure 3-25 schematically illustrates a typical cover system used for tailings waste, which can be quite reactive. This cover would be applied following a number of years during which the tailings settled, increasing their density and lowering the water content.

The climate is another major factor when selecting closure cover systems. Consideration is given to annual precipitation and potential evapotranspiration rates. In arid climates, a thick soil layer and vegetated cover may be adequate to prevent water infiltration into underlying material. In contrast, wetter climates like Virginia may require a low-permeability synthetic liner, a compacted clay layer, organic covers (a layer with chemically reducing conditions to remove oxygen), or other type of barrier system to be installed between the waste material and the upper soil and vegetation layers to adequately prevent deeper infiltration of water and oxygen (see Figure 3-26). This generally requires a cover material capable of shielding the acid-producing material and supporting plant cover. Dry cover systems are not static and may need to be monitored during postclosure. Complex multilayer covers may be compromised by tree roots that penetrate the cover layers, especially if the roots rip through the layers when trees are toppled during storm events.

The decision to cover a disturbed area with solid material at closure ("dry cover") or to allow a "water cover" system to form is evaluated based on the site-specific waste management goals and the hydrologic, geotechnical, and geochemical conditions of the disposal area and the potential fill. In wet climates such as Virginia, wet closure or "water cover" systems may be considered instead of dry closure for some facilities. Wet closure methods are more common for the disposal of tailings than waste rock piles, unless the waste rock has been placed as backfill into pits or underground voids. Wet closure could include a permanent pond or wetland feature within an impoundment or other disposal area, or the in-pit storage of waste rock or tailings beneath a pit lake. The goal of this method is to keep the waste rock, highwall or underground mine surfaces, and/or tailings in a saturated, low-oxygen environment. This approach may require stratification of the water body (e.g., Flite, 2006), which can be

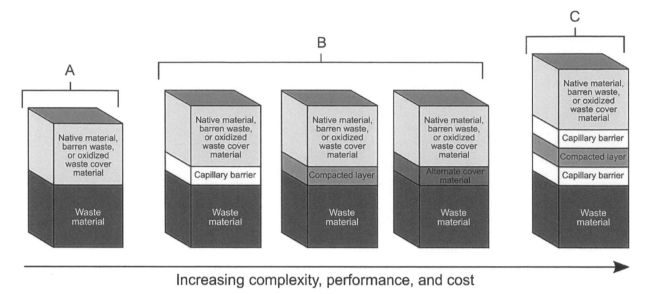

Increasing complexity, performance, and cost

FIGURE 3-26 Sample soil cover designs. (A) The base method of reclamation is a waste material capped by unreactive material. More complex variations are where (B) capillary barriers, compacted layers, alternate layers, or (C) a combination of them are added over the waste material.
SOURCE: Image modified from Price (2009).

challenging from a long-term management perspective. In addition, the long-term management of water-retaining facilities is difficult with the occurrence of extreme weather events. Long-term geotechnical stability is also a significant factor that should be considered for facilities with wet closure systems, particularly if a tailings facility is located upstream of humans or important ecological resources. Wet closure methods for tailings impoundments are not consistent with current best practices for geotechnical stability, so a thorough analysis of site-specific conditions and facility designs should be conducted prior to implementing such a closure system.

Water Management During Reclamation

Similar to during mine operations, water quantity management during reclamation and closure may include dewatering of pits, tailings storage facilities or underground mines to control water levels and flow paths. However, long-term/post-closure requirements for dewatering should be avoided when possible because of the ongoing costs (see Box 3-4). Operators should instead rely on passive or low-maintenance engineered systems to manage water quantity, whenever possible. Water quantity management during reclamation may also include the reestablishment of flow paths/channels for springs, seeps, and surface water features that may have been diverted around mine disturbance features during operations. This reestablishment of flow paths may include liners, riprap, or other methods/designs to construct and protect stream banks and/or preclude erosion of slopes. Other methods of stabilization may include the installation of gabions (rock- or dirt-filled cages), concrete, shotcrete (pneumatically applied concrete), and geotextiles.

The potential need to treat contaminated water is an important part of water quality management during reclamation. In some cases, water treatment may only be needed for a short time after operations have ended, as a temporary requirement of closure because there is only a finite amount of blasting agents or reagents within waste materials and these chemicals will flush, disperse, and degrade with time. In other cases, like acid rock drainage or the leaching of metals and metalloids, the contaminants are contained within host rock or waste materials and may be mobilized slowly, so treatment may be needed for longer periods of time. Treatment for these pollutants is necessary until acceptable hydrologic and geochemical steady-state conditions can be achieved, which could take

---

BOX 3-4
Failed Reclamation of Brewer Gold Mine Superfund Site, South Carolina

The most recent mining activities at Brewer Gold Mine commenced in 1987, but planning for closure and reclamation does not appear to have been evaluated until 1992 (Candice Teichert, personal communication, 2022). During reclamation of the Brewer Gold Mine, an estimated 120 million gallons of acid water was removed from both the Brewer and B-6 pits. A temporary water treatment plant was built to handle the water from the pits, which was discharged to Little Fork Creek under a National Pollutant Discharge Elimination System (NPDES) permit issued by the South Carolina Department of Health and Environmental Control. The Brewer pit was then backfilled with rinsed heap leach tailings, wastewater treatment sludge, demolition debris, and waste rock, and the B-6 pit was backfilled with sulfide-rich waste rock. A geosynthetic liner was installed over the pit area and a limestone-filled drainage conduit was constructed within the backfill to passively adjust the pH of the water while it drained from the pits. Unfortunately, the bedrock of the pits was highly fractured and seepage in multiple places has prevented the water from rising to a level where it would flow through the limestone-filled drainage conduit. In 1999, Brewer Gold Mine and its corporate owner Costain Holdings abandoned the site. In 2005, the U.S. Environmental Protection Agency (EPA) placed the site on the Superfund program's National Priorities List (NPL) due to the acidic seeps that were being released from the B-6 and Brewer pits to the nearby Little Fork Creek. EPA installed an extraction well to keep pit water level low enough in the Brewer and B-6 pits to stop the seepage of this acidic and metal-laden water to Little Fork Creek. This system is actively maintained to this day with an average ongoing annual cost of $1.18 million (EPA, 2021d). Inadequate planning for closure, poor characterization of groundwater hydrogeology and design of a passive limestone channel, and inadequate bonding were likely reasons why this site ended up on the Superfund program's NPL.

---

years, decades, or even centuries. Water treatment methods are designed to ensure that water quality standards are attained. The potential need for closure and post-closure water treatment should be determined early in the mine life cycle, so that it is factored into the designs and costs of the operation and the financial assurance provided to regulatory agencies.

Various methods may be used for active water treatment depending on the contaminant(s) of concern (described above), like filtration, adsorption, ion exchange, or the use of chemical reagents or biologic activity to promote precipitation, coagulation, and/or degradation of the contaminants. The operation and reclamation plans for mines should be designed to avoid long-term active water treatment as much as possible because of the ongoing costs and environmental risks. Other options may include passive or low-maintenance treatment systems, but these require careful site-specific planning. Constructed wetlands (see Figure 3-27) have been successfully deployed to control ARD at both coal and metal mines, particularly for sites with relatively low flow rates (5–20 gallons/min; EPA, 2015; Hassan et al., 2021; Perry and Kleinmann, 1991). These systems may pretreat drainage under oxidizing conditions, to encourage precipitation of iron as Fe^{3+}. Metals may also be removed by sorption on organic acids. Care is needed at the planning stage to ensure that constructed wetlands can, indeed, comply with all applicable water quality standards. Although some wetland designs include limestone to chemically neutralize acid that may be present, acid is removed principally in wetlands by the action of sulfate-reducing microorganisms in reducing conditions, which are maintained with organic media such as peat, used mushroom compost, etc. The action essentially reverses the oxidation reaction responsible for the formation of ARD, consuming acid and forming sulfide. If conditions are sufficiently acidic, sulfide forms gaseous hydrogen sulfide (with its characteristic "rotten egg" smell). As the pH increases, the sulfide precipitates heavy metal ions as sulfides, which are retained in the anaerobic sediments at the base of the ponds.

Removal of Facilities, Equipment, and Structures

The decommissioning and removal of facilities, equipment, and structures requires the appropriate handling and recycling of materials and equipment. These materials can include equipment with resale or scrap value, potentially hazardous materials, utility systems, and consumable products like fuel, lubricants, and reagents. Disposal of wastes can be in on-site or off-site repositories. There may be potential for some buildings or features to

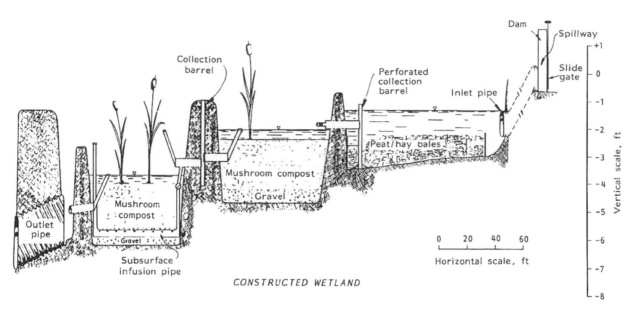

FIGURE 3-27 Depiction of a *Typha* (cattail)-dominated constructed wetland, showing a series of gravity-fed ponds in which sulfate is reduced to sulfide under anoxic conditions using sulfate-reducing bacteria.
SOURCE: Image from Perry and Kleinmann (1991).

remain or be relocated intact, depending on the approved postmine land use or other terms established with the local government or landowner.

Closure of Portals

Reclamation for underground mines include closing or securing entrances to underground workings in order to prevent access by the public, as well as to manage the potential release of water from underground excavations due to rising groundwater elevations or during periods of flooding. This might include installing a secure gate or similar barrier across the portal opening, or may include using backfill (e.g., mine waste such as waste rock or tailings) or other engineered materials such as cemented rock fill. Bulkheads and hydraulic plugs may be necessary to isolate water within the mine workings, unless such discharge is necessary for water management and treatment. If there is no need to maintain underground access or allow the release of water, the ground surface around the portal may be closed using backfill and graded with suitable fill material and a stable slope can be formed over the former portal opening and entry pad.

Revegetation and Habitat Reestablishment

The goals of revegetation will depend on the specific postmine land use. Generally, a diverse mixture of self-sustaining species is established across recontoured surfaces through seeding or direct planting. In some cases, boulders or other features like logs or similar debris may be placed on recontoured surfaces to provide additional important habitat components. In Virginia, the postmine land use may also include productive agricultural land, and there are requirements described in Chapter 5 for acceptable crop production rates to demonstrate that the land use has been achieved. While non-native species may have specific applications, native plant species are preferred because they provide the best support for the established local ecosystems. Some aspects of mining generate features that may benefit some wildlife species, such as cracks in an open pit highwall for roosting species, topographic variation in the reclaimed surface, and vegetation diversity. These aspects may contribute to habitat establishment for wildlife species that are already present at the site, in addition to species that may not have been present prior to mining.

TABLE 3-2 Commonly Measured Properties Required by Postclosure Monitoring Plans

Water quantity measurements	Groundwater levels
	Pit lake levels
	Underground mine pool levels
	Piezometric surfacea within dumps
	Piezometric surface within tailings
	Flow from springs
	Flow from surface water bodies
Water quality measurements	Groundwater
	Pit lake
	Underground mine pool
	Treated water discharge
	Piezometric surface within dumps
	Piezometric surface within tailings
	Springs
	Surface water bodies
Geotechnical features	Erosion, animal burrows, and large vegetation
	Slope stability
	Subsidence
	Ground movement
Vegetation success	Diversity
	Density
	Microhabitats
	Productivity metrics
Wildlife success	Individual count of mammals, bats, birds, macroinvertebrate
	Habitat assessment for mammals, bats, birds, macroinvertebrate

a Piezometric surface is the imaginary level to which groundwater rises under hydrostatic pressure.

Postclosure Monitoring

A postclosure monitoring plan details the sites, observation methods, schedules, and sample types required for measuring reclamation success, based on the unique characteristics of the site, the surrounding areas, and the postmining land use. Ultimately, the results of postclosure monitoring will be used to determine the adequacy of completed reclamation and whether financial assurances can be released to the project operator, as discussed further in Chapter 5. Plans may require the monitoring of many different properties, as shown in Table 3-2.

Environmental Risks Associated with Reclamation, Closure, and Long-Term Stewardship

There are a number of potential impacts during reclamation and closure that are similar to the impacts of mine operation. These include the production of fugitive dust during grading, emissions from vehicles, and the production of waste by-products from water treatment as described in operational water management.

The extent of ongoing impacts following reclamation and closure depends on the quality of site characterization and careful project design prior to operations. Active, long-term, or perpetual water management and treatment scenarios have high costs and elevated risk for environmental impacts. Preventing perpetual long-term water management and treatment and other active care relies on thorough site characterization and careful project designs prior to mining and diligent monitoring and regulatory compliance during operations. These scenarios

---

BOX 3-5
Case Study of the McLaughlin Mine Remediation

McLaughlin Mine, located about 100 km north of San Francisco, was operated by Homestake Mining Company from 1985 to 2004, producing 3.5 million oz of gold. During operation, acid-producing waste was identified and mined separately. This material was encapsulated on all sides with low-permeability clays and buffered by acid-neutralizing materials and the tailings storage facilities were built to withstand a 1,000-year, 72-hour storm (Krauss, 2002; Schoenberger, 2016). Following closure and reclamation, most of the mine area was established as a University of California, Davis, nature reserve and environmental research station. However, 1,200 acres of the 6,430-acre site remained within the care of the mine. This area includes the tailings storage facility, where reclamation remains ongoing. This ongoing reclamation is necessary, in part, because updated regulations precluded implementation of the originally permitted reclamation plan, which would have channeled surface runoff through the tailings storage facility to the original watercourses. The revised plan established differing moisture regions, with grassland on the upper, dryer zones, and marshland behind berms in the lower regions of the tailings (Benchmark Resources, 2022). The tailings have now been fenced off and are being dewatered and covered with 1–2 feet of compacted soil. Cattails, tules, willows, and cottonwood have been reestablished in the marshland. No toxic discharge has been reported in the surrounding area and the reclamation has generally been considered a success (Schoenberger, 2016).

---

can rarely be prevented or reversed once mining has been completed. In contrast, fully successful reclamation projects, sometimes referred to as "walk-away" sites, require no additional maintenance after final reclamation and closure. It is rare that an entire mine can meet those conditions, as certain components of a mine site may need ongoing maintenance and passive care (International Finance Corporation, 2007). Tailings facilities and pit lakes, in particular, may require some degree of care postclosure (The Mining Association of Canada, 2017; Vandenberg et al., 2022; see Box 3-5). However, fully successful pit lake closure has been demonstrated at certain sites such as the Canmore Creek Open Pit Coal Mine in Alberta, Canada (Stephenson and Castendyk, 2019).

MINING METHODS AT COMPARABLE DEPOSITS

In Chapter 2, several gold deposits around the world were suggested as potentially analogous to the Virginia gold deposits based on their geologic, mineralogical, hydrologic, and climatic characteristics. Here, the committee highlights mine sites that are thought to broadly reflect the scale and type of mining operations that could occur in Virginia, while cautioning that engineering design is influenced by many factors beyond those considered here, including site-specific considerations such as climate, proximity to population centers, current use of the site, and local infrastructure.

The currently active Kensington Mine in southeast Alaska broadly reflects the underground mining and processing methods that could be utilized in Virginia. As discussed in Chapter 2, gold production and mineral resources at the Kensington Mine are about one to two orders of magnitude larger than those of the known deposits in Virginia and are found in several spatially associated ore bodies. Should gold mining occur in Virginia in the future, it is likely the operator would similarly seek to develop several closely spaced ore bodies using a common processing infrastructure. All mining at the Kensington site is underground and utilizes a long-hole stoping method with backfill (see the "Mining Methods" section), where open stopes are backfilled with a combination of cemented paste fill, Cemented RockFill,[3] and waste fill. Conventional underground equipment and mining methods are employed with three portals to access the underground workings, one mill to process the ore, and a single tailings pond for waste storage. The processing operations use a flotation mill to recover a pyritic gold concentrate from sulfide-bearing rock. Ore is segregated by grade and blended before being fed to crushing, ball milling, and flotation. The final cleaner concentrate is thickened and filtered to approximately 10 percent moisture, then shipped to Europe and Asia for final processing (Pascoe et al., 2022). Because processing at the mine site stops with a pyrite-rich concentrate, an off-site processing facility bears the responsibility of preventing any potentially adverse environmental risks associated with the final processing of concentrate to gold metal.

---

[3] Cemented RockFill is a mixture of waste rock mixed with cement.

As discussed in Chapter 2, open pit mining could also be an economic and technically viable approach for some Virginia deposits, especially to access the near-surface oxidized ores. When considering an analogue for potential surface mining in Virginia, many stakeholders reference the recent open pit mining in South Carolina (see Box 3-1). Many of these examples are larger than what is likely to occur in Virginia. For example, the planned open pit at the Haile Mine for 2031 is estimated to be 2,500 m (8,200 feet) long, 1,250 m (4,100 feet) wide, and 370 m (1,214 feet) deep (SRK Consulting, 2020). Owing to the style and geometry of mineralization in Virginia, operators might be more likely to develop multiple, adjacent, smaller open pits that utilize shared mining infrastructure to extract near-surface ore from the linear gold-quartz veins. This style of surface mining may be more comparable to the Harvard Pit at the Jamestown Mine in the Mother Lode district of California (Savage et al., 2009), which started as small open pits to recover near-surface gold, followed by underground mining to access deeper ores. For reference, the size of the open pit at the Jamestown Mine in California is 823 m (2,700 feet) long, 243 m (800 feet) wide, and 152 m (500 feet) deep (SMGB, 2007; see Figure 3-28). The committee cannot state with certainty that a large mine comparable to that being developed at Haile would not be developed in

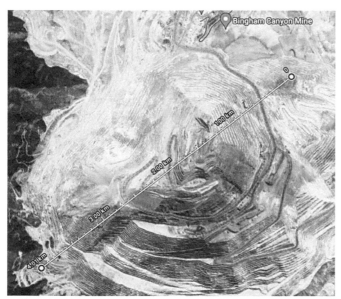

A. Bingham Canyon Mine, Utah
Pit dimensions: ~ 4 km diameter x 1.21 km deep

B. Haile Mine, South Carolina
Final pit dimensions (2031): ~ 2.5 km long x 1.25 km wide, 370 m deep

C. Fort Knox Mine, Alaska
Pit dimensions: ~ 2 km long x 0.8 km wide, Depth not available

D. Willis Mountain Kyanite Mine, Virginia
Open cut dimensions: ~ 1.37 km x ~0.05 km

E. Harvard open pit, Jamestown, CA
Pit dimensions: ~ 800 m long x 240 m wide, 150 m deep

F. Royal Mountain, King Mine, CA (North Pit)
Pit dimensions: ~ 600 m long x 200 m wide, depth not available

├─────── 1 km ───────┤

FIGURE 3-28 Compilation of images of various open pit mines (all images at the same scale), including (A) the open pit at the Bingham Canyon copper mine in Utah with a diameter of ~4 km (2.5 miles); (B) the Haile Gold Mine in South Carolina that currently consists of three open pits and that will be merged to form one large 2.5 km × 1.5 km pit by 2031; (C) the Fort Knox Gold Mine located northeast of Fairbanks, Alaska, that is approximately 2 km × 0.8 km; (D) the Willis Mountain Kyanite Mine in Buckingham County that is about 1.4 km × 0.05 km; and (E-F) the Harvard pit and the north pit at the Royal Mountain King Gold Mine in the Mother Lode district of California that are 0.8 km × 0.24 km and 0.6 km × 0.2 km, respectively. SOURCES: Images from Google Earth. Pit size data from Mining People International (2016); Sims (2018); SRK Consulting (2020); USGS (2007b, 2022b); SMGB (2007); Chaffee and Sutley (1994).

Virginia at some time in the future, although the probability is low based on the known geometries of the ore zones in Virginia deposits.

SUMMARY OF FINDINGS

Best practices for modern gold mining incorporate a life-cycle approach, which requires a robust analysis of site-specific environmental, economic, and social impacts of gold mining, processing, and long-term stewardship prior to permitting a proposed gold mine. This analysis includes careful site characterization to ensure appropriate planning for water management, waste management, and reclamation. It also includes stakeholder involvement in the early planning process, which may influence engineering design and promote specific mining and reclamation strategies.

The most probable mining methods for commercial development of gold ores in Virginia are small, elongated, open pit mines and underground mining. Processing could include crushing, grinding, flotation, cyanidation, adsorption and elution, electrolysis, and smelting. Many of the mining and processing methods will be determined based on site-specific conditions. However, due to the shape and small size of the known gold deposits in Virginia (see Chapter 2), mining would most likely occur via one or more smaller open pits for near-surface deposits, and via underground methods for deeper vein deposits. These methods would lead to smaller areal surface disruption compared to gold mines in the western United States and the Haile Gold Mine in nearby South Carolina. If underground mining is used, subsurface storage of tailings and waste rock could reduce surface storage and related impacts to air and water quality. Processing methods for gold ores in Virginia would likely include some on-site methods (e.g., crushing, grinding, flotation), as well as other methods that could occur on-site or off-site (e.g., cyanidation, adsorption and elution, electrolysis, smelting). Due to the small size of gold deposits likely to be discovered and mined in Virginia, it may be more economical for companies to ship ore or pyrite concentrates off-site for the later stages of processing.

Modern engineering methods and best practices for mining, processing, waste management, and reclamation can significantly reduce impacts to air and water quality. Nevertheless, impacts cannot be completely eliminated, and the remaining impacts may still be of concern to communities. History is filled with examples of engineering failures at mine sites caused by human error, incomplete understanding of site conditions, changing site conditions due to climate change, or compounding design flaws (e.g., spills, impoundment breaches, failed reclamation). The gold mining and regulatory community should continue to learn from mistakes so that best practices, codes, and regulatory oversight can continue to be advanced.

4

Potential Impacts to Health and the Environment

Gold mines are industrial operations that can have significant impacts on the surrounding environment and local communities. At each stage in the life cycle of a mining operation—including exploration, development, mining, processing, closure, and reclamation—a variety of chemical and physical hazards can be encountered and adverse impacts to the landscape, ecosystems, and local communities can occur. As indicated in the introductory chapter, the committee interpreted the Statement of Task to confer an emphasis on human health and ecological concerns, rather than on the potential negative or positive societal and economic impacts associated with mining. Thus, these societal and economic impacts are not considered here, even though they are important to consider when completing site-specific environmental assessments prior to permitting.

To ensure a thorough consideration of the potential impacts of gold mining on public health and the environment, the committee focused on the impacts specified in the Statement of Task ("air and water quality"; see Box 1-3) and sought relevant information widely. This information included the concerns voiced by community members and stakeholders in public meetings or through written comments (see Box 4-1). Additionally, as there are few studies of environmental and health impacts specific to gold mining in Virginia, the committee considered case studies in other geographic settings that could be relevant to the Commonwealth. The committee evaluated this comprehensive catalog of potential impacts in the context of the Statement of Task (see Box 1-3), the geological and environmental context of Virginia (see Chapter 2), and the availability of reliable engineering controls to mitigate the risks (see Chapter 3) in order to identify specific impacts that deserved more detailed discussion. This chapter first summarizes some of the broad environmental impacts that could occur due to gold mining and processing in Virginia and then discusses individual impacts that can affect human health and the environment in more detail.

BROAD DISCUSSION OF IMPACTS

Although the study task emphasizes effects to public health, air, and water quality, the committee acknowledges that other concerns have been raised about potential impacts of commercial gold mining operations. For example, industrial-scale mining can disrupt the rural character of a region. Physical conversion of a site into a mine and associated facilities using heavy equipment can alter the viewscape and soundscape that are inherent parts of rural history and culture. Such disruptions, especially if sizable, can have undesirable consequences for property values, the future attractiveness of the region to people who value the character of rural areas, and the mental health of affected local communities.

BOX 4-1
Concerns from Citizens of Buckingham County

In multiple open forums and in written comments, the committee heard public concerns that were expressed by dozens of community members, several community and environmental organizations, and business owners in areas of possible future mining in Buckingham County, Virginia. Although this report is concerned with potential impacts of gold mining in the entire Commonwealth of Virginia, most of the public input came from the citizens of Buckingham County and nearby areas owing to the more imminent concern there as a result of active exploration for gold deposits. Overarching concerns from these citizens included dewatering of aquifers and the effects on well water supply; pollution of local groundwater and surface water, including impacts to drinking water supplies, the James River, and the Chesapeake Bay; detrimental impacts to local fish, wildlife, and livestock; air pollution; adverse impacts to livelihoods of local residents; the inability to pass wealth and property on to future generations; and the loss of the rural character and lifestyle that is core to their community's identity and values. Community members expressed a strong connection to rural life, natural environments, and environmental stewardship.

Many citizens who spoke with the committee come from families that have lived in the area for many generations or had moved to the area because of their desire to live in a rural community in close proximity to natural landscapes. Citizens were troubled by the possibility that an industry could come to the area for a relatively short period of time, extract its resources, disrupt the local rural character, close the mine, and potentially leave long-lasting impacts behind. These comments provide important context for understanding the concerns of citizens about the potential impacts of gold mining on their communities, even if the committee cannot evaluate many of these site-specific issues without more data on the deposit and proposed mining, processing, waste management, and reclamation plans.

Physical conversion of land into gold mining operations also destroys or degrades natural habitat for flora and fauna, which may lead to decreased biodiversity. Virginia is home to an extraordinary diversity of plants and animals, and has several regions and streams that are recognized biodiversity hotspots (Roble, 2022). Across the Commonwealth, dozens of species are threatened or endangered and vulnerable to mining activities, to include bats, birds, amphibians, turtles, and freshwater fishes and mussels (Roble, 2022). Disturbance to these and other species can occur through the removal of trees and other vegetation, removal of topsoil overburden that releases organic carbon and nitrogen, installation of access roads, blasting and excavation of soil and rock, redistribution of water on-site, and transport of solutes and chemicals (e.g., metals, nitrates) in surface water and groundwater. Such adverse effects on habitat can affect local species diversity, but can also extend to migratory species, such as neotropical migrating bird species, that may rely on these habitats in Virginia for seasonal breeding activities or stopovers during longer-distance migrations. One prevalent impact of mining to natural habitat is the loss of soil and subsequent sediment and nutrient (e.g., nitrogen) loading into wetlands and waterways, because the removal of soils is necessary to allow construction of open pits, roads, facilities, ponds, tailings storage facilities, and waste rock piles. In some cases, the original soil may be lost if not appropriately salvaged prior to mining or stockpiled and maintained during operations. Even if soil material is salvaged for future use, re-creating the physical properties, microbial communities, and nutrient status of these original soils may not be feasible, even during land reclamation. As discussed in Chapter 3, open pit mines in Virginia are likely to be quite small, so these physical impacts to habitat would likely be spatially limited compared to larger mining operations in other portions of the United States.

Although the committee does not cover these broad impacts in more detail below, they remain important considerations in the siting and development of a mine. The following sections discuss in more detail some of the most likely impacts of concern, as well as some that are unlikely to occur but were raised by the study's charge and concerned citizens. These include waterborne and sediment-associated contaminants and nutrients (acid rock drainage [ARD], metals, cyanide, nitrogen), tailings dam failures, water table depression, air emissions, and cumulative health effects. The discussion of these impacts draws from examples and lessons learned from other locations in the United States and abroad.

ACID ROCK DRAINAGE

Reactive sulfide minerals can occur in the ore, on the walls of open pits or underground workings, or within the waste rock and tailings generated by mining and processing. When mining exposes sulfide-rich minerals, such as pyrite, to air and water, these minerals oxidize to form sulfuric acid and dissolved iron (see Boxes 2-1 and 2-3). This process is autocatalytic, meaning that once sulfide oxidation occurs, it tends to occur faster and is difficult to stop (see Box 2-3). Unless there is sufficient alkaline content (e.g., carbonates) within the mineralized zone or adjacent host rock to neutralize the acidity produced from sulfide oxidation, the resulting drainage water can create acidic runoff (referred to as acid rock drainage or acid mine drainage, hereafter ARD). ARD has low pH (often ranging between 2.0 and 5.0) and can contain high concentrations of sulfates and iron. Additionally, the acidic solution typically mobilizes a wide range of metals, such as lead, cadmium, copper, zinc, and other elements, from the ore minerals and associated host rocks (metals and their associated health effects are discussed in the next section). Thus, ARD is a complex mixture of elements in a low-pH solution.

The acid-generating potential of mines differs broadly depending on site-specific characteristics. Most of the recorded gold-quartz vein deposits in Virginia are relatively low in pyrite and have carbonate minerals (e.g., calcite and ankerite) that may neutralize some acid. Therefore, such deposits are not likely to be a high risk for generating extensive ARD. However, gold deposits in Virginia that are located in or in close proximity to massive sulfide deposits could pose higher risks of producing ARD if this material is exposed or disturbed during excavation (Hammarstrom et al., 2006; see Figures 2-3, 2-4, and 2-5). There is some evidence that ARD has been problematic for some mines in Virginia. For example, drainage from the historical Vaucluse Mine (see Chapter 2) has been reported to be extremely acidic (Virginia Energy, 2022e) and the Virginia Department of Energy (Virginia Energy) reported a brief period of ARD discharge from the active kyanite mine near Dillwyn in Buckingham County from February to April 2016 (Virginia Energy, 2022c). Conversely, waters associated with the low-sulfide Greenwood Gold Mine in Prince William County show pH values of 5.9 and 6.1 with no evidence of ARD (Seal et al., 1998). Without comprehensive, site-specific acid-base accounting and kinetic geochemical testing of relevant geologic materials, it is not possible to make a definitive assessment of the likelihood of ARD occurring in Virginia gold mines broadly. Thus, a robust site-specific analysis (e.g., quantity and reactivity of the pyrite exposed, presence of co-occurring minerals, bacterial activity) is necessary to determine the acid-generating potential of a particular deposit and its surroundings. Documents that describe best practices related to the prediction and treatment of acid rock drainage and metal leaching are described in Chapter 3.

If present, ARD can be one of the most persistent and significant environmental problems associated with mining of sulfide-bearing deposits, including gold deposits. Because sulfide oxidation is autocatalytic, mines can continue to generate ARD long after mining operations cease unless appropriate precautions are incorporated into the mine design during operations and upon mine closure (see Chapter 3). For example, in Johannesburg, South Africa, tailings dumps of crushed rock from former gold mining operations have produced ARD for decades, polluting both ground- and surface water with dissolved metals, low pH, and salinity (Naicker et al., 2003). Similarly, the high-sulfide Summitville Mine in Colorado released extensive ARD for years, resulting in a cleanup process that has taken more than three decades (USGS, 1995). Closer to Virginia, the high-sulfide Brewer Gold Mine in South Carolina was designated as a Superfund Site due to its ARD production; the site has continued to produce large quantities of ARD since it was abandoned in 1999 (see Box 3-4).

ARD is extremely toxic to plant and animal life due to its acidity, high specific conductance, and high concentrations of heavy metals and other elements. The low pH of ARD is directly toxic to many animals, especially aquatic life (Fromm, 1980; Haines, 1981). Most freshwater fauna are intolerant of low pH because it can disrupt respiration, osmoregulation, growth, and reproduction of many species of invertebrates and fish (Fromm, 1980; Haines, 1981). Environmental impacts tend to occur when ARD contaminates streams and wetlands, either through direct surface runoff from mine sites, from acidic seeps, or from subsurface water that has hydraulic connectivity to surface waters (Johnson et al., 2017; McCarthy, 2011; Tutu et al., 2008). The acidity is eventually attenuated through a combination of neutralization and dilution within the groundwater system or downstream surface water.

The high level of dissolved ions in ARD can increase the specific conductance and/or salinity of receiving waters to levels that are inhospitable for many freshwater organisms (Cañedo-Argüelles et al., 2013; Pond et al., 2008).

For example, groundwater polluted by ARD from South African gold mines that ultimately enters perennial streams can have specific conductance as high as ~4,000 microsiemens per centimeter (µS/cm) (Tutu et al., 2008), an order of magnitude higher than that known to be detrimental to much freshwater life (EPA, 2011a). Likewise, seeps of ARD from the inactive Minnesota gold and silver mine in Colorado have specific conductance that fluctuates daily, seasonally, and after rainfall events between ~1,500 and 2,500 µS/cm (Johnston et al., 2017). The seeps contribute to contamination of a nearby headwater stream (Lion Creek), causing the conductivity in the stream to rise to seasonal highs (~800 µS/cm) sufficient to harm many sensitive freshwater fauna. High conductivity/salinity disrupts osmoregulation and ionoregulation in freshwater fauna, which can result in a myriad of adverse sublethal effects and death in some instances (Griffith, 2017; Reid et al., 2019).

Finally, elevated concentrations of dissolved metals and other elements are common in ARD and have a wide array of adverse effects on organisms and ecosystems. Often some combination of these elements will co-occur in ARD, and cumulative exposure is likely. Ecological health risks and exposure pathways for key metals of concern are discussed in the next section.

Collectively, low pH, high dissolved metals, and high conductivity/salinity can depress populations of aquatic organisms at all levels of the food web (including plants) and, as a result, entire aquatic communities can be decimated by ARD. This in turn has consequences for ecosystem-level processes like primary production and nutrient cycling. Unlike some of the other toxic constituents potentially released from gold mining that can be relatively short lived (e.g., cyanide), release of ARD containing metals and sulfates has long-lasting toxic effects. Although low-pH discharge can naturally attenuate in some circumstances based on local geochemical conditions, many constituents of ARD do not degrade (e.g., dissolved metals), but can precipitate or be transformed to other forms that can be more or less toxic to plants and animals. Thus, economically costly interventions (e.g., constructed wetlands, phytoremediation, neutralization with limestone) are typically needed to mitigate sources of ARD and remediate habitats impacted by ARD to prevent continual long-term damage. For example, it currently costs $1.18 million per year to treat ARD that is still being generated from tailings at the Brewer Gold Superfund Site in South Carolina (EPA, 2021d).

In addition to harming the environment, ARD can impact drinking water that is sourced from the local aquifer or from downstream surface water intakes. Toxic metals dissolved in acid rock drainage can pose serious risks to human health, as discussed in the next section. Additionally, ARD can cause aesthetic impacts such as elevated concentrations of iron in drinking water that generates an unpleasant flavor and that can stain clothing and household surfaces. Likewise, elevated sulfur compounds may lead to unpalatable taste or odor in the water, with the potential for gastrointestinal impacts (EPA, 2022n).

METALS AND METALLOIDS

Gold mining can be associated with the mobilization of numerous metals and metalloids,[1] all hereafter referred to as metals. Although many metals can be solubilized and transported by ARD, others may be mobilized and released in the absence of ARD (Ashley, 2002; Ashley and Savage, 2001). There is considerable uncertainty involved in estimating the potential for the release of metals associated with gold mining across the Commonwealth. This is due to the variable spatial relationship of the primary quartz-hosted gold deposits with massive sulfide deposits that may have higher concentrations of metals of concern (see Chapter 2). With a few exceptions, the primary mechanisms and processes that can introduce metals into the environment are tailings dam failures, discharge of wastewater contaminated with metals, and the settling of metalliferous fugitive dust (see the "Air Emissions" section) into soils, wetlands, and surface waters (Donkor et al., 2005; Entwistle et al., 2019; Grimalt et al., 1999). Alternatively, metal mobilization could also occur following the disturbance of historical mine waste or materials—for example, the remobilization of mercury used for gold amalgamation in the past. Best practices for controlling erosion, mining-influenced water, and fugitive dust, as well as managing tailing storage facilities, are described in Chapter 3.

Effective waste and water management of mines can greatly reduce the release of metals to the surrounding environment. For example, following the abandonment of the Brewer Mine site in South Carolina in 1999,

---

[1] A metalloid is an element with properties that are intermediate between those of metals and solid nonmetals or semiconductors.

stream sampling between 2002 and 2004 identified aluminum, barium, cadmium, cobalt, copper, iron, manganese, mercury, silver, zinc, and cyanide mean concentrations that were all above water quality standards (EPA, 2021d). However, following capture and treatment of the acid-producing seeps by a pump installed by the U.S. Environmental Protection Agency (EPA), the concentrations of metals were brought below water quality standards (EPA, 2021d; see Box 3-4). EPA reported that the site would be contributing 2,248 pounds/day of metals to the nearby creek if the pit water was not captured and treated (EPA, 2014b).

Metals can exist in multiple oxidation states including metallic (valence zero), inorganic (charged cations combined with a variety of anions), and organic (e.g., methylmercury, tetraethyl lead, arsenobetaine, organotin compounds). The oxidation state of metals affects their fate, transport, and toxicity. Metals may sorb to sediments or precipitate in downstream wetlands and streams, generating contaminated sediments, or be distributed in floodplain soils following high-flow events. Depending on the chemical species of the element and local biogeochemical conditions, these precipitated metals can sometimes have low bioavailability and low toxicity. In other cases, bioavailable forms of metals deposited in sediments and soils can reach high concentrations that are toxic to benthic and soil organisms, respectively. Some organic forms, such as methylmercury, can bioaccumulate, while other organic forms, such as arsenobetaine and arsenocholine, are not bioaccumulative and are relatively nontoxic. Several metals can be solubilized during the cyanide leaching process and some metal–cyanide complexes are stable and can persist in tailings ponds.

Key Metals of Concern for Plants and Animals

Many metals are essential trace elements necessary for health of plants and animals, including copper, selenium, and zinc, but can become toxic at high concentrations or in complex mixtures. The primary routes of ecological exposures are through ingestion of metal-containing surface water, sediments, or food chain transfer, as well as across the gill epithelium in aquatic species (Clements et al., 2021). Table 4-1 lists those metals that may be associated with potential gold mining in Virginia; have plausible environmental exposure pathways to invertebrates, fish, and wildlife; and are inherently toxic, especially to aquatic fauna.

Key Metals of Concern for Human Health

The committee's approach to screening the primary metals of concern for human health associated with gold mining in Virginia is outlined in Figure 4-1. First, the committee identified metals of potential concern that were mentioned in the scientific literature associated with gold mining worldwide. Then, the committee reviewed the scientific literature on the exposure, epidemiology, and toxicology of these metals. This literature review included assessments by EPA, the Agency for Toxic Substances and Disease Registry, and the International Agency for Research on Cancer (IARC) of the World Health Organization. Metals that are essential trace elements with low inherent human toxicity (e.g., copper, chromium-III, selenium, zinc) and those with widespread environmental ubiquity or limited evidence of human health impacts (e.g., aluminum, boron, cobalt, nickel, silver, vanadium) were deprioritized for potential human health impacts.

The geologic literature from Virginia, including Tables 2-2, 2-3, and 2-4, was then reviewed to determine if the remaining metals may be present in Virginia's gold mining areas (see Chapter 2 and Figure 4-1). Cadmium, lead, and thallium were noted in high concentrations in water samples immediately downstream of mined massive sulfide deposits (see Table 2-4), and as discussed in Chapter 2, similar deposits could be disturbed during mining of the low-sulfide, gold-quartz veins. Although antimony, arsenic, and mercury have not been noted in high concentrations in Virginia ores (see Table 2-2), nor in downstream mine-influenced water (see Tables 2-3 and 2-4), the committee thought it was important to consider these elements given the limited data, their occasional association with low-sulfide, gold-quartz vein deposits (Ashley, 2002), and/or their presence as legacy contaminants in historical mine sites. Other elements, like uranium, were deprioritized as they are not commonly associated with low-sulfide, gold-quartz vein deposits and the committee did not find evidence for elevated contents in the gold-bearing rocks of Virginia, nor in downstream mine-influenced water (Owens and Peters, 2018; Owens et al., 2013; Pavlides et al., 1982; Seal et al., 2002). Additionally, hexavalent chromium (Cr[VI]) was considered highly

TABLE 4-1 The Potential Adverse Effects of Selected Metals in Plants and Animals (Excluding Humans)

Element	Potential Adverse Effects in Plants and Animals
Aluminum	Toxic at high aqueous concentrations to aquatic animals, particularly fish, amphibians, and aquatic invertebrates. Disrupts osmo- and ionoregulation after both acute and chronic exposures. Particularly problematic in low-pH surface waters (Rosseland et al., 1990; Sparling and Lowe, 1996).
Arsenic	Acutely toxic to aquatic vertebrates and invertebrates, but lower concentration exposure and chronic toxicity are more common and can be associated with some accumulation in tissues, adversely affecting growth, reproduction, and survival. Can also disrupt food webs at low concentrations by influencing lower trophic levels like phytoplankton (Eisler, 2004; Sanders et al., 2019).
Cadmium	Adversely affects growth, reproduction, early development, and survival in fish, wildlife, and invertebrates. Can cause cancer. Freshwater aquatic species are generally more sensitive than birds and mammals. Can accumulate in some tissues (Eisler, 1985).
Copper	An essential element for both plants and animals. At high dissolved concentrations, copper is toxic to freshwater invertebrates and fish, but toxicity is greatly influenced by water chemistry (e.g., hardness, pH, alkalinity, etc.). Excessive copper can disrupt the nervous system, enzymes, and blood chemistry, ultimately impairing growth, reproduction, and survival. High concentrations in soils are toxic to plants (Rehman et al., 2019; Santore et al., 2001).
Lead	Highly toxic to plants and animals. Wildlife consuming excessive lead experience adverse neurological effects that can lead to death. Affects other tissues including kidneys and reproductive organs (Assi et al., 2016).
Mercury	Highly toxic to aquatic and terrestrial animals and can bioaccumulate and biomagnify in food webs, especially in its methylated form. Interferes with the nervous system, cardiovascular system, and reproduction (Eagles-Smith et al., 2018).
Selenium	Essential at low dietary concentrations, but toxic at slightly higher concentrations. Selenium is highly bioaccumulative but does not biomagnify in food webs. Egg-laying animals (fish, birds, and invertebrates) are particularly at risk of toxicity because selenium concentrates in eggs and is a potent teratogen (Janz et al., 2010).
Thallium	Highly toxic to terrestrial and aquatic animals and plants and can accumulate in tissues. In vertebrates it can induce reproductive abnormalities and metabolic disorders. In birds it can cause embryonic developmental abnormalities. In mammals, hair loss is a common symptom of sublethal exposure (Peter and Viraraghavan, 2005; USACHPPM, 2007).
Zinc	An essential element for enzymes in both plants and animals, but excessive zinc can be toxic. In mammals and birds zinc toxicity primarily affects the pancreas and bone. In fish, zinc disrupts gill tissue which can cause acute or chronic toxicity depending on aqueous concentrations. High concentrations in soils are toxic to plants (Eisler, 1993).

NOTE: Metals such as cobalt, nickel, and vanadium are excluded because they are unlikely to be released from gold mining in Virginia in sufficient quantities to elicit toxicity (see Chapter 2).

unlikely to be mobilized from Virginia mines given the low concentration of chromium in the host rock and the lack of a mechanism for the oxidation of trivalent chromium (Cr^{3+}) to Cr^{6+} in ARD.[2]

Through these steps, the committee identified antimony (Sb), arsenic (As), cadmium (Cd), lead (Pb), mercury (Hg), and thallium (Tl) as being priority metals of potential concern for human health due to their documented or potential association with Virginia ores or mine sites and their potential for toxicity. The primary routes of human exposure to these metals are through the ingestion of contaminated surface water, groundwater, biota, or crops. Contaminated soils can pose a health hazard if ingested, which is a pathway generally limited to children. The human health impacts of these metals are summarized in Table 4-2 and discussed in the following sections.

---

[2] At pH 2, ARD is in equilibrium with Cr^{3+} at lower redox potentials (1.07 Volts [V]) compared to the standard hydrogen electrode (SHE); Pourbaix, 1966), while Cr^{6+} predominates at higher potentials. The redox potential of ARD is controlled by the availability of dissolved oxygen and the redox equilibrium between Fe^{2+} and Fe^{3+}, for which the standard redox potential is 0.771 V SHE (Pourbaix, 1966). Accordingly, there is no apparent mechanism for oxidation of Cr^{3+} to Cr^{6+} in ARD.

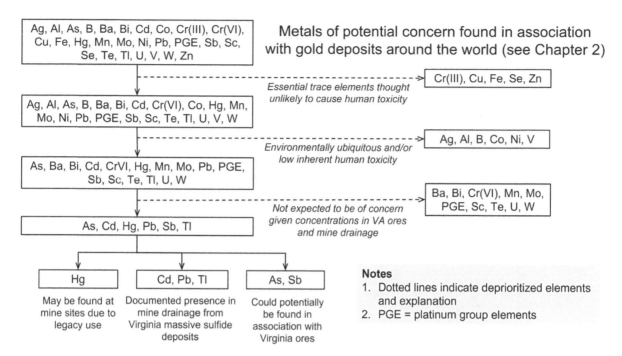

FIGURE 4-1 Schematic for the prioritization of metals. Initially all metals of concern found in association with gold deposits around the world were considered. Those metals that are essential trace elements and unlikely to cause human toxicity, those that are ubiquitous or have low inherent human toxicity, and those that are not expected to be present in relevant concentrations in Virginia ores and mine drainage were deprioritized. The remaining elements—antimony (Sb), arsenic (As), cadmium (Cd), lead (Pb), and thallium (Tl)—were identified as being of potential concern due to their documented presence in mine drainage from Virginia massive sulfide deposits or their potential association with Virginia gold-quartz vein deposits (Ashley, 2002). Mercury (Hg) may also be a concern in contaminated mine sites where it was used for processing in the past.

Antimony

Antimony can be a trace to minor constituent in sulfides, like pyrite (USGS, 2017a). While elevated concentrations of antimony have not been identified in downstream mine-influenced water in Virginia (see Tables 2-3 and 2-4), it has been documented in low-sulfide, gold-quartz vein deposits that are similar to those found in Virginia (Ashley, 2002). Additionally, antimony might be a concern if massive sulfide deposits—sometimes located near the low-sulfide, gold-quartz vein deposits—are disturbed during mining. EPA has established 0.006 mg/L antimony in drinking water as both the Maximum Contaminant Level Goal (MCLG) and the Maximum Contaminant Level (MCL; 40 CFR § 141.62).

The general U.S. population has exposure to low levels of antimony in food and water. An increasing number of human epidemiologic studies have been published in the past 15 years assessing the potential relationships between a range of health outcomes and various antimony concentrations in urine (ATSDR, 2019). The most consistent finding observed in more than one study was an association with high blood pressure (ATSDR, 2019). Antimony is predominantly found in the pentavalent oxidation state in water, but it can also be found in the trivalent oxidation state. Trivalent antimony is classified as probably carcinogenic to humans (Group 2A), whereas pentavalent antimony has been evaluated as not classifiable as to its carcinogenicity (Group 3; IARC, 2022).

Arsenic

Arsenic is a major element in the sulfide mineral arsenopyrite (FeAsS) and a trace to minor constituent in pyrite (Schellenbach and Krekeler, 2012). Arsenopyrite is rare in most gold deposits in Virginia (Pardee and Park, 1948), and elevated concentrations of arsenic have not been identified in downstream mine-influenced water in

TABLE 4-2 The Potential Adverse Effects of Selected Metals in Humans

Element	Potential Human Health Effects
Antimony	Lower doses have been correlated with increased blood pressure (ATSDR, 2019). Health effects at higher doses unlikely to be relevant to gold mining include respiratory effects, gastrointestinal symptoms, joint and muscle pain, neurodevelopmental effects, risk of diabetes, and electrocardiogram changes. Trivalent antimony is classified as probably carcinogenic to humans (Group 2A), whereas pentavalent antimony has been evaluated as not classifiable as to its carcinogenicity (Group 3; IARC, 2022).
Arsenic	Exposure is linked to circulatory system, skin, and neurologic effects; lung, bladder, and skin cancer; type 2 diabetes; and other adverse outcomes (ATSDR, 2007). Arsenic causes lung, bladder, and skin cancer and is classified as a Group 1 carcinogen (IARC, 2012a).
Cadmium	Associated with neurodevelopmental toxicity in children and renal toxicity in children and adults (ATSDR, 2012). Cadmium causes lung cancer and is classified as a Group 1 carcinogen (IARC, 2012b).
Lead	Associated with a range of health effects in fetuses, children, and adults. Lead can have adverse effects in almost every organ system, with particular effects on brain health and development. Lead bioaccumulates, is deposited in bone, and can be slowly released over time to affect various organs (ATSDR, 2020). Inorganic lead compounds are classified as probably carcinogenic to humans (Group 2A), whereas organic lead compounds have been evaluated as not classifiable as to their carcinogenicity (Group 3; IARC, 2006).
Mercury	Exposure leads to neurologic, cognitive, and neurodevelopmental effects, and there is some evidence for cardiovascular effects. Exposure is of greatest concern for pregnant women and their fetuses. Methylmercury bioaccumulates in fish and is a risk after ingestion (ATSDR, 2022). Methylmercury has been classified as possibly carcinogenic to humans (Group 2B), whereas inorganic mercury compounds have been evaluated as not classifiable as to their carcinogenicity (Group 3; IARC, 1993).
Thallium	Low doses for longer duration are associated with obesity, impaired thyroid function, autism spectrum disorders, adverse pregnancy outcomes, measures of oxidative stress, gestational diabetes, and others (Campanella et al., 2019).

Virginia (see Table 2-4). Nevertheless, the committee considered arsenic since it is documented to occur in some low-sulfide ore deposits (Ashley, 2002; Schellenbach and Krekeler, 2012) and might be a concern if nearby massive sulfide deposits are disturbed during mining. It is also possible that arsenic could be mobilized in aquifers (Alpers, 2017; Peters and Blum, 2003; Verplanck et al., 2008) if mining triggers changes in hydrologic conditions. For example, increases in dissolved oxygen in groundwater due to water table changes can oxidize arsenopyrite and pyrite and release arsenic into solution as As^{3+} and As^{5+} species. Alternatively, arsenic sorbed to Fe-Mn oxyhydroxide minerals in aquifer materials can be released into solution by reductive dissolution of the Fe-Mn minerals (Fendorf and Kocar, 2009; Peters and Blum, 2003). Compilations of arsenic data from well water in Virginia show that 15 to 23 percent of well water samples have arsenic concentrations over 5 µg/L, but most well water samples with elevated arsenic were found in the Culpeper Basin and other Triassic sedimentary basins that do not host gold (VanDerwerker et al., 2018). These limited data leave considerable uncertainty regarding the possible mobilization of arsenic in aquifers following water table changes in the gold-bearing regions of Virginia. EPA has established 0 mg/L arsenic in drinking water as the MCLG and 0.010 mg/L as the MCL (40 CFR § 141.62).

The most common route for human exposure to arsenic is through drinking water containing geologically derived arsenic—it has been estimated that more than 2 million Americans use drinking water wells with arsenic levels in excess of drinking water limits (USGS, 2019). Arsenic is tasteless, odorless, and colorless and has been linked to circulatory system, skin, and neurological effects; type 2 diabetes; and other adverse outcomes (ATSDR, 2007). Arsenic causes lung, bladder, and skin cancer and is classified by EPA and the IARC as a known human carcinogen (Group 1; IARC, 2012a). Long-term ingestion of arsenic has been associated with increased risk of heart disease, skin abnormalities, adverse pregnancy outcomes, and diabetes (ATSDR, 2007, 2020; Bräuner et al., 2014; Ettinger et al., 2009; Farzan et al., 2015a,b; Navas-Acien et al., 2008).

Cadmium

Cadmium can be hosted at minor to trace concentrations in sphalerite, chalcopyrite, galena, and pyrite (Schwartz, 2000). These sulfides are commonly found in massive sulfide deposits, such as those that are sometimes near Virginia's low-sulfide gold deposits. In South Carolina, the Haile Gold Mine has been fined for cadmium discharges to surface waters (The State, 2021), and elevated cadmium concentrations have been observed in mine-influenced water downstream of Virginia's massive sulfide deposits (see Table 2-4). EPA has established 0.005 mg/L cadmium in drinking water as both the MCLG and the MCL (40 CFR § 141.62).

Absorption of significant levels of cadmium from water sources can result in a number of adverse health outcomes. Cadmium is associated with neurodevelopmental toxicity in children, has a long retention time in the kidney, and is a known cause of renal toxicity in children and adults as a function of cumulative dose (ATSDR, 2012; Satarug, 2018). Cadmium is often found with other metals in well water, some of which are also renal toxicants (e.g., lead; Rehman et al., 2018). This explains why some authors are cautious in solely attributing observed chronic renal effects to cadmium (Butler-Dawson et al., 2022; Herath et al., 2018; Kaur et al., 2020; Wasana et al., 2016). Cadmium also causes lung cancer and is classified as a Group 1 carcinogen (IARC, 2012b).

Lead

Galena (PbS) is reported as a common trace mineral in many Virginia gold deposits (Schellenbach and Krekeler, 2012), including the London and Virginia Mine, Moss Mine, and Vaucluse Mine gold deposits (see Chapter 2). In addition, elevated lead concentrations have been observed in mine-influenced water downstream of Virginia's massive sulfide deposits (see Table 2-4), which are sometimes located in the vicinity of low-sulfide, gold-quartz veins in Virginia. Finally, lead is sometimes used as an additive to assist in the gold dissolution process (Kyle et al., 2011, 2012). EPA has established 0 mg/L lead in drinking water as the MCLG and 0.015 mg/L as the MCL action level, which is the level at which additional steps must be taken (40 CFR § 141.62). The U.S. Centers for Disease Control and Prevention has repeatedly lowered the threshold at which blood lead levels are considered to be of concern in children from 60 μg/dL[3] to 5 μg/dL over the past 40 years, although it is generally agreed that there is no safe level of lead exposure for children (AAP, 2021).

Lead is a human toxicant with well-documented health effects in fetuses, children, and adults. Toxicity can be found in almost every organ system, including the central nervous system and peripheral nervous system, as well as the reproductive, cardiovascular, hematopoietic, gastrointestinal, and musculoskeletal systems. Inorganic lead compounds are classified as probably carcinogenic to humans (Group 2A), whereas organic lead compounds have been evaluated as not classifiable as to their carcinogenicity (Group 3; IARC, 2006).

Lead poisonings from gold mining have resulted in tragic events internationally. Lead exposure due to artisanal gold mining in northern Nigeria (Lo et al., 2012; Tirima et al., 2016) was the largest known occurrence of lead poisoning in history (CDC, 2016). This setting was unique in that the gold-containing ore contained a vein with more than 10 percent lead, and occupational and public health safeguards were lacking. Given the more robust regulations in the United States, the committee concluded that the unique features of this extreme example are not relevant to gold mining in Virginia.

Mercury

Mercury is unique among the metals considered by the committee in that it has both a natural source from sulfide ores (including some ores mined for gold, such as those at the McLaughlin Mine in California) and an anthropogenic source from the historical use of it for amalgamation of gold. Although the mercury content of gold deposits in Virginia is expected to be low (see Chapter 2), the limited number of analyses leaves significant uncertainty in estimating the concentrations of mercury. In addition, mercury was widely used in Virginia in the 1800s to amalgamate gold at mine sites. Large quantities of mercury were often lost during the gold mining process, and previous gold mining areas and downstream rivers are often highly contaminated with mercury (e.g., in the Sierra Nevada foothills, California; Saiki et al., 2010). Because metallic mercury is relatively stable in the environment, it can be found in high concentrations in stream sediments and soils hundreds of years after mining activities have ceased (see Box 1-1). Sampling of the distribution and occurrence of mercury at historical gold

---

[3] A deciliter (dL) is one-tenth of a liter.

mining sites in Virginia is limited. Hammarstrom et al. (2006), however, reported up to 1.5 mg/kg of mercury in the pond sediment at the site of the Mitchell Gold Mine. Up to 40 mg/kg of mercury in soil and 3.7 mg/kg of mercury in sediment was reported near the Greenwood Gold Mine (Seal and Hammarstrom, 2002; Seal et al., 1998). Elevated mercury concentrations in stream sediments have also been reported near the Vaucluse Mine (Virginia Energy, 2022e). If a new mine were established on the site of a historic mine where mercury was used to amalgamate gold, the legacy mercury could be excavated and re-released into surface waters, unless it is fully captured and removed for processing (see Box 3-2).

The general U.S. population is often exposed to mercury from the consumption of fish, other seafood, and rice, and via dental amalgams. Based on 2016 data from the National Health and Nutrition Examination Survey, the general U.S. population has been estimated to have a geometric mean total blood mercury level of 0.81 µg/L. EPA has established 0.002 mg/L mercury in drinking water as both the MCLG and the MCL (40 CFR § 141.62), but drinking water is generally considered a minor source of mercury exposure (WHO, 2005).

Mercury can be found in many forms, including metallic/elemental mercury (Hg^0); oxidized, inorganic divalent mercury (Hg^{2+}); and organic, mono-methyl mercury (MMHg; Morel et al., 1998). The production of MMHg from Hg^{2+} occurs in low-oxygen environments (wetlands and lake/river bottom sediments) mainly by sulfate- and iron-reducing bacteria (Morel et al., 1998). In aquatic systems, MMHg is taken in by algae and subsequently transferred up the food web to zooplankton, small forage fish, and finally large predatory fish in lakes and rivers. Thus, it is often larger/older fish feeding at high trophic levels that have the highest levels of MMHg. Consumption of high-trophic-level fish caught either for sport or as a needed source of protein (most often by people with low incomes) can lead to unsafe levels of exposure to MMHg. Numerous rivers and lakes in and downstream of historically gold-producing counties in Virginia are under fish consumption advisories for mercury. Water bodies with fish consumption advisories that are in historically gold-producing counties include Lake Gordonsville in Louisa County; Nottoway River in Dinwiddie County; Motts Run Reservoir in Spotsylvania County; Dan River in Pittsylvania, Halifax, and Mecklenburg Counties; Roanoke River in Pittsylvania, Campbell, Halifax, Charlotte, and Mecklenburg Counties; and Kerr Reservoir and Lake Gaston in Mecklenburg County (VDH, 2022a). For some of these water bodies, the point source of mercury is from industrial operations, but in others, the source is unknown but is likely a combination of legacy mines, industrial inputs, and atmospheric deposition. For example, samples collected by the Virginia Department of Environmental Quality from streams at the Vaucluse Mine site indicated that mercury levels in fish tissue (up to 0.47 mg/kg) were above background levels and very close to the current action levels for a fish consumption advisory in Virginia (0.5 mg/kg). The report concluded that mercury from the historical mine site was entering the aquatic food chain (Holmes, 2022). Although fish consumption advisories due to mercury can help protect public health, local communities that rely on fish for sustenance lose valuable protein from their diet when fish is unsafe to eat. Additionally, the sport fishing industry in Virginia is estimated to generate $1.3 billion annually with 800,000 anglers participating each year (Virginia DWR, 2022), highlighting the potential economic consequences of fish advisories caused by mercury pollution.

Mercury is a potent neurotoxicant in each of its environmental forms. Although Hg^0 and Hg^{2+} can be hazardous, toxic levels are generally limited to occupational exposures. MMHg is, however, more toxic than other forms of mercury and, as described above, is strongly bioaccumulated and biomagnified by about tenfold in concentration for each trophic level. Studies of MMHg report consistent neurologic, cognitive, and neurodevelopmental effects; some evidence for cardiovascular effects; and the possibility of other developmental effects (e.g., structural malformations). Many of these effects are of greatest concern for pregnant women and their fetuses, although people can potentially be adversely affected at any point in the lifespan. Animal studies also raise concern about renal effects (ATSDR, 2022). Methylmercury has been classified as possibly carcinogenic to humans (Group 2B), whereas inorganic mercury compounds have been evaluated as not classifiable as to their carcinogenicity (Group 3; IARC, 1993).

Thallium

Thallium can be hosted as a trace metal in pyrite, galena, and sphalerite and, therefore, could be a potential metal of concern if nearby massive sulfide deposits are disturbed during the mining of Virginia low-sulfide, gold-quartz veins. The Haile Gold Mine was fined for discharging thallium into surface waters (The State, 2021) and one sample of mine-influenced water in Virginia had elevated thallium concentrations (see Table 2-4). Additionally, thallium may be elevated in the host rock adjacent to low-sulfide, gold-quartz veins deposits (Ashley, 2002), such as the deposits in Virginia's gold-pyrite belt and the Virgilina district. EPA has established 0.0005 mg/L thallium in drinking water as the MCLG and 0.002 mg/L as the MCL (40 CFR § 141.62).

Thallium has two primary oxidation states, Tl^+ and Tl^{3+}; both are toxic but Tl^{3+} is likely more so (Rickwood et al., 2015; Zhuang and Song, 2021). Tl^+ is the most common species in surface waters. In the body, Tl^+ competes with potassium (K^+) and is widely distributed, including to heart and brain cells. Because of its similarity to potassium, thallium concentrates in tissues with high potassium concentrations and can inhibit potassium-dependent processes. Of increasing concern is the toxicity of thallium at lower doses for longer durations, as can be found in the consumption of drinking water containing thallium through natural or anthropogenic contamination (Biagioni et al., 2017; Campanella et al., 2019). A growing epidemiologic literature has associated increased levels of thallium in urine and blood with a number of adverse health outcomes, including obesity, impaired thyroid function, autism spectrum disorders, adverse pregnancy outcomes, measures of oxidative stress, gestational diabetes, and others (Campanella et al., 2019). Thallium contamination of drinking water has been highlighted as an emerging environmental health issue that requires more attention. A growing number of studies in the past 10 years have identified putative health effects at contaminant levels far below the current EPA MCL (Campanella et al., 2019).

CYANIDE

Since the late 1800s, cyanide leaching has been one of the primary mechanisms for recovering gold from ore. Today, it has completely replaced the use of mercury in gold mining both in the United States and in other high-income countries. Cyanide, primarily in the form of dilute sodium cyanide solutions, is typically applied to mined and crushed ore using either tank or heap leaching techniques (see Chapter 3). Gold is then removed from the resultant gold-bearing solutions using zinc or activated carbon, and the remaining cyanide solution is recycled to leaching. Any waste materials containing cyanide typically undergo cyanide destruction treatment during operation or prior to final mine closure (EPA, 1994c). Although some alternatives to cyanide leaching have been developed, none are as widely available, efficient, or as economical as cyanide-based methods. However, cyanide is extremely toxic and must be managed carefully to avoid harm to human health and the ecosystem (see Box 3-3). Accidental releases of cyanide from gold mining into the environment have occasionally harmed humans and resulted in mass mortality of fish and other wildlife (Cleven and Van Bruggen, 2000; Donato et al., 2017; Eisler et al., 1999; Moran, 1998, 1999). Despite successful use and improved management of cyanide at mines (described in Chapter 3), its potential to cause considerable harm if mismanaged understandably makes it one of the most significant concerns of the public.

Cyanide is extremely toxic to humans, fish, wildlife, invertebrates, and to a lesser extent other life such as certain aquatic plants and algae. In animals, cyanide blocks oxidative energy metabolism by disrupting a critical enzyme (cytochrome oxidase), which then deprives cells of energy, results in calcium imbalances, and ultimately causes cell death (Solomonson, 1981). As a result, cyanide toxicity often manifests as disruptions to the cardiovascular and nervous systems (Borowitz et al., 2005) because heart and brain tissue are particularly reliant on oxygen and energy for proper function and are also susceptible to changes in electrical activity important in cellular signaling. Symptoms of cardiovascular disruptions following cyanide poisoning include slowed heart rate, abnormal heart rhythms, and heart failure (Borowitz et al., 2005). Neurotoxic effects of cyanide can present as behavioral abnormalities, seizures, impaired vision, and loss of consciousness. Other manifestations can include vomiting and shortness of breath, and, at high doses, death within minutes (Borowitz et al., 2005).

Cyanide spills pose acute risks to human health and the environment, but minimal long-term risks because cyanide does not bioaccumulate in animal tissues and tends to break down in the environment quickly. These are among the reasons that cyanide has replaced mercury amalgamation as the preferred method of gold extraction in many places around the world (Veiga and Meech, 1999). Cyanide does not accumulate in animal tissues because low doses are readily detoxified and metabolized by animals and acute exposure to high doses are fatal, making transfer via the food chain negligible in most situations (Eisler, 1991). Free cyanide (HCN and CN^-) is its most toxic form (Gensemer et al., 2006), but this form naturally breaks down over time by photodegradation, chemical oxidation, volatilization, and microbial processes (Dzombak et al., 2005; Ebbs et al., 2005). Cyanide also readily binds to a variety of metals (e.g., zinc, cadmium, iron, copper, mercury, cobalt) to form relatively nontoxic metal-locyanide complexes,[4] which can persist for longer periods of time but are slow to release toxic, free cyanide into solution (Borowitz et al., 2005; Dzombak et al., 2006). Cyanide is not typically bioavailable in sediments and soils (Gensemer et al., 2006) and does not persist for extended periods of time in these environmental media (Eisler et al., 1999). Thus, when cyanide is accidentally released from mining operations, it does not typically persist in soils and sediments for extended periods of time, and degrades in water within days to weeks (Eisler et al., 1999). However, these degradation processes can be slow in holding ponds due to the high concentrations of cyanide and water chemistry in these settings (Simovic and Snodgrass, 1985), so gold mines often employ a variety of chemical (e.g., alkaline chlorination), ultraviolet light, and microbial treatments to accelerate the degradation of cyanide on-site.

Cyanide toxicity among human populations near gold mining operations is rare. Cyanide can be toxic to humans via direct inhalation and contact, but ingestion of cyanide-contaminated water is the primary exposure route (e.g., Pannier, 2020). When not stored in carefully controlled basic solutions, cyanide can convert to hydrogen cyanide gas, which can be inhaled and is extremely toxic to humans (The Canadian Press, 2016; International Cyanide Management Code, 2022; Peiyue, 2021). However, this impact is primarily an occupational concern and can be managed using best practices, such as the International Cyanide Code (see Chapter 3). In contrast to rare gaseous exposures, aqueous cyanide exposure has occasionally resulted from mishandling of cyanide at gold mines, such as a large spill from a truck carrying NaCN in Kyrgystan in 1998 that resulted in contamination of surface drinking water, which produced conflicting reports regarding fatalities and thousands of illnesses (Cleven and Van Bruggen, 2000; Moran, 1998, 1999). In addition, probable low-level aqueous exposures to cyanide in communities living near gold mines have been linked to headaches, dizziness, eye irritation, and skin irritation in Malaysia, and in some cases these symptoms were associated with biomarkers of exposure to cyanide (i.e., urinary thiocyanate; Hassan et al., 2015). The committee could not identify any publications that described accidental release of cyanide from modern gold mining that affected drinking water in the United States, but the potential exists in circumstances where communities rely on surface water. For example, the cyanide release at the Brewer Mine in South Carolina was prevented from contaminating local drinking water supplies by the rapid response of local authorities (Jim McLain, personal communication, 2022).

In contrast to rare human exposures, the potential for ecological impacts is far greater if proper precautions are not in place. Historically, wildlife exposures to cyanide have occurred on-site in extraction and tailings ponds, as well as small pools of cyanide solution on top of ore heaps (Henny et al., 1994). In addition, accidental release of cyanide from mine sites can have catastrophic consequences for downstream ecological communities. The vast majority of documented cases of cyanide poisoning of fish and wildlife linked with mining activities (both on and off mine sites) involve acute aqueous exposure.

Prior to the relatively widespread adoption of international best practices such as those outlined in the International Cyanide Code (see Chapter 3), ponds containing cyanide-bearing leach solutions often attracted wildlife to bathe, drink, or reproduce. At gold mines in the western United States, diverse species of birds, amphibians, reptiles, and mammals have been found at cyanide gold leaching ponds (Clark and Hothem, 1991; Griffiths et al., 2014). Birds are particularly vulnerable, as they are attracted to even small pools of open water. For example, birds comprised about 90 percent of the dead wildlife found near cyanide leach ponds near gold mines in California, Nevada, and Arizona (Clark and Hothem, 1991). Thousands of bird deaths involving waterfowl and migratory species in Nevada were attributed to birds drinking, bathing, and resting in ponds at gold mines containing high cyanide concentrations (Henny et al., 1994; Hill and Henry, 1996). Observations at these sites indicate that some birds die on-site quickly, but others fly off-site after swimming and drinking cyanide-polluted water, suggesting that on-site counts of dead birds may underestimate the actual impact of improperly managed cyanide ponds (Henny et al., 1994). The risk to birds extends beyond cyanide ponds, as the small pools of cyanide solution that may form on top of heaps at mine sites also attract birds and cause mortality (Donato et al., 2007; Henny et al., 1994).

Fortunately, adoption of best practices that are typically required during the mine permitting stage has minimized many of these problems. Common best practices include cyanide treatment to decrease concentrations

---

[4] Including weak acid dissociable cyanide complexes and strong acid dissociable cyanide complexes.

(often to <50 mg/L weak acid dissociable cyanide) prior to release into surface impoundments to decrease the risk of acute toxicity to wildlife. Likewise, a variety of deterrents (e.g., exclusion netting, pond covers, floating balls, noise/light) have been developed to deter wildlife from accessing surface water at these facilities. Similar risks of exposure would need to be carefully managed in the gold pyrite belt of Virginia, especially in light of its high biodiversity and population densities of wildlife. In Virginia, amphibians, migratory songbirds, waterfowl and waterbirds, and bats are among the groups of wildlife that depend on open surface water and should be deterred from using ponds containing cyanide. Because the toxicity of cyanide is not well studied in many of these species but some of them are known to be highly sensitive to environmental pollutants (e.g., amphibians), deterrents are critical for minimizing possible on-site exposure to cyanide.

In addition to on-site exposures, accidental releases of cyanide have resulted in catastrophic consequences for downstream ecosystems (e.g., streams and wetlands; see Box 3-3). In general, aquatic animals are particularly sensitive to cyanide poisoning because they are often completely immersed in water, even exposing their sensitive respiratory structures (i.e., gills) to dissolved cyanide. Fish kills are often the most conspicuous effect of cyanide release because fish are particularly sensitive to cyanide toxicity (Eisler, 1991). Additionally, dead fish are more easily observed than other species such as birds that can move off-site before dying and aquatic invertebrates that are simply less conspicuous. For example, cyanide release from a mine in Canada killed more than 20,000 steelhead trout (Leduc et al., 1982).

In addition to acute mortality, long-term exposure to sublethal levels of cyanide can have consequences for fish and freshwater communities. Most notably, nonlethal exposure to cyanide has long been known to impair fish reproduction (Leduc, 1981, 1984; Leduc et al., 1982; Lesniak and Ruby, 1982; Ruby et al., 1986). Other sublethal effects of cyanide on fish include behavioral abnormalities, poor swimming performance, and reduced growth, all of which have implications for survival (Eisler et al., 1999). These long-term sublethal impacts are less common than acute mortality events given that cyanide tends to break down in the environment quickly. Instead, the long-term effects of cyanide on the environment likely relate primarily to the pace of ecological recovery processes. For example, cyanide released from a gold mine in Japan following an earthquake killed all biota in a stream for approximately 10 kilometers. However, within days cyanide was no longer detectable in the water and within 6 months local plant, invertebrate, and fish species were recolonizing the impacted region of the stream (Yasuno et al., 1981). At the Brewer Mine in South Carolina, taxa richness and abundance of aquatic invertebrates were reduced for months downstream of the point of cyanide release but other signs of recovery were beginning to become evident months after the spill (Shealy Environmental Services Inc., 1991; see Box 3-3).

NITROGEN

As discussed in Chapter 3, operators often use a mixture of ammonium nitrate and fuel oil for blasting during mining. Proper detonation will ensure the blasting product is wholly consumed to produce gases such as CO_2, N_2, and H_2O (Martel et al. 2004). However, nonideal blasting practices (e.g., wet conditions) may produce more toxic gases such as CO, NO, and NO_2, and estimates for the mass of explosive nitrogen remaining after detonation ranges from 0.2 percent for near-ideal conditions to up to 28 percent in nonideal conditions (Bailey et al., 2013; Brochu, 2010; Morin and Hutt, 2009; Pommen, 1983). Residual nitrogen compounds and undetonated ammonium nitrate may occur on the surfaces of the host rock (in pit or underground mines) and the blasted rock, and it may be processed as ore or disposed as waste material. Undetonated ammonium nitrate and the ammonium ion (NH_4^+) are readily soluble in water and could be further mobilized by runoff, infiltrating water, or process solutions. Without implementing appropriate strategies for the management and treatment of this water, poor hydrologic containment may lead to loading of nitrogen in surface runoff and groundwater discharge mostly as nitrate species (NO_3^-), but also as ammonia (NH_3) and to a lesser extent nitrite (NO_2^-) (Brochu, 2010).

Depending on the size of a mining operation and the frequency of blasting, hundreds to tens of thousands of kilograms of ammonium nitrate may be used at a site. This can lead to a substantial amount of nitrogen-laden effluent that can exceed water quality criteria. Several mines around the United States—including the Buckhorn Mine in Washington state and the Jamestown, McLaughlin, and Royal Mountain King Mines in California—have received violations from the discharge of excessive nitrogen that range from 25 to 600 mg/L nitrate (5.6–135.5 mg/L nitrate as nitrogen) and 10 to 40 mg/L ammonia (Brochu, 2010; Maest, 2022). Concentrations will be most elevated proximal to the mine site, as movement downstream or down-gradient will result in attenuation and dilution of the nitrogen levels. Virginia has set the surface water quality criteria for the protection of human health as 10 mg/L for nitrate as nitrogen (9VAC25-260-140) and has also set the groundwater standards for the Piedmont and Blue Ridge regions as 5 mg/L nitrate and 0.025 mg/L for nitrite and ammonia (9VAC25-280-50). Although there are no surface water quality criteria for nitrate for the protection of aquatic life, acute and chronic water quality criteria have been established for ammonia in order to protect freshwater mussel species and the early life stages of fish (9VAC25-260-155). These acute and chronic water quality criteria for ammonia are determined by site-specific pH and temperature conditions, with values ranging from 0.27 to 51 mg/L and 0.08 to 4.9 mg/L, respectively (9VAC25-260-155).

Excessive nitrogenous compounds in water can pose health risks to humans and the environment. Although ammonium ion toxicity is low, it is readily converted to nitrate in aquatic systems (Camargo et al., 2005), which can affect aquatic life or pose a health hazard to humans (Brochu, 2010). For example, high nitrates in drinking water can induce production of methemoglobin, a form of hemoglobin that cannot effectively transport and release oxygen to tissues, and infants less than 6 months of age are particularly at risk for this condition. When methemoglobin is produced in high quantities it can lead to methemoglobinemia, a syndrome of inadequate tissue oxygenation. Public health actions and the water quality criteria have been established by EPA (10 mg/L nitrate as nitrogen) to protect infants, the most sensitive population (EPA, 2022e; Minnesota Department of Health, 2018).

In addition, the presence of excessive nitrate in surface water can promote algal blooms, growth of harmful cyanobacteria, and eutrophication. This is especially true when excessive nitrogen occurs in conjunction with an excess of other elements such as phosphorous and iron (Wurtsbaugh and Horne, 1983; Xiao et al., 2021a), the latter of which can occur in very high concentrations in mining effluent and runoff. Because iron is often a limiting factor for the growth of phytoplankton, its release in conjunction with excessive nitrogen can accelerate algal growth. Eutrophication of surface waters depletes dissolved oxygen and over time also decreases pH, which can both be lethal to invertebrates and fish, sometimes resulting in anoxic zones (also known as "dead zones") and fish kills. Nitrogen loading from mining poses concerns to aquatic habitats near mining sites but also potentially contributes to loads that have consequences for more distant habitats, such as the Chesapeake Bay. A multi-state effort is under way to restore the habitats of the Chesapeake Bay, with a major focus on reducing loads of nitrogen, phosphorus, and sediment to improve conditions for the bay's aquatic life. Total maximum daily loads of nitrogen, phosphorus, and sediment for each state and each watershed have been established by EPA to reach the restoration goals (see also Chapter 5), and Virginia has worked aggressively to reduce its nutrient loads to meet the restoration targets (see Figure 4-2).

TAILINGS STORAGE FACILITIES FAILURE AND TAILINGS RELEASE

Some gold mining operations produce large amounts of slurry effluents, called tailings (Adler and Rascher, 2007). Tailings, which can contain a wide range of metals, are often stored in impoundments behind perimeter dikes (i.e., in tailings storage facilities [TSFs]). Although numerous best practices designed to safely retain these materials are presented in Chapter 3, TSFs can occasionally fail, releasing toxic materials downstream into streams and rivers with negative effects on natural ecosystems and on human health. These events can lead to acute danger (e.g., fatalities, injury, destruction of property). For example, the failure of two iron ore TSFs in Brazil in 2015 and 2019 resulted in numerous immediate fatalities (Vergilio et al., 2020).

Although the acute effects of tailings dam failures are well documented, there is significantly less information in the scientific literature regarding the potential chronic environmental impacts on ecosystems and human populations. The most significant chronic environmental impact of tailings dam failures from gold mines is the release of metals, which can be dissolved in surface water runoff, sorbed to sediment particles, or dispersed by wind (Barcelos et al., 2020; Fashola et al., 2016). Metals can be a serious health issue because they persist in the environment and thus can pose long-term effects on ecosystems (Singh et al., 2011). Metal-rich sediments can be lethal to stream invertebrates and vertebrates at each level of the food chain (Vergilio et al., 2020), and metals that are deposited in soils adjacent to rivers can become incorporated into plants and crops, which can lead to negative

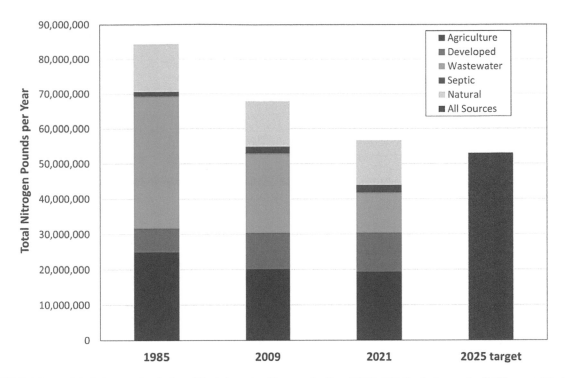

FIGURE 4-2 Modeled nitrogen loads from Virginia to the Chesapeake Bay (1985–2021) relative to the 2025 target. The loads are simulated using the Chesapeake Assessment and Scenario Tool version 2019 and jurisdiction-reported data on wastewater discharges.
SOURCE: Data from Chesapeake Bay Program (2022).

health effects if those plants or crops are consumed by humans (Barcelos et al., 2020). A key factor that influences the magnitude of impacts from tailings dam failures is the bioavailability of the toxic metals in the tailings. The minerals that toxic metals are included within or sorbed to affect their availability to humans and wildlife (Barcelos et al., 2020). One case study relevant to Virginia is the Valzinco massive sulfide deposit in Spotsylvania County, which was mined intermittently until 1945. At the time of its reclamation in 2001, tailings had moved up to 2.5 km downstream of the failed tailings dam, contaminating the water with toxic metals (Hammarstrom et al., 2006; see Table 2-4). Another relevant case study is the tailings dam failure at the Mount Polley copper and gold mine in British Columbia, Canada. Analytical studies concluded that metals in the fine sediments deposited from the tailings breach were bioavailable and potentially toxic to invertebrates several years after the event (Pyle et al., 2022; see Box 4-2).

WATER TABLE DEPRESSION

The practice of dewatering a mine by pumping the water from the bottom of the pit or underground workings can affect the groundwater table (see Chapter 3). Groundwater wells near open pit or underground mines may run dry depending on the well depths, distance from the mine being dewatered, and the hydrogeologic properties of the local aquifers (see Figure 4-3). Mine dewatering in a low-permeability aquifer could create steep cones of depression in the water table that would affect residents living relatively close to the mine site. Dewatering in high-permeability or highly fractured aquifers, by contrast, could result in more extensive, but less steeply depressed, areas of drawdown that could be unequally distributed based on the orientation of bedrock fractures. The Piedmont and Blue Ridge regions of Virginia are composed of crystalline rock aquifers (86 percent), Early Mesozoic basins aquifers (9 percent), and low-permeability carbonate rock aquifers (3 percent; see Figure 4-4).

BOX 4-2
Other Case Studies of the Chronic Environmental Impacts of Tailings Release

Besides Pyle et al. (2022), which describes the bioavailability of metals following the tailings dam failure at the Mount Polley copper and gold mine, there are few case studies that report the chronic environmental effects of TSF failures at gold mines. Below, two studies are described that report the chronic ecological effects of TSF failures, but that are unrelated to gold mining. The first relates to a failure at a zinc/silver/lead/copper mine in Spain and the second relates to an iron ore mine in Brazil.

The 1998 mine tailings spill of 4 million cubic meters of acidic water and 2 million cubic meters of mud containing metals at the Aznalcóllar zinc, silver, lead, and copper mine in southwest Spain resulted in widespread distribution of zinc, lead, arsenic, copper, antimony, cobalt, thallium, bismuth, cadmium, silver, mercury, and selenium (Grimalt et al., 1999), much of it upstream of Doñana National Park, a critical habitat for migratory birds and other wildlife. In follow-up studies, underlying soil was found to contain a number of metals in an accumulation zone up to 30 centimeters deep (Kraus and Wiegand, 2006). The committee was unable to locate any published studies that identified human health impacts from environmental exposures from the Aznalcóllar spill, but metals such as arsenic, lead, and cadmium were elevated in tissues of terrestrial wildlife several years later in the areas impacted by the spill (Fletcher et al., 2006).

Two major tailings dam failures occurred in Brazil, one in 2015 and the other in 2019. The first, the Fundão Dam at the Germano iron ore mine in Bento Rodrigues in Minas Gerais State, released more than 40 million cubic meters of tailings, contaminating over 668 kilometers of surface waters. The second, when Dam B1 failed at the Córrego do Feijão iron ore mine in Brumadinho in Minas Gerais State, released 12 million cubic meters. Both incidents resulted in numerous immediate fatalities and a wide distribution of metals in the surrounding environment. Toxicological tests from the Brumadinho Dam rupture demonstrated that the contaminated soil and sediments could affect different trophic levels, from algae to microcrustaceans and fish. They also demonstrated that metals were accumulated in the muscle tissue of fish following the event (Vergilio et al., 2020).

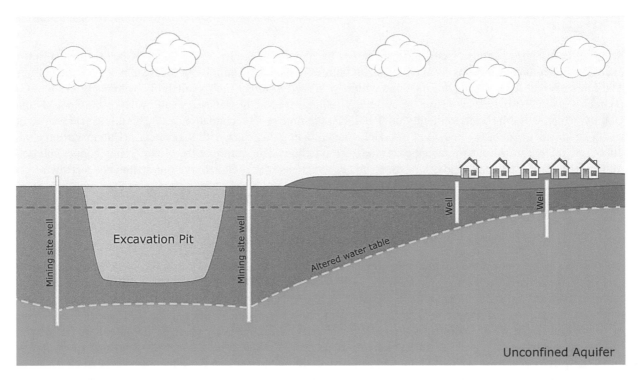

FIGURE 4-3 Pumping at an open pit mine in high-permeability homogeneous aquifers leads some wells near the mine to run dry due to groundwater table drawdown. The darker blue dotted line represents the water table prior to being altered by mining.

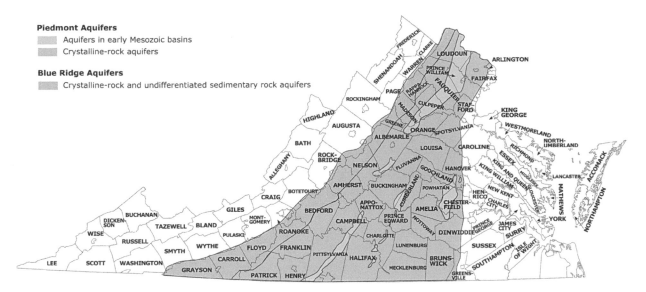

FIGURE 4-4 Rock types associated with aquifers in Virginia's Piedmont and Blue Ridge regions.
SOURCE: Image modified from Trapp and Horn (1997).

Groundwater in the crystalline rock and low-permeability carbonate rock moves along steeply angled joints and faults. Because bedrock fractures often have preferred directions of orientation, groundwater will flow more readily along those orientations (see Figure 4-5), making it more difficult to anticipate the exact area and magnitude of the groundwater drawdown (Cohen et al., 2007).

Following the termination of pumping, the water table will begin to recover. The rate of recovery depends on the recharge rate and aquifer permeability. Recharge rate is highly variable in Virginia's Piedmont and Blue Ridge regions (e.g., varying from 4 to 28 inches per year in Bedford County; Cohen et al., 2007) and is determined by precipitation, runoff, and thickness of the unconsolidated material overlying the bedrock. The unconsolidated material is thicker in the Piedmont region than in the Blue Ridge region, which leads to faster aquifer recharge in the Piedmont (Trapp and Horn, 1997). However, even a temporary lack of well water for household use and irrigation may require installation of new wells or the transport of water to properties near the mine. If a new well

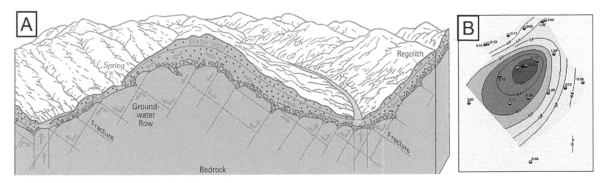

FIGURE 4-5 Cross-section of crystalline rock aquifer in Virginia. (A) Groundwater-saturated bedrock fractures often have a preferred orientation. (B) Contours of equal water-level decline after pumping shows that preferred orientation of the fractures may lead to greater water-level decline parallel to the preferred direction of fracturing.
SOURCE: Images from Trapp and Horn (1997).

is not provided, depression of the water table may lead to increased cost of living and decreased property values and quality of life for residents.

Rivers, lakes, and springs have a close relationship with groundwater. Dewatering a mine can reduce surface water flows if these surface waters intersect the cone of depression of the water table and if the near-surface saturated zones are well connected with the aquifer being dewatered (and not separated by a confining layer such as clay or shale). If mine-related water withdrawals are large and significantly impact the flow volume of the stream, downstream water users and the ecosystem could be affected. Major impacts to surface water are not expected given the relatively limited size of mines expected in Virginia, although site-specific analysis would be needed for any proposed mine to evaluate potential impacts.

AIR EMISSIONS

Various air pollutants can be generated from mining activities (see Chapter 3). Some of these agents are hazardous air pollutants known to cause cancer or other serious health impacts (e.g., mercury, certain species of volatile organic compounds [VOCs]), whereas others are common air pollutants called criteria air pollutants (e.g., particulate matter, carbon monoxide [CO], sulfur dioxide [SO_2], nitrogen oxides [NO_x], ozone [O_3]). Mining activities do not result in direct emissions of ozone but it can be produced through photochemical reactions in the presence of ozone precursors emitted from mines (namely, VOCs, CO, NO_x).

Historically, the most important impacts for mining-associated air emissions have been occupational exposures to certain kinds of particles that cause a large set of occupational lung diseases. These are generally interstitial lung diseases, and include examples such as asbestosis, coal workers' pneumoconiosis (black lung disease), and silicosis. Inhalational exposure to dusts containing high concentrations of elements such as aluminum, antimony, iron, and barium, or minerals such as graphite, kaolin, mica, and talc, can also cause pneumoconiosis (NIOSH, 2022). However, these minerals and elements are not found in high enough concentrations in the Virginia gold deposits that they would be a concern for the nonoccupational communities in Virginia. There are federal regulations and best practices to limit workplace exposure to dust, but the occupational impacts are not addressed in this report.

Fugitive dust may be emitted from mine sites from drilling, blasting, ore crushing, roasting, smelting, hauling and moving of materials, operation of machines and vehicles on unpaved roads, and storage and disposal of waste. The dust produced from many of these operations tends to contain relatively large particles that settle out of the air quickly and do not penetrate far into the respiratory system (Entwistle et al., 2019). But if not controlled, these dusts can be hazardous, especially if they contain high concentrations of potentially toxic elements, such as the metals described in the "Metals and Metalloids" section. In fact, studies in Chile have reported that residential proximity to large gold or copper open pit mining was associated with a higher prevalence of respiratory diseases among children (Herrera et al., 2016, 2018). However, fugitive dust can generally be limited through best practices on mine sites (see Chapter 3) and is typically less of a concern in the United States than in low- and middle-income countries where dust emissions are less regulated (Entwistle et al., 2019). Hence the spatial and temporal scales of the impacts of fugitive dust from gold mining in Virginia would be fairly limited. Nevertheless, when fugitive dust is not properly controlled on mine sites it can be a significant concern to nearby communities and can adversely affect public health.

Another source of air pollutant from gold mines that may impact air quality and public health beyond the mine site is the exhaust from fuel-burning vehicles and machines. Combustion of fossil fuels, in particular diesel, leads to emissions of gases and vapors, including CO, NO_x, and VOCs, as well as fine particulate matter that comprises elemental and organic carbon, ash, sulfate, and metals (IARC, 2014). Diesel exhaust is a Group 1 carcinogen that can cause lung cancer and bladder cancer (IARC, 2014). The impact of diesel exhaust will be proportional to the truck traffic and heavy equipment operation at the site. Given that future gold mining operations in Virginia are likely to be limited in size, the impacts of diesel emissions on surrounding communities may be limited.

Other activities on a mine site may also produce air emissions. For example, the processing of ores, including the high-temperature combustion or heating processes such as roasting and smelting, can release nitrogen oxides, sulfur dioxide, and mercury. The amounts of mercury compounds produced are very site specific and dependent on the ore composition and the mining processes used. In 2018, the processing plant at Haile Gold Mine in South Carolina exceeded compliance levels for mercury (40 CFR Pt. 63 Subpart EEEEEE, 2022). In response to this, Haile operators installed a mercury abatement control device system and the mine has not exceeded mercury compliance levels since (USACE, 2022). The source of the mercury was unclear in this case (Morton, 2020). Although the committee does not have any evidence of elevated mercury in Virginia gold deposits, there are limited data, which leaves significant uncertainty. In addition, there could be significant amounts of mercury near old gold mine sites where mercury was once used during gold processing (see Chapter 2). If mercury from contaminated historical mine sites was inadvertently brought into the processing stream, it could lead to atmospheric emissions. Most atmospheric mercury is in the metallic gaseous form (about 90 percent or greater), but atmospheric redox reactions can convert mercury between different forms (Horowitz et al., 2017). Metallic mercury has a long atmospheric lifetime (around 1 year), enabling it to be transported far downwind (Horowitz et al., 2017) before redepositing to Earth's surface. The redeposited mercury has further implications for soil and water quality.

Effective control strategies and techniques have been developed for various mining-associated emissions. Examples include spraying water or dust suppressant on roads and waste piles, reducing open surfaces, enclosing ore-crushing areas, switching from internal combustion engines to other power sources, and applying air pollution control systems (scrubbers) for specific air pollutants. There are very effective (with 90 percent or greater controlling efficiency) air pollution control systems for all the major air pollutants, including particulate matter, sulfur dioxide, nitrogen oxides, VOCs, and mercury. Such systems are widely used in stationary sources such as power plants, oil refineries, and manufacturing settings and similar approaches have been used in gold mining operations at Haile Gold Mine in South Carolina (USACE, 2022).

The impacts on air quality from various sources of air pollution depend not only on the emission fluxes but also on the background or baseline air quality (Sillman et al., 1990; Wu et al., 2009). The committee reviewed air quality data from Virginia for the past 5 years (EPA, 2021a). All the criteria air pollutants were found to be in compliance (levels lower than the National Ambient Air Quality Standards [NAAQS]) for gold-bearing regions in Virginia, but there were multiple counties with fine-particle particulate matter ($PM_{2.5}$) levels close to the NAAQS (annual average of 12 $\mu g/m^3$ and 24-hour average of 35 $\mu g/m^3$).

CUMULATIVE HEALTH IMPACTS FROM COMBINED EXPOSURES

Recently, there has been growing recognition that human populations and ecosystems are exposed to multiple stressors in combinations that can interact in a dynamic way to produce a range of outcomes. In 2003, EPA developed a long-term initiative to evaluate combined risks of adverse effects on human health or ecosystems from multiple environmental stressors (Callahan and Sexton, 2007). Stressors that may impact human or ecological health include not just chemical toxicants but any combination of chemical, biological, physical, and psychosocial hazards. Importantly, special attention has been given to evaluating how chemical and nonchemical stressors can interact to increase the risk of adverse health outcomes (Sexton, 2012). Progress on EPA's cumulative risk assessment efforts has been somewhat slow because studies of concurrent exposure to multiple hazards are more methodologically challenging, time-consuming, and costly to conduct. For cumulative ecologic risk assessment, the additional complexities are often so significant that many of the available studies are only qualitative or semiquantitative (EPA, 1998a).

Several kinds of complexity are introduced by the cumulative risk assessment process. These include time- and spatial-related aspects of exposures (e.g., concurrent exposure, serial exposure, past exposure in critical time period combined with current exposure), vulnerability of exposed populations (i.e., based on biology, exposures, underlying health, and recovery), identification of subgroups with exposures of special concern (e.g., higher exposures based on occupation or behaviors), and characterization of interactions between psychosocial stress (i.e., that could arise from poverty, inadequate housing, street crime, discrimination, unemployment, and other sources of stress) and other hazardous exposures (Callahan and Sexton, 2007; EPA, 2003; Gallagher et al., 2015). When populations are exposed to more than one hazard at a time, effects can exhibit a range of risks that may equate to a sum of the individual risks or exceed that sum. Studies of multiple hazards are increasingly appearing in the peer-reviewed scientific literature and highlight that concurrent exposure to multiple pollutants and exposure to chemical and nonchemical stressors can increase the risk of adverse health outcomes (Bobb et al., 2015; Domingo-Relloso et al., 2019; Green et al., 2015; Iakovides et al., 2021; Lee et al., 2021; Meza-Montenegro et al., 2012; Park et al., 2017; Peters et al., 2014; Sanders et al., 2019; Wang et al., 2018, 2020a,b, 2021; Xiao et al., 2021b; Zhou et al., 2019). Studies like these have direct relevance to gold mining in Virginia because they suggest that the toxicant exposures arising from gold mining operations could differentially impact the health of populations that experience concurrent exposure to nonchemical stressors that affect psychological or physical health.

Gold mining operations present multiple hazards to communities, including direct exposure to pollutants, increased stress, and changing perceptions of community conditions. These different hazards can interact to increase the risk of adverse health outcomes. For example, a study in a community with small-scale and industrial-scale gold mining in Ghana evaluated perceived stress, salivary cortisol as a biomarker of stress, personal noise exposure, and heart rate, documenting that communities with gold mining can experience multiple, often additive, exposures that can contribute to such health outcomes as hearing loss and cardiovascular disease (Green et al., 2015). In addition, these mixed exposures are occurring in populations with differences in social vulnerability (Emmett, 2021), underlying health vulnerability, behavioral vulnerabilities, and individual susceptibility to these factors. Because of this, the impacts of degraded water quality from mining on nearby populations are best interpreted in light of a variety of cultural, social, and economic vulnerabilities (French et al., 2017).

The County Health Rankings system is one tool for considering the concurrent stressors that may already exist in Virginian communities (see Figure 4-6). This tool evaluates health outcomes according to premature

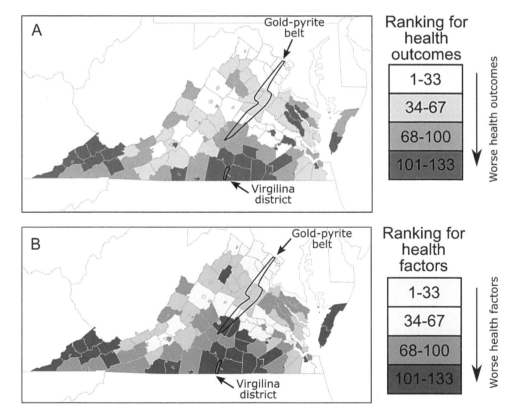

FIGURE 4-6 Health outcome and factor rankings for all counties in Virginia with higher rankings (darker shading) being worse for health outcomes and factors. Black outlined areas are the gold pyrite belt and the Virgilina district, and the yellow dot is the recent exploratory drilling in Buckingham County. (A) Health outcomes include premature death, poor or fair health, poor physical health days, poor mental health days, and low birthweight. (B) Health factors include access to clinical care, local economics, and prevalence of behaviors that can affect health. Some of the southern regions of the gold-pyrite belt and the Virgilina district have worse health outcomes and health factors than the northern part of the belt. For example, Buckingham County was ranked in the second worst quartile for health outcomes and the lowest quartile for health factors.
SOURCE: Modified map from The University of Wisconsin Population Health Institute (2022).

death, poor or fair health, poor physical health days, poor mental health days, and low birthweight. It also evaluates health factors using nine indicators for health behaviors (e.g., adult smoking, physical inactivity, teen births), seven indicators for clinical care (e.g., uninsured, mental health providers, flu vaccinations), nine indicators for social and economic factors (e.g., high school completion, children in poverty, income inequality, violent crime, injury deaths), and five indicators for physical environment (e.g., particulate matter air pollution, drinking water violations, severe housing problems).

The impacts of mining also tend to be distributed unevenly across landscapes and communities, based on proximity to the mine site, characteristics of the physical environment, socioeconomic and political structure, and demographic factors. Some communities are relatively successful in exercising self-determination with regard to proposed mineral mining projects, for example by ensuring that mineral developments reflect community priorities. Other communities with less socioeconomic or political influence, especially those who live near the mine, tend to bear the brunt of the negative impacts of mining, but the positive impacts accrue to others, including investors, shareholders, and users of the end products manufactured from mined materials (Dunbar et al., 2020). This poses well-established environmental justice issues (Kivinen et al., 2020). As defined by EPA, environmental justice ensures that all communities have the same degree of environmental protections and equal access to the decision-making processes that shape the environments in which they live (EPA, 1994a). Environmental justice is at the intersection of environmentalism (protect and improve the environment) and justice (fairness among members of society) and fundamentally addresses the fact that the health impacts of environmental degradation are unevenly distributed across society. Hence, environmental justice efforts address such issues as the disproportionally high distribution of industrial activities in poor and minority communities, the heightened impacts of pollution in these communities, the lower prevalence of environmental amenities in such communities (e.g., green spaces, healthy food), and the complicating factor that such places also are characterized by worse access to health care, worse underlying health, and higher prevalence of adverse health behaviors that make these populations more vulnerable to the hazardous agents (Diez Roux and Mair, 2010; Hilmers et al., 2012; Hynes and Lopez, 2007; Olvera Alvarez et al., 2018). Mining operations have long been a focus of environmental justice concerns (Aydin et al., 2017; Lewis et al., 2017; Liévanos et al., 2018; Morrice and Colagiuri, 2013). For example, the impacts of acid rock drainage are different in Indigenous communities than in non-Indigenous communities (Clausen et al., 2015).

CONCLUSIONS AND RECOMMENDATION

This chapter outlines the potential human health and ecological impacts from gold mining in Virginia based on a review of the impacts of gold mining at U.S. and international sites and on the concerns expressed by community members during the information-gathering activities for this study. As little commercial gold mining has occurred in Virginia in the past 70 years, there is limited information about the impacts from historical gold mining in Virginia or the constituents of concern in the remaining gold deposits. The committee therefore could not predict site-specific impacts from gold mining in Virginia, but instead evaluated the impacts reported at other gold mining sites in the context of the environmental, geologic, and social conditions of the gold-bearing regions of the Commonwealth. The committee used the best available scientific and technical information to draw the following conclusions and recommendation.

Remobilization of Legacy Contaminants

Remobilization of legacy mercury from mining operations that take place at historically mined sites poses a significant risk to human health and the environment. Mercury is no longer used for the processing of gold in the United States, but it was used at historical gold mines in Virginia. As a result, considerable legacy mercury may exist in surface waters, soil, and mine waste at previously mined sites. These areas may still harbor unmined gold deposits and unrecovered gold in historic waste material, and future gold mining operations could remobilize this legacy mercury unless appropriate extraction and processing circuits are implemented to capture the mercury. Because of mercury's high toxicity, careful characterization for mercury is essential at all potential mine sites in order to protect environmental and human health.

Impacts to Water Quality

Acid rock drainage (ARD) is among the most important potential environmental impacts of concern and poses a substantial risk if massive sulfides are disturbed during gold mining operations and if proper engineering controls are not in place. ARD can persist long after mining has ended and can cause acidity, high salinity, and elevated concentrations of toxic metals in surface water and groundwater if appropriate engineering controls are not in place. Many gold deposits in Virginia are not directly associated with large quantities of sulfide-containing minerals, reducing the likelihood of extensive ARD associated with mining. However, if adjacent massive sulfide deposits or sulfide-bearing country rock are disturbed and if appropriate engineering controls are not applied, ARD could adversely impact sensitive freshwater fauna in nearby streams and wetlands, resulting in substantial remediation costs. Site-specific characterization, engineering controls, and monitoring throughout the life cycle of gold mines are important to minimize and mitigate ARD that could negatively impact surface water and ecological communities.

Site-specific geologic conditions determine whether metals could be released from gold mining operations in sufficient quantities to pose human health threats to surrounding communities. The primary elements of concern for human health that could be released from Virginia gold deposits or from nearby rocks disturbed during mining include antimony, arsenic, cadmium, lead, mercury, and thallium. Most Virginia gold deposits occur in low-sulfide, gold-quartz veins and the few reliable geochemical data that are available for these deposits show low concentrations of metals of concern in discharge waters. However, some gold deposits in Virginia are located in close proximity to massive sulfide deposits, which have higher concentrations of pyrite and higher risk of toxic metal discharge, leaving considerable uncertainty in predicting risk across the state. Therefore, any future efforts to mine gold deposits in Virginia should be accompanied by detailed studies to characterize the mineralogy, metal content, and geochemistry of each deposit and its surrounding rock. Site-specific characterization, water quality management, and monitoring throughout the life cycle of gold mines will be important to minimize and mitigate the release of metals that could negatively impact surface water and groundwater quality.

Mining can increase nitrate loading to local waterways, which can contribute to eutrophication of local surface waters. Although best practices for blasting activities can limit nitrogen loading of surface water and groundwater (see Chapter 3), incomplete combustion of ammonium nitrate and fuel oil explosives under wet, nonideal conditions may result in nitrate-laden mine-influenced water that can exceed water quality criteria. If this water is not appropriately managed and it reaches local surface waters without significant dilution, depleted dissolved oxygen and reduced pH due to eutrophication may result, which can be lethal to invertebrates and fish. Mining could also contribute to the total loading of nitrogen to more distant habitats (e.g., the Chesapeake Bay), although the relative contributions to the total loads are expected to be small. Elevated nitrate in drinking water can also be harmful to human populations, but these higher concentrations are likely only possible in groundwater in the immediate vicinity of the mine site and can be prevented with best practices for blasting activities.

Open impoundments that contain cyanide pose acute toxicity risks to wildlife unless proper management and deterrents are in place. Wildlife species are attracted to virtually any kind of surface water body, natural or constructed, including waste and treatment impoundments. In the arid western United States, there have been numerous acute toxicity events affecting wildlife (especially birds) at cyanide impoundments in gold mining sites, although there have been fewer reports documenting these toxicity events following the establishment of modern best practices for cyanide management. Although surface water is plentiful in Virginia, the Commonwealth hosts diverse and abundant wildlife species that are dependent on access to open surface water. Unless best practices (e.g., deterrent systems, cyanide destruct systems) or alternative methods (e.g., enclosed tank leaching) are used, wildlife acute toxicity events could occur at open impoundments containing cyanide.

Impacts to Air Quality

The committee did not find evidence to indicate that gold mining in Virginia would significantly degrade air quality if appropriate engineering controls were in place. Fugitive dust produced from excavation activities, heavy equipment, and mine road traffic can be a nuisance that impacts the quality of life of affected neighbors. In addition, toxic fine particles and gaseous pollutants generated from fuel combustion and gold processing can be hazardous if released, because of their greater respiratory impacts and longer atmospheric transport distance. Given the likely small scale of future commercial gold mining in Virginia that would lead to limited heavy equipment operation and traffic, and the technological advancements in recent decades that allow for effective dust suppression and control of hazardous air pollutants, the impacts of air pollutants on surrounding communities are expected to be limited.

Rare But Catastrophic Events

Catastrophic failures of gold mine tailings dams and cyanide solution containment structures are low-likelihood but high-consequence events that have caused significant impacts where they have occurred. Tailings dam failures can lead to acute danger (e.g., fatalities, injury, destruction of property) as well as long-term ecological effects that are caused by the dispersal of toxic metal-containing mine wastes in rivers and floodplains. The magnitude of the long-term ecological effects depends on the scale of the spill, bioavailability of the contaminants, and effectiveness of cleanup efforts. In contrast, cyanide spill events do not pose long-term risks because cyanide degrades in the surface environment relatively quickly. However, because of cyanide's high acute toxicity, accidental spills have caused mass mortality events of aquatic life and pose an acute human health risk where water affected by the spill is used as a drinking water supply. If tailings and cyanide containment structures are not designed to accommodate seismic, high-precipitation, and flooding events, then the likelihood of these potential high-consequence events will increase. This is especially pertinent in light of the potential for increased frequency and severity of precipitation events due to climate change.

Impacts to Water Quantity

Drawdown of the water table associated with the dewatering of an open pit or underground mine could impact local groundwater users, depending on aquifer conditions and the proximity of wells to the mine site. Unless drawdown effects are appropriately mitigated, these impacts could significantly affect the quality of life and the cost of living for residents near the mine site who rely on groundwater supplies. Rigorous site characterization and modeling is needed to estimate the level and geographic span of groundwater impacts and to evaluate whether alternative sources of water or new wells need to be provided to local citizens. Public engagement and participation during permitting is essential if alternative sources of water or new wells may need to be provided.

Cumulative Risk

Robust analyses of the potential impacts of mining consider cumulative health risks. Human populations are exposed to multiple hazard types, including biological, physical, chemical, psychological, and social (e.g., poverty, discrimination, unemployment, limited access to health care). These hazards can occur through different exposure settings (e.g., environmental, occupational) and multiple media (e.g., air, water, soil). Different hazard types, especially chemical and nonchemical stressors, can interact to affect human health in complex and dynamic ways. These multiple, sometimes synergistic, stressors can lead to asymmetric impacts within and between communities, and historically underresourced and underrepresented populations are often most affected.

***

In light of the general impacts of gold mining in Virginia that are outlined in the conclusions above, robust site- and project-specific analyses are necessary in order to assess potential impacts and determine what mining operation procedures will be most protective of human and ecological health.

RECOMMENDATION: To minimize impacts to human health and the environment, the Virginia General Assembly and state agencies should ensure that robust site- and project-specific analyses of impacts are completed prior to the permitting of a gold mining project.

5

Virginia's Regulatory Framework

Mining projects in the United States, including gold projects, require numerous permits and approvals issued by a combination of federal, state, and local government agencies that are designed to help protect public health and the environment, among other goals (see Table 5-1). The numbers and types of required permits vary according to the size, type, location, and other specifics of a project, but it is common for mining projects to require dozens of permits and authorizations. In contrast to coal mining, there is no overarching federal regulatory program for gold mining that applies to all lands, regardless of ownership.[1] Gold mine projects typically need permits under federal environmental regulatory programs like the Clean Water Act and the Clean Air Act; state permits that address mine design, operation, reclamation, closure, and financial assurances; and, potentially, local permits that may cover a variety of local concerns including transportation, noise, timing of certain activities, as well as many other issues of concern (§ 45.2-1227 of the Code of Virginia; see Box 5-1). The interplay of the different authorities and permitting requirements can be confusing to the public, especially when different permits each have separate public notice and comment requirements.

In addition to the different roles of federal, state, and local governments in regulating gold mines, different laws can be triggered if mining were to be proposed on private, state, or federal land. For example, certain federal and state requirements, like the requirements for an environmental review process, may not be triggered by a proposed mining activity on land that is privately owned, which includes the majority of land in Virginia, as noted in Chapter 1.

FEDERAL ENVIRONMENTAL REVIEWS

In Virginia, a formal review of environmental impacts would not be required to issue permits for a project on private lands unless a major federal action (e.g., certain federally issued permits or authorizations) were involved and triggered the National Environmental Policy Act (NEPA). Examples of major federal actions that might occur during gold mining operations include permitting from the U.S. Army Corps of Engineers (USACE) for discharges of dredged or fill material into waters of the United States (33 CFR § 322.3), or the permitting of mining operations that would occur on federal lands (see Box 5-2). If federal permits or approvals associated

---

[1] The Bureau of Land Management (BLM) and the U.S. Forest Service have regulations that govern the conduct of mining related activities on the lands they administer (43 CFR 3809 [BLM] and 36 CFR 228 [U.S. Forest Service]).

TABLE 5-1 Various Agencies and Their Role in Permitting and Regulating Gold Mines in Virginia

State/Federal	Agency	Role in Regulation
Federal	U.S. Army Corps of Engineers	Discharge of dredged or fill material into waters of the United States, including jurisdictional wetlands. National Environmental Policy Act compliance and consultations under the Endangered Species Act and National Historic Preservation Act
	U.S. Fish & Wildlife Service	Endangered Species Act consultation
	U.S. Environmental Protection Agency	Clean Water Act Section 404 application, review of draft Environmental Impact Statement, and Underground Injection Control well permitting
	Bureau of Alcohol, Tobacco, Firearms and Explosives	Transportation, use, and storage of explosives
	Mine Safety and Health Administration	Mine safety and health (occupational)
Virginia	Virginia Energy, Mineral Mining Program	Mineral mining (nonfuel) operations and reclamation
	Department of Environmental Quality, Air Programs	Air pollution emissions and permitting
	Department of Environmental Quality, Water Programs and State Water Control Board	Point-source discharges to waters, underground or surface petroleum storage tanks, groundwater withdrawal, water rights authorization, groundwater management areas
	Department of Environmental Quality, Waste Programs	Treatment, storage, disposal, or transportation of solid and hazardous waste; reclamation of nonhazardous wastes
	Chesapeake Bay Local Assistance Department	Impacts on the Chesapeake Bay
	Department of Conservation and Recreation	Impact on recreation resources and unique habitats, non-point source water pollution, stormwater management
	Department of Historic Resources	Protection of historic structures and archaeological resources
	Department of Agriculture and Consumer Services	Endangered plants and insects
	Department of Health	Protection of public or private water supply, drinking water quality, disposal of biosolids
	Department of Game and Inland Fisheries	Effect on fish and endangered animals
	Department of Forestry	Impact on state forests
	Virginia Marine Resources Commission	Construction and disturbances in waterways and wetlands, activities affecting state-owned subaqueous lands
	Virginia Department of Transportation	Entrance and access to public highways from mineral extraction sites, use of public highways by trucks, highway right-of-ways

SOURCE: Table modified from Virginia DMME (2007).

with a gold mining project are deemed to be a major federal action, then NEPA requires either an environmental impact statement (EIS), environmental assessment (EA), or confirmation that a categorical exclusion[2] applies to the action. The Virginia Department of Environmental Quality (DEQ) Office of Environmental Impact Review coordinates the review of any federal EA and EIS documents developed under NEPA (§ 10.1-1183 of the Code of Virginia; 40 CFR Part 1500–1508). NEPA also facilitates input from other governmental agencies that have jurisdiction by law or special expertise, including other federal agencies, state and local agencies, and tribes, by allowing and encouraging them to formally participate throughout the NEPA process as "Cooperating Agencies."

---

[2] A categorical exclusion is a type of action that has been determined to not have a significant effect on the human environment. Normally, neither an environmental assessment nor an environmental impact statement is required for these actions. A categorical exclusion would likely only occur for minor disturbances to earth, air, or water, like the construction of minor access roads and streets (33 CFR § 230.9).

BOX 5-1
Federal, State, and Local Oversight in Regulation of Gold Mines

In U.S. environmental law, there is a long-standing commitment to "cooperative federalism," meaning that regulatory authority is shared by the state and the federal government. While the federal government sets mandatory minimum standards, it can delegate to the state implementation of these standards. Many U.S. environmental laws, such as the Clean Water Act (CWA), the Clean Air Act (CAA), the Safe Drinking Water Act (SDWA), and the Resource Conservation and Recovery Act (RCRA), adopt this system, and states are given the authority to create their own programs to implement these laws. States must enforce the federal standards as minimums, but they also have the discretion to be more (but not less) protective of the environment if they so choose (Elliott and Esty, 2021). Virginia is delegated to run most programs under the CAA, CWA, SDWA, and RCRA (Troutman Sanders LLP, 2008, pp. 1–2). It does not have delegation under Section 404 of the CWA. As a result, a project that requires the discharge of dredged or fill material into waters of the United States, including jurisdictional wetlands, would require a CWA 404 permit from the U.S. Army Corps of Engineers.

In addition to those federal programs delegated to the state, in certain instances local governments can adopt their own ordinances. The Code of Virginia states that "Any locality may establish standards and adopt regulations for mineral mining, so long as such standards and regulations are no less stringent than those adopted by the Director [of Virginia Energy]" (§ 45.2-1227 of the Code of Virginia).

NEPA was one of the first laws that established a broad national framework for protecting the environment (42 USC § 4321 et seq.) and established a "look before you leap" approach for permitting actions. The efforts to develop EA and EIS documents involve a thorough examination of the existing conditions or baseline information for a wide range of resources and an assessment of potential effects under the proposed action, as well as alternative hypothetical scenarios. NEPA requires federal agencies to consider the potential environmental effects on natural resources, as well as social, cultural, and economic resources. It also requires that the federal agencies inform the public about their decision making process (40 CFR § 1502.3; Council on Environmental Quality, 2021; EPA, 2021b). The gathering of this type of information can be very useful for state permitting processes as illustrated in an example from the Haile Mine in South Carolina (see Box 5-3). The NEPA analysis may "serve as a framework" to meet other requirements, such as those associated with the National Historic Preservation Act, the Endangered Species Act, the Environmental Justice Executive Order, and other federal, state, tribal, and local laws and regulations (VDOT, 2022). NEPA, however, does not require that an environmentally preferable alternative is selected or that adverse environmental effects are prohibited.

BOX 5-2
Waters of the United States

The Clean Water Act (CWA) applies to "navigable waters," which are defined as "waters of the United States" (see CWA section 502(7)). The term "waters of the United States" (or WOTUS) is therefore an important concept because it dictates whether certain activities are covered by this law. The CWA gives the U.S. Environmental Protection Agency (EPA) and the U.S. Department of the Army discretion to define "waters of the United States" (CRS, 2022; EPA, 2021b). However, defining what is—and is not—WOTUS is complicated and controversial. For example, it is unclear whether wetlands that are not navigable but are next to navigable water or wetlands that are on private property a mile from the closest navigable stream would be defined as WOTUS. If these are defined as WOTUS, then CWA permits must be obtained to fill, dredge, or emit pollutants to them.

In recent years, the U.S. courts have stepped in to examine the meaning of this term. At present, there is a case before the U.S. Supreme Court (SCOTUS Blog, 2022) that will review whether certain wetlands are covered under the WOTUS definition. The decision in this case will impact the applicability of the CWA to wetlands. Any potential gold mining project that impacts wetlands could, therefore, be affected by this decision.

BOX 5-3
NEPA EIS Aids Permitting Process for a South Carolina Gold Mine

An environmental impact statement (EIS) for the Haile Gold Mine in South Carolina was completed in 2014 and a supplemental EIS for a permit modification was completed 2022. South Carolina does not have state-level requirements to conduct an environmental assessment (EA) or EIS for gold mining permits, but the U.S. Army Corps of Engineers initiated an EIS in this case because the proposed operations would require a Clean Water Act (CWA) 404 permit to impact wetlands, constituting a major federal action. In the absence of this federal action, it seems unlikely that an EIS would have been conducted prior to issuing the gold mining permit. Jeremy Eddy, with the South Carolina Department of Health and Environmental Control, indicated that the EIS was helpful for the South Carolina permitting processes because it provided the regulatory agency more resources and information than would otherwise be available (Jeremy Eddy, personal communication, 2022). The benefits of an environmental impacts analysis can include:

- Providing baseline information for environmental resources and evaluating project details for potential impacts ("look before you leap" approach to permitting);
- Evaluating technical considerations (e.g., Failure Modes Effects Analysis), identifying mitigation for adverse impacts, and comparing potential impacts for project alternatives; and
- Engaging with citizens and stakeholders during scoping and document development.

NEPA has many procedural requirements. These include publication of a "Notice of Intent," a scoping process, multiple public notice and comment opportunities, a description of the affected environment, evaluations of environmental impacts including cumulative impacts, and an analysis of alternatives that must include the "No Action Alternative" (Council on Environmental Quality, 2021). The distinctions between an EA and an EIS are described below.

Environmental Assessment

If a categorical exclusion does not apply to a proposed action, an EA may be completed. This assessment determines whether the action may cause significant environmental effects and generally includes a brief discussion of the need for the action and alternatives to the proposed action, the environmental impacts of the proposed action and alternatives, and documentation of the agencies and people consulted. Based on the conclusions of the EA, the applicable federal agency may issue a Finding of No Significant Impact, which discusses why the agency has concluded that there would be no significant environmental impact. If it is determined that the impact will be significant, then an EIS must be prepared. Project applicants may choose to skip over an EA and proceed directly to an EIS when significant environmental impacts are expected from a project's development.

Environmental Impact Statement

An EIS must be prepared if the proposed major federal action will significantly affect the human environment.[3] The EIS is much more detailed and rigorous than the EA. The agency must first publish a Notice of Intent in the *Federal Register*, which describes how the public can be involved. This begins the scoping process, where the agency and the public define the issues and potential alternatives. The agency drafts the EIS and makes it available for public review and comment for a minimum of 45 days. After the comment period closes, the agency must consider all substantive comments and conduct further analyses if necessary. The agency then publishes the final EIS, which begins a 30-day "wait period" or "no action period," before making a final decision. This decision is documented through the issuance of the Record of Decision, which explains the agency's decision.

---

[3] Human environment means "the natural and physical environment and the relationship of people with that environment" (40 CFR § 1508.14).

STATE ENVIRONMENTAL REVIEWS

A federal review of environmental impacts would not be required to issue permits for a project on private lands in Virginia unless a federally issued permit or authorization were deemed to be a major federal action. If, however, mining is proposed to occur on state-owned land, the project proponent is responsible for preparing an EIS and submitting it to Virginia DEQ in a 1-year timeframe (Virginia DMME, 2007). This state process is known as a Virginia Environmental Impact Report (VA EIR; § 2.2-1157 of the Code of Virginia) and can be loosely compared with environmental review documents prepared under NEPA, although there are several important distinctions (see Table 5-2).

Given that only a small percentage of mining projects are carried out on state-owned land, VA EIRs are rare for mining proposals. They are more commonly completed for the construction of state facilities (§ 10.1-1188 of the Code of Virginia). The State Minerals Management Plan outlines the requirements for leasing and extraction of minerals on state-owned lands (Virginia DMME, 2007). The application of the VA EIR to proposed mineral leases and mining projects requires additional baseline information compared to other non-mining projects. It also includes more public engagement opportunities than the VA EIR procedures described for other major state projects unrelated to mining. The VA EIR is used to assist the state in making a determination whether or not to issue a lease for the use of state-owned lands for the proposed activity. The document will include the items shown in Table 5-3, as applicable (Virginia DMME, 2007). Project proponents may request that all or part of the VA EIR be waived after a public hearing if the project (1) does not affect the surface land owned by the state, (2) does not adversely affect surface or groundwater beneath state-owned land, and (3) complies with all other requirements for environmental protection.

In summary, a proposed gold mining activity would only trigger the VA EIR process if mining would occur on state-owned lands. In comparison, other states have their own requirements to complete an evaluation of environmental impacts for major permitting actions or state decisions. These evaluations of environmental impacts are conducted whether or not federal land partners or permitting agencies are involved, and are not limited to projects on state-owned land (Montana, MEPA; Washington, SEPA; California, CEQA). There are 15 other states that have NEPA-like planning requirements (Council on Environmental Quality, 2021).

IMPLICATIONS OF FEDERAL REGULATIONS FOR GOLD MINING

As noted above, the implementation of most programs under federal environmental laws including the Clean Air Act, the Clean Water Act, and the Resource Conservation and Recovery Act are delegated to Virginia. Below is an overview of those regulations and their implications for gold mining in Virginia.

The Clean Air Act

The Clean Air Act (CAA) is the major federal environmental law that regulates "general" or ubiquitous air pollution and air emissions at specific sources, like gold mining sites. The State Air Pollution Control Board (the "Air Board") and the director of the Virginia DEQ have shared authority under the CAA (§ 10.1 of the Code of Virginia). For the purposes of regulating the potential impacts of gold mining and processing on air quality, CAA regulatory tools include implementation of (1) the National Ambient Air Quality Standards (NAAQS), (2) New Source Performance Standards (NSPS), and (3) National Emission Standards for Hazardous Air Pollutants (NESHAP;

TABLE 5-2 Major Differences Between EIS and EA Produced Under NEPA and the VA EIR Process

	NEPA EA/EIS	VA EIR
Notice of intent	Yes	No
Scoping period	Yes	No
Public draft document prepared for public review?	Yes	No
Final document part of the public record?	Yes	Yes (if conducted under the State Mineral Management Plan)

SOURCES: Council on Environmental Quality (2020); Virginia DMME (2007).

TABLE 5-3 Typical Components of a VA EIR for Mining on State-Owned Lands

Components of EIR	Details
1. Purpose and need for proposed activities	
2. Description of the baseline settings for environmental factors	Physical conditions: topography, timber and other vegetation, geology, soils, hydrology, flood potential, climate, and air quality
	Biological conditions: terrestrial and aquatic ecosystems, wetlands, and threatened or endangered species
	Socioeconomic conditions: location, size, and distribution of existing population and labor force; existing land uses, community facilities, and transportation infrastructure; and historical, archaeological, recreational, or scenic sites
3. Description of the proposed actions and alternatives	Site access and preparation; conduct of exploration, extraction, and related activities; and deactivation of activities and land reclamation
4. Description of potential impacts to environmental factors (above), from the proposed activities, methods, or plans	Polluting substances which may be employed or may result from the operation and the plan for use, reuse, recycling, or disposal of all substances
	The nature, size, and expected duration of operations that will produce adverse noise levels or be visible from any public roadways, use areas, or viewpoint
	The location, length, and width of all roadways that would be constructed, or the anticipated use, upgrades, or repairs required for existing roadways
	Areas requiring the clearing of timber, brush, or undergrowth and the value of the timber, total forest cover, and disposition of proceeds
	Ground-disturbing activities that may occur (like excavation, drilling, and mining facilities), especially in areas where the disturbance may adversely affect teams, other waterways, or roadways
	Nature, size, and location of all areas in which the contour of the land may be altered, and the plans for restoring the affected land according to reclamation
	Utility, petroleum, or gas transmission lines, including associated construction and maintenance of right-of-way, and monitoring plans for leaks or breaks
5. Description of the mitigations to minimize the adverse impact of proposed activities	
6. Description of any irreversible environmental changes that would occur as a result	
7. List of local, state, and federal permits which are applicable to the proposed operations	
8. An executive summary of the EIS report	

SOURCE: Modified from Virginia DMME (2007).

EPA Region 10, 2003). In addition, the Clean Air Act has a "Good Neighbor" provision (42 USC § 7410 (a)(2)(d) (i)) that requires the U.S. Environmental Protection Agency (EPA) and individual states to address interstate transport of air pollution. Since 2015, the Cross-State Air Pollution Rule has required 28 states in the eastern United States, including Virginia, to reduce SO_2 and NO_x emissions from power plants that may affect downwind states' ability to attain and maintain the NAAQS.

National Ambient Air Quality Standards

The central feature for regulating general air pollution under the CAA is the development of criteria documents that summarize the scientific information relevant to particular pollutants. Based on these documents, EPA establishes the NAAQS for pollutants deemed "criteria pollutants," which are minimum standards that are implemented by the states to limit emissions of these pollutants into the air from point sources. Criteria pollutants currently include NO_2, SO_2, CO, ozone, lead, and particle pollution[4] (EPA Region 10, 2003). Areas that are in compliance with these minimum standards are classified as "attainment areas," whereas "nonattainment areas" are not in compliance. Virginia has seven air quality control regions (9VAC5-20-200) for purposes of classifying attainment and nonattainment areas (42 USC § 7407). Depending on the area in question, a facility evaluation could include both attainment and nonattainment, because certain places are out of attainment for one or more NAAQS, but considered "clean" for other NAAQS.

States issue New Source Review (NSR) permits for major air pollution sources[5] according to the classification of the area. These permits include the Prevention of Significant Deterioration permit, which applies to attainment areas and prohibits activities that would make air significantly dirtier in clean areas, as well as Nonattainment NSR permits, which can be issued in areas that are not meeting NAAQS (Elliott and Esty, 2021). These permits impose different requirements, ranging from the most stringent requirements for Lowest Achievable Emission Rate in nonattainment areas to Best Available Control Technology in attainment areas.

A NSR permit for smaller sources is called a minor source permit.[6] The regulation of minor sources is generally left entirely to the states (EPA Region 10, 2003) and minor sources may have requirements that are easier to meet than those for major sources. In South Carolina, the Haile Gold Mine's Supplemental Environmental Impact Statement (SEIS) estimates that the current operations for Haile Gold Mine do not meet major source thresholds, but that the proposed expansion would increase NO_x emissions and possibly lead to exceedances of these thresholds and the NAAQS. Any future gold mining that occurs in Virginia would likely be on a smaller scale than activities currently occurring in South Carolina (see Chapter 2) and would likely not meet the threshold for a major source. However, as explained in the NESHAP section (40 CFR § 63.11640), operations that meet the criteria of a gold processing plant are required to apply for a Title V permit, which is a federal program designed to standardize the permitting for major sources of emissions across the country.

New Source Performance Standards

The CAA also authorizes control of air emissions from specific operations that can be directly regulated through NSPS. Virginia has adopted NSPS regulations that mirror the federal regulations (9VAC5-50; 40 CFR Part 60). If a processing plant were to be developed on a mine site in Virginia, the NSPS for metallic mineral processing plants would apply (40 CFR Subpart LL). NSPS includes standards for opacity and particulate matter, as well as source testing for determining the direct emissions.

National Emission Standards for Hazardous Air Pollutants

The CAA also authorizes control of specific hazardous air pollutants (HAPs) through the NESHAP. Virginia has also adopted EPA's NESHAP requirements (9VAC5-60), which require gold mines to comply with emission standards for certain HAPs (40 CFR § 63.1(b)). Hazardous air pollutants that may apply to gold mining include cyanide, arsenic, cadmium, chromium, cobalt, lead, mercury, and selenium (EPA, 2022g). Certain source categories, including processing plants for gold mines,[7] have to comply with additional standards (40 CFR Part 63 Subpart EEEEEEE), which is mostly concerned with limiting emissions of mercury.

---

[4] Particle pollution is described by the size of the particulate matter (PM). $PM_{2.5}$ are particles with diameters that are generally 2.5 micrometers and smaller; PM_{10} are particles with diameters that are generally 10 micrometers and smaller.

[5] Major sources are facilities that may emit higher levels of pollutants than the major source threshold levels, which vary by pollutant and source category. In attainment areas, a major source is defined as having the potential to emit more than 100 tons/year of any nonhazardous pollutant, or more than 10 tons/year of any hazardous air pollutants (EPA, 2022r).

[6] Minor sources are any sources that are not major sources.

[7] These regulations define gold mining and processing as "any industrial facility engaged in the processing of gold mine ore that uses any of the following processes: Roasting operations, autoclaves, carbon kilns, preg tanks, electrowinning, mercury retorts, or melt furnaces. Laboratories (see CAA section 112(c)(7)), individual prospectors, and very small pilot scale mining operations that processes or produces less than 100 pounds of concentrate per year are not a gold mine ore processing and production facility. A facility that produces primarily metals other than gold, such as copper, lead, zinc, or nickel (where these metals other than gold comprise 95 percent or more of the total metal production) that may also recover some gold as a byproduct is not a gold mine ore processing and production facility. Those facilities whereby 95 percent or more of total mass of metals produced are metals other than gold, whether final metal production is on-site or off-site, are not part of the gold mine ore processing and production source category" (40 CFR § 63.11651).

A major source of HAPs is defined as one that has the potential to emit 10 tons or more of one HAP, or 25 tons or more of a combination of HAPs, per year. The SEIS for the Haile Gold Mine in South Carolina estimates that emissions of HAP are less than the federal major source thresholds (USACE, 2022). Any gold mining that occurs in Virginia will likely be on a smaller scale than that currently occurring at Haile and there is no evidence that gold ores in Virginia have elevated mercury content (see Chapter 2). This suggests that future gold mines in Virginia are unlikely to reach the criteria for a major source. Nevertheless, the NESHAP for gold processing plants has a requirement that the permittee obtain a Title V permit (40 CFR Part 70; 40 CFR Part 71), even if the activity does not meet the threshold for a major source (40 CFR § 63.11640). Thus, even though the Haile Gold Mine currently does not meet the threshold for major source emissions, it is permitted under Title V major source (Mareesa Singleton, personal communication, 2022). As a result, any future gold mines in Virginia that have on-site processing plants would be permitted under a Title V major source permit, which has extensive requirements for monitoring and reporting (EPA, 2021e).

The Clean Water Act

The Clean Water Act (CWA) regulates pollution flow into "navigable waters" including rivers, streams, and other bodies of water primarily through effluent limitations on point sources, such as outflows from industrial facilities. These effluent limitations are placed as conditions in permits. They are determined based on the water quality criteria applicable to the receiving water as well as industry-specific and technology-based criteria. The regulatory tool through which these effluent limitations are imposed is a permit under the National Pollution Discharge Elimination System (NPDES) for discharges into surface waters. The Virginia DEQ and the Virginia State Water Control Board (the "Water Board") have shared authority in implementing and administering these regulations (Troutman Sanders LLP, 2008).

The CWA also contains provisions that attempt to address non-point sources that do not come from a defined outfall, such as runoff of sediment. EPA requires states to identify water bodies where effluent standards have not been sufficient to clean up surface waters, and establish total maximum daily loads (TMDLs) for these waters. NPDES permits are then tightened up to meet TMDLs for these water bodies.

Finally, EPA and USACE share authority under CWA Section 404 to control discharges of dredged and fill material into waters of the United States, including jurisdictional wetlands. Under this program, people who seek to discharge fill or dredged material to waters of the United States, including wetlands, must obtain a CWA 404 permit from USACE. Most permits under this section of the CWA require that adverse impacts to the wetlands be minimized, that compensatory mitigation be undertaken, or that fees be paid to support wetlands protection (Elliott and Esty, 2021).

Water Quality Standards and Criteria

The Virginia Water Board has adopted surface water quality standards that have been approved by EPA (see Table 5-4) and reviews these standards at least once every 3 years. Virginia has also adopted the federal antidegradation provisions, which require that waters whose quality is better than established standards must be protected and maintained. Certain water bodies, designated as Tier 3 waters (with exceptional water quality), are singled out for added protection (9VAC25-260-30(a)(3)). However, the Water Board can allow a change that would lower water quality when that change is needed for economic or social development (9VAC25-260-30(a)(2)).

The Water Board has also established enforceable standards and nonmandatory criteria for groundwater. These include an antidegradation policy for groundwater (9VAC25-280-30), enforceable groundwater standards that are specific to the full state, and nonenforceable criteria applicable to individual physiographic provinces[8] (see Table 5-5).

---

[8] While not mandatory, criteria provide guidance for preventing groundwater pollution.

TABLE 5-4 Virginia's Surface Water Quality Criteria for Protection of Freshwater Aquatic Life and Human Health for Chemicals of Concern to This Study

Contaminant	Freshwater Aquatic Life		Human Health	
	Acute	Chronic	Public Water Supply	All Other Surface Waters
Ammonia (µg/L)	Dependent on pH, temperature, and biota	Dependent on pH, temperature, and biota		
Antimony (µg/L)			5.6	640
Arsenic (µg/L)	340[a]	150[a]	10[b]	
Cadmium (µg/L)	Freshwater values are a function of total hardness	Freshwater values are a function of total hardness	5[b]	
Copper (µg/L)	Freshwater values are a function of total hardness	Freshwater values are a function of total hardness	1,300[b]	
Free cyanide (µg/L)	22[a]	5.2[a]	4	400
Lead (µg/L)	Freshwater values are a function of total hardness	Freshwater values are a function of total hardness	15[b]	
Mercury (µg/L)	1.4[a]	0.77[a]		
Nitrate as N (µg/L)			10,000	
pH in nontidal waters	6.0–9.0	6.0–9.0	6.0–9.0	6.0–9.0
Selenium (µg/L)[α]	20	5.0	170	4,200
Sulfate (µg/L)[β]			250,000	
Thallium (µg/L)			0.24	0.47
Total Dissolved Solids (µg/L)[β]			500,000	
Zinc (µg/L)	Freshwater values are a function of total hardness	Freshwater values are a function of total hardness	7,400	26,000

NOTES: Regulations require that surface water conditions must not be acutely or chronically toxic for freshwater aquatic life except as allowed in mixing zones. The definition of a mixing zone is a "limited area or volume of water where initial dilution of a discharge takes place and where numeric water quality criteria can be exceeded but designated uses in the waterbody on the whole are maintained and lethality is prevented" (9VAC25-260-5). "Acute" toxicity is an adverse effect that occurs shortly after exposure, and "chronic" toxicity is that which is irreversible or progressive.

[α] Freshwater criteria expressed as total recoverable.

[β] Criterion to maintain acceptable taste, odor, or aesthetic quality of drinking water.

[a] Equivalent to the National Recommended Water Quality Criteria (EPA, 2022j).

[b] Equivalent to National Primary Drinking Water Regulations (40 CFR § 141.62).

SOURCES: 9VAC25-260-140; 9VAC25-260-155; EPA (2022f).

NPDES/VPDES Permits

Virginia DEQ administers the federal NPDES program as the Virginia Pollutant Discharge Elimination System (VPDES). A VPDES permit is required for every discharge into "state waters," defined as all surface or groundwater that is wholly or partially within or bordering the Commonwealth, or within its jurisdiction (§ 62.1-44.3 of the Code of Virginia, 2022). This definition would include "waters of the United States," plus additional Virginia surface or groundwaters that do not meet that definition (see Box 5-2). The categories of discharges that are likely from mineral mines such as gold mines are process wastewater,[9] mine drainage,[10] and industrial stormwater[11] (EPA Region 10, 2003).

---

[9] Process wastewater is "any water which, during manufacturing or processing, comes into direct contact with or results from the production or use of any waste material, intermediate product, finished product, byproduct, or waste product" (40 CFR § 122.22).

[10] Mine drainage is "any water drained, pumped, or siphoned from a mine" (40 CFR § 400.132).

[11] Industrial stormwater is "the discharge from any conveyance which is used for collecting and conveying storm water and which is directly related to manufacturing, processing or raw materials storage areas at an industrial plant" (40 CFR § 122.26).

TABLE 5-5 Statewide and Province-Specific Groundwater Standards and Nonenforceable Criteria

Constituent	Concentration
Arsenic	50 µg/L
Cadmium	0.4 µg/L
Copper	1000 µg/L
Cyanide	5 µg/L
Lead	50 µg/L
Mercury	0.05 µg/L
Selenium	10 µg/L
Zinc	50 µg/L
pH[a]	5.5–8.5
Ammonia[a]	25 µg/L
Nitrite[a]	25 µg/L
Nitrate[a]	5,000 µg/L
Alkalinity[b]	10,000–200,000 µg/L
TDS[b]	250,000 µg/L
Sulfate[b]	25,000 µg/L
Iron[b]	300 µg/L
Manganese[b]	50 µg/L

NOTE: TDS = total dissolved solid.

[a] Groundwater standards only applicable to the Piedmont and Blue Ridge regions.

[b] Nonenforceable groundwater criteria.

SOURCES: 9VAC25-280-40; 9VAC25-280-50; 9VAC25-280-70.

When an operator applies for an individual NPDES permit, they must first determine the applicable gold mining–specific Technology-Based Effluent Limits (40 CFR Part 440 Subpart J; 40 CFR Part 440 Subpart M; see Table 5-6). This establishes standards for metal contaminants, pH, and total suspended solids standards according to best practicable control technology or best available technology (40 CFR § 440.104). These limitations apply to process wastewater and mine drainage, including potential discharges from tailings piles, but not stormwater (EPA, 2011b). The discharge of process wastewater to WOTUS is generally prohibited, but an exception is provided in areas where the precipitation is greater than annual evaporation. In practice, this provision means that in

TABLE 5-6 New Source Performance Standards for the Mining and Processing of Gold According to Best Available Demonstrated Technology

	Effluent Limitations	
	Maximum for 1 Day	Average for 30 Consecutive Days
Total Suspended Solids	30,000 µg/L	2,000 µg/L
Copper	300 µg/L	150 µg/L
Zinc	1,500 µg/L	7,500 µg/L
Lead	600 µg/L	300 µg/L
Mercury	2 µg/L	1 µg/L
Cadmium	100 µg/L	50 µg/L
pH	within 6.0 to 9.0	within 6.0 to 9.0

SOURCE: 40 CFR § 440.102.

Virginia, where average annual precipitation is almost always going to exceed evaporation, treated wastewaters would likely be discharged into surface waters (40 CFR § 440.103(c); 40 CFR § 440.103(d)).

Following the determination of gold mining–specific Technology-Based Effluent Limits, the permit writer then determines discharge limits for the facility that are protective of state water quality standards (see Table 5-4). The permit writer must compare the Technology-Based Effluent Limits with effluent requirements necessary to ensure attainment of the state water quality standards and choose the more stringent of the two (EPA, 2011b).

Consistent with federal oversight and guidance (40 CFR § 131.13), many states (including Virginia) allow for the use of mixing zones where aquatic life criteria may be exceeded within a specifically defined zone of a receiving water body (see Table 5-7). The mixing zone allows for dilution and instream mixing to attenuate the pollutant discharges within this prescribed area. States have various methods to determine the allowable size of mixing zones and often limit mixing zone widths, cross-sectional areas, and flow volumes and lengths (EPA, 2014a). EPA guidance states, "The area or volume of an individual mixing zone or group of mixing zones should be as small as practicable so that it does not interfere with the designated uses or with the established community of aquatic life in the segment for which the uses are designated" (EPA, 2014a). Because low flows in the receiving water provide less dilution of effluent discharges, EPA (2014a) requires mixing zones be determined so that they ensure protection of the applicable criteria under low-flow conditions (EPA, 2014a).

Virginia DEQ allows instream mixing when setting effluent limits for any toxic impacts, including whole effluent toxicity and temperature (James Golden, personal communication, 2022). These limits must not prevent the movement or cause serious harm to passing and drifting aquatic organisms through the water body (9VAC25-260-20). Additionally, no mixing zone can be used as a substitute for treatment required by the CWA and other state and federal laws (9VAC25-260-20) and they are not allowed for wetlands, swamps, marshes, lakes, or ponds (9VAC25-260-20). The mixing zone standard does not require protection for organisms that permanently reside within a mixing zone, but additional consideration must be given if there are critical beneficial uses of the stream or sensitive resident species that require special protection. The mixing zone cannot not be utilized if there is a rare and endangered species within reasonable proximity, unless it is demonstrated that the specific parameters will not result in adverse impacts on that species (Virginia DEQ, 2000).

EPA guidance states that bioaccumulative pollutants may not be appropriate for mixing zones and recommends that state and tribal policies do not allow mixing zones for discharges of bioaccumulative pollutants (EPA, 2014a). This is because bioaccumulative pollutants may cause significant risks to human health and non-human biota and their persistence in sediments, water, or biota may adversely affect the water body. Some states like Alaska have requirements that prohibit the bioaccumulation of pollutants to significantly adverse levels (see Table 5-7), but Virginia DEQ does not have any written policies requiring special consideration for mixing zones with bioaccumulative substances during their permitting (Allan Brockenbrough, personal communication, 2022; Virginia DEQ, 2000). Some examples of bioaccumulative pollutants include arsenic, lead, mercury, cadmium, selenium, and copper, all of which are discussed in Chapter 4 as potential pollutants from future gold mines in Virginia.

TMDLs and Non-Point Sources

Section 304(l) of the CWA (33 USC § 1314) requires that Virginia create a list of water bodies for which water quality standards have not been achieved and establish TMDLs for these waters (9VAC25-720-20). In this way, the use of TMDLs represents a "watershed approach," which differs from the NPDES approach of controlling pollution from an outflow or point source (Elliott and Esty, 2021). As of 2006, 644 stream segments had TMDLs and another 1,200 stream segments needed TMDLs (9VAC25-720). If gold mining activities impacted waters for which a TMDL applied, it is possible that additional regulatory requirements could be added to a gold mine's VDPES permit (Virginia DEQ, 2022a,c).

In 2010, EPA established the Chesapeake Bay TMDL, which set limits on the nutrients (e.g., nitrogen and phosphorus) and sediment that can flow into the Chesapeake Bay (EPA, 2022b). Bay jurisdictions developed Watershed Implementation Plans (WIPs) in order to meet the needed pollution reductions by 2025. Virginia's most recent WIP Plan—Phase III—established state basin planning targets shown in Table 5-8 (Virginia DEQ, 2022b; see Figure 5-1). As described in Chapters 2 and 4, mining operations can increase sediment and nitrogen loading

TABLE 5-7 Comparison of Mixing Zone Requirements in Selected States

State	Definitions	Size	Parameters
Virginia	**Mixing zone:** The area where chronic criteria can be exceeded, but acute criteria must not be exceeded. **Allocated impact zone:** The area within a mixing zone where acute criteria can be exceeded (9VAC25-260-20).	**Mixing zones:** • Width must be less than one-half of the width of the receiving watercourse. • May not constitute more than one-third of the area of any cross-section of the receiving watercourse. • Length must be less than five times the width of the receiving watercourse. **Allocated impact zones:** • Shall be sized to prevent lethality to passing and drifting aquatic organisms. • No required size, but internal policy and EPA guidance recommends a size that is smaller than 10% of the distance to the boundary of the mixing zone, 50 times the discharge length scale, and 5 times the local water depth (James Golden, personal communication, 2022; 9VAC25-260-20).	Effluent limits for any toxic impact, including whole effluent toxicity (WET) and temperature (James Golden, personal communication, 2022). Current guidance does not have any special consideration for mixing zones and bioaccumulative substances (Virginia DEQ, 2000).
South Carolina	**Mixing zone:** The area where chronic toxicity limit can be exceeded, but acute toxicity limit must not be exceeded. **Zone of initial dilution:** The area within a mixing zone where acute toxicity limit can be exceeded (S.C. Code Regs. § 61-68.E).	The size of the mixing zone shall be minimized, as determined by the Department, and shall be based on applicable critical flow conditions. **Recommended chronic mixing zones:** • Width of one-half of the river width • Length of twice the river width **Recommended acute mixing zones:** • Width of one-tenth of the river width • Length of one-third of the river width (S.C. Code Regs. § 61-68.E)	Mixing zones are only applied to toxicity (WET) and thermal limitations, not to individual parameters such as metals (Byron Amick, personal communication, 2022).
Alaska	**Mixing zones:** The area where chronic aquatic life criteria can be exceeded. The pollutants discharged will not exceed acute aquatic life criteria at and beyond the boundaries of a smaller initial mixing zone surrounding. **Initial mixing/acute zone:** The area where acute aquatic life criteria may be exceeded.	**Mixing zones:** • Size will be as small as practicable. **Initial mixing/acute zone:** One of the following must be used: • The initial discharge velocity is 3 m/s or greater; and the mixing zone is no larger in any direction than 50 times the discharge length scale. • Size is smaller than 10% of the distance to the boundary of the mixing zone, 50 times the discharge length scale, and 5 times the local water depth. • A drifting organism reaches the acute mixing zone boundary (i.e., the zone in which aquatic life criteria are exceeded) in 15 minutes or less. • A drifting organism does not receive harmful exposure when evaluated by a valid toxicological analysis approved by the department (18 AAC § 70.240).	The pollutants discharged will not • bioaccumulate, bioconcentrate, or persist above natural levels in sediments, water, or biota to significantly adverse levels; • present an unacceptable risk to human health from carcinogenic, mutagenic, teratogenic, or other effects; • settle to form objectionable deposits; • produce floating debris, oil, scum, and other material in concentrations that form nuisances; • result in undesirable or nuisance aquatic life; or • produce objectionable color, taste, or odor in aquatic resources harvested from the area for human consumption (18 AAC § 70.240).

TABLE 5-7 Continued

State	Definitions	Size	Parameters
Montana	**Mixing zones:** The area where chronic aquatic life standards can be exceeded. Acute aquatic life standards for any parameter may not be exceeded in any portion of the mixing zone unless DEQ specifically finds that allowing minimal initial dilution will not threaten or impair existing beneficial uses. **Minimal initial dilution:** Area where acute criteria may be exceeded if DEQ finds that it will not threaten or impair existing beneficial uses (ARM 17.30.507).	Mixing zones are required to have the smallest practicable size, a minimum practicable effect on water uses, and definable boundaries (75-5-301(4), MCA). **Mixing zone:** • Length downstream must be less than one-half mixing width distance or extend downstream more than ten times the stream width, whichever is more restrictive (The stream width and discharge limitations are considered at the 7Q10 low flow, or seasonal 14Q5 in conjunction with base numeric nutrient standards in DEQ-12A, ARM 17.30.516). **Minimal initial dilution:** No size restrictions given.	Specific parameters not excluded (ARM 17.30.505). The department shall assess biological, chemical, and physical characteristics of the receiving water and the nature of the pollutant (toxic, carcinogen, bioconcentration; ARM 17.30.700).

of waterways after the movement of soils, and these impacts may be compounded following the migration of nitrates from the use of blasting agents. As a result, if a gold mine were to be developed within a Chesapeake Bay watershed, additional regulatory requirements might be added to VPDES permits.

Wetlands Permitting (CWA 404)

USACE and EPA share authority to regulate the dredging and filling of WOTUS, including wetlands (Troutman Sanders LLP, 2008; Virginia DEQ, 2019). Dredging and filling permits include provisions for mitigating wetlands loss and compensating impacts so that there is no net loss of existing wetlands acreage or functionality (§ 62.1-44.15:20-62.1-44.15:21.1 of the Code of Virginia). The CWA also allows USACE to issue general permits for activities with minimal impact, which would not constitute a major federal action that triggers the NEPA process. General permits are applicable to any project causing less than 0.5 acre of impacts, and certain other small projects (§ 62.1-44.15:21 of the Code of Virginia). According to the Virginia Water Protection permit guide, if a gold mining "project meets the eligibility criteria and conditions within the general permit, the activity can typically be authorized by the [USACE] under one of these general permits within 45 days of application and without further sister agency or public comment" (Virginia DEQ, 2019). Virginia DEQ has issued blanket permits for some activities that qualify under USACE's nationwide and regional permit program (Virginia DEQ, 2019).

TABLE 5-8 State Basin Planning Targets with Basin-to-Basin and Nitrogen: Phosphorus Exchanges

State Basin	Nitrogen (million pounds/year)	Phosphorus (million pounds/year)	Sediment (million pounds/year)
Eastern Shore	1.83	0.152	473.3
Potomac River Basin	16.51	1.823	1,929.7
Rappahannock River Basin	7.09	0.819	1,505.1
York River Basin	5.71	0.548	949.1
James River Basin	21.81	2.241	2,015.2

SOURCE: Table modified from Linker et al. (2019).

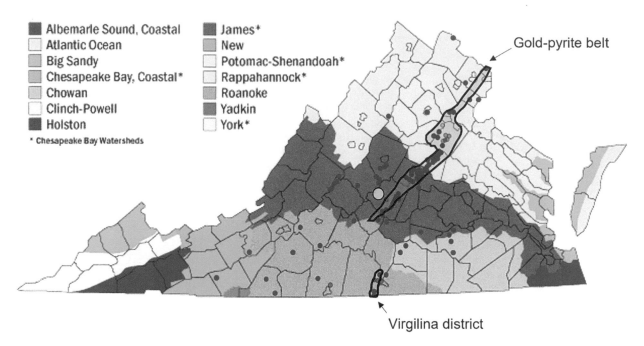

FIGURE 5-1 Major state watersheds in Virginia. Overlain on the map are historic gold mines (red dots) and the gold-pyrite belt and Virginia District outlined in black. The large yellow circle denotes the location of Aston Bay's exploration property in Buckingham County.
SOURCE: Modified from Virginia Department of Conservation and Recreation, Soil and Water Conservation Programs.

The Resource Conservation and Recovery Act

The Resource Conservation and Recovery Act (RCRA) primarily regulates waste handling. Generally, the regulations fall on both the generators of the waste and the facilities that treat, store, and dispose of the waste (40 CFR Part 264/265, Subpart A–E). It applies broadly to many types of discarded materials, which are defined in the statute as "solid wastes." This term, as used in the RCRA law, is counterintuitive; it includes both gaseous and liquid wastes as well as solid materials (Elliott and Esty, 2021). Under RCRA, a "solid waste" will be classified as a hazardous waste (and subject to much more stringent regulation) under two situations. First, EPA can specifically list a category of waste from an industrial or production process. These types of waste are known as "listed wastes." Second, a waste can exhibit one or more of four characteristics: corrosivity, ignitability, reactivity, and toxicity. These types of waste are known as "characteristic wastes."

Solid waste from the mining and processing of ores and minerals is generally exempt from regulation as listed wastes under RCRA Subtitle C. This exemption, called the "Mining Waste Exclusion" or the "Bevill Amendment," was added to RCRA by law in 1980 (EPA Region 10, 2003). Mining wastes, and several other categories of excluded wastes, are known as "special wastes." This provision precluded EPA from regulating these special wastes until the agency performed a study. These steps have been taken (EPA, 2022h), and as of this report's time of writing, most extraction (e.g., waste rock) and processing wastes (e.g., tailings, spent ore) from mineral mining have been excluded from federal hazardous waste regulations under Subtitle C of the RCRA (EPA, 2022m), except for spent furnace dust and slag (EPA, 1998b), which are both produced during smelting—the final stage for the processing of gold. Additional wastes from gold mining and processing could be subject to RCRA if they are determined to be characteristic wastes under the statute. For example, some precipitated wastes from water treatment facilities, for example those at Brewer Gold Mine in South Carolina, do not pass toxicity limits and would be treated as hazardous waste under RCRA (Jim McLain, personal communication, 2022).

Safe Drinking Water Act

The Safe Drinking Water Act (SDWA) protects the quality of drinking water. This law focuses on all waters actually or potentially designated for drinking use, whether from above ground or underground sources (EPA, 2022o). The SDWA covers six categories of contaminants: micro-organisms, radionuclides, inorganic chemicals, organic chemicals, disinfectants and disinfection by-products. At present, EPA regulates 91 contaminants (Elliott and Esty, 2021).

Drinking Water Standards

The SDWA authorizes the EPA to set enforceable national primary drinking water standards. Public water systems are responsible for ensuring that contaminants in tap water do not exceed these standards. These regulatory levels are based on Maximum Contaminant Level Goals (MCLGs), which are human exposure limits that protect against the hazards of these contaminants with an adequate margin of safety. Using these MCLGs, EPA sets its regulatory levels based on Maximum Contaminant Levels (MCLs; see Tables 2-3 and 2-4 and Chapter 4), which are set as close to the MCLGs as possible after considering technology limits and costs (Elliott and Esty, 2021).

Underground Injection Wells

The Underground Injection Control (UIC) well program is authorized by the SDWA (40 CFR Parts 144–148). The regulations outline 6 classes of wells, two of which might be associated with mining—Class III and Class V. Class III wells utilize fluids to extract minerals in situ, which has not be commercially deployed for gold mining (Guthrie, 2020). In contrast, Class V wells are potentially pertinent to gold mining in that they involve the disposal of mining waste fluids and materials in deep wells above drinking water sources (40 CFR section 146.5). According to the EPA, most Class V wells are associated with storm water drainage and large capacity septic systems (EPA, 2022i), but the regulations are also relevant to the injection of tailings or other mining waste products underground (EPA, 1999). Specifically, the regulations pertain to both conventional drilled wells that place slurries/solids in underground mines, but also piping systems within mine shafts that are utilized for the same purpose (EPA, 1999). Mine shafts can also be considered mine backfill wells under UIC regulations (EPA, 1999), if the depth of the shaft is greater than the largest surface dimension (40 CFR 144.3). Examples and potential methods for underground mine backfill are described in more detail in Chapter 3.

Virginia does not have primacy for its UIC program. Instead, EPA administers UIC permitting, monitoring, and enforcement in Virginia (40 CFR Part VV sections 147.2350-2352; EPA, 2022q). Class V wells are authorized by rule, which means that Virginia operations may not require a permit if an operator complies with certain requirements (EPA, 2022f), including if they:

- Submit inventory information to EPA and verify that they are authorized to inject. EPA will review the information to be sure that the well will not endanger a drinking water source.
- Operate the wells in a way that does not endanger drinking water sources as defined by EPA.
- Properly close their Class V well when it is no longer being used so that the movement of any contaminated fluids into drinking water sources is prevented.

After reviewing this information, EPA could determine that an individual permit is necessary to prevent contamination of a drinking water source.

VIRGINIA'S MINERAL MINING LAWS, REGULATIONS, AND GUIDANCE

The Mineral Mining Program within the Virginia Department of Energy (Virginia Energy) is introduced in Chapter 1 and is expanded upon here. The codes and regulations that are administered by the Mineral Mining Program reflect the history of the program and address two broad areas: (1) occupational safety and health and (2) mine operations and reclamation (see Chapter 1). Virginia's Mineral Mine Safety Act (§ 45.2-1100 et seq. of

TABLE 5-9 Codes, Regulations, Guidance Documents, and Policies Relevant to This Study

Document	Title	Part/Agency	Chapter
Code of Virginia	Title 45.2: Mines, Minerals and Energy	Part A. Mineral Mines Generally	Chapter 12. Permits for Certain Mining Operations; Reclamation of Land (§§ 45.2-1200 to 45.2-1243 of the Code of Virginia)
			Chapter 13. Mineral Mining Retaining Dams; Adjacent Owners (§§ 45.2-1300 to 45.2-1304 of the Code of Virginia)
		Part B. Underground Mineral Mines	Chapter 14. Requirements Applicable to Underground Mineral Mines (§§ 45.2-1400 to 45.2-1405 of the Code of Virginia)
		Part C. Surface Mineral Mines	Chapter 15. Requirements Applicable to Surface Mineral Mines (§§ 45.2-1500 to 45.2-1505 of the Code of Virginia)
Virginia Administrative Code	Title 4: Conservation and Natural Resources	Agency 25. Department of Energy	Chapter 31. Reclamation Regulations for Mineral Mining (4VAC25-31-10 to 4VAC25-31-570 of the code of Virginia; 4VAC25-31-10 to 4VAC25-31-570)
Division of Mineral Mining Manual	The Mineral Mine Operator's Manual	n.a.	n.a.
	Enforcement Policy and Procedures Manual	n.a.	n.a.

SOURCE: The Code of Virginia, Virginia Administrative Code, Division of Mineral Mining.

the Code of Virginia) provides requirements that are similar to or expand upon those administered by the U.S. Department of Labor's Mine Safety and Health Administration (MSHA). The codes and regulation that are applicable to this study's Statement of Task are shown in Table 5-9 along with other policy and guidance documents. This includes the Mineral Mine Operator's Manual (Virginia DMME, 2011) provided by Virginia Energy, which is a technical guidance document to assist operators in complying with the Reclamation Regulations for Mineral Mining (4VAC25-31 et seq.). Although not a directly enforceable document, the Operator's Manual contains forms, guidelines, and support materials to assist users in implementing the enforceable standards. Additionally, the committee obtained a copy of the Division of Mineral Mining Enforcement Policy and Procedures document from Virginia Energy, some of which is publicly available on Virginia.gov (2022).

Definitions, Exemptions, and Permitting Categories

The framework in Virginia for permitted mining activities is established with general definitions that identify the activities categorized as mineral mining, which includes gold mining, and the different levels of permitting required for such activities. There are multiple exemptions for excavation projects and the Director of Virginia Energy can consider the length of time or duration of the activity and whether it is a one-time activity when considering whether an activity is exempt (4VAC25-31-70).

It is unlawful for any operator to begin mineral mining,[12] without having first obtained a mine permit and safety license from the Mineral Mining Program (§§ 45.2-1124 and 45.2-1205 of the Code of Virginia). A separate permit and license need to be secured for each operation (§§ 45.2-1124 and 45.2-1205 of the Code of Virginia). However, the Director of Virginia Energy may combine noncontiguous areas into a single permit if the areas are close to each other and part of the same operation (4VAC25-31-80).

---

[12] Mining is the "breaking or disturbing of the surface soil or rock in order to facilitate or accomplish the extraction or removal of minerals or any activity constituting all or part of a process for the extraction or removal of minerals so as to make them suitable for commercial, industrial, or construction use" (§ 45.2-1200 of the Code of Virginia). Mineral is the "ore, rock, and any other solid homogeneous crystalline chemical element or compound that results from the inorganic processes of nature other than coal" (§ 45.2-1200 of the Code of Virginia).

TABLE 5-10 Permitting Categories for Mineral Mining Activities in Virginia

Mining Activity Category	Definition of Activity	Required Components for Permitting			
		Mining Permit?	Financial Assurance?	Public Notification?	Public Hearing or Meeting?
Exploration (drilling)	Searching, prospecting, exploring, or investigating for minerals by drilling (4VAC25-31-70; § 45.2-1200 of the Code of Virginia).	No	No	No	No
Exploration (other surface disturbance)	Searching, prospecting, exploring, or investigating for minerals through other surface disturbance (§ 45.2-1101 of the Code of Virginia).	Yes	Yes	Initial notice only. Not required for future permit modifications[a]	If requested within 10 days of initial notice[a]
Restricted Mining Permit	Less than one acre of land disturbance and removal of less than 500 tons of minerals at any site (4VAC25-31-200; § 45.2-1203 of the Code of Virginia).	Yes	No	Initial notice only. Not required for future permit modifications[a]	If requested within 10 days of initial notice[a]
Mining Permit	All other activities for the extraction or removal of minerals, or any activity constituting the process of extraction or removal of minerals, to make them suitable for commercial, industrial, or construction use. Does not include deep mining that does not affect the surface (§ 45.2-1101 of the Code of Virginia; § 45.2-1200 of the Code of Virginia).	Yes	Yes	Initial notice only. Not required for future permit modifications[a]	If requested within 10 days of initial notice[a]

[a] Prior to submitting an application to the Mineral Mining Program, permit applicants must provide a notice of intent to "property owners within 1,000 feet of the permit boundary, the chief administrative official of the local political subdivision" (county or city), and "all public utilities on or within 500 feet of permit boundary" (4VAC25-31-170; § 45.2-1210 of the Code of Virginia). Additional details about notifications are provided later in this chapter.

There are subcategories for mineral mining permits, based on the scale of disturbance and nature of the activity, as summarized in Table 5-10. Depending on the phase of project development and the size of disturbance, gold mining activity could potentially fit into different subcategories. A "General Mining Permit" governs sand or sand and gravel operations that disturb a total area of less than 10 acres (Virginia DMME, 2011). Operations specific to gold development would not be permitted under this subcategory, even if free or placer gold were to be discovered in a sand and gravel mine, which later modified its operations to collect gold. This modification would require additional steps, including the application for a regular mining permit, because the operation would exceed the terms of the General Mining Permit. Because it is not applicable to gold mining, the General Mining Permit is not discussed in detail here. The exempt activities and restricted mining permits that are applicable to gold mining in Virginia are described below.

Exploration

The definition of mineral mining does not include searching, prospecting, exploring, or investigating for minerals by drilling (4VAC25-31-70; § 45.2-1200 of the Code of Virginia) and as a result such drilling activities are exempt from regulation. The surface disturbances associated with such drilling operations (e.g., roads, drill pads, sumps for water or cuttings) are also exempt from mine permitting, although these activities may be subject to local requirements and could require permits for controlling erosion, sediment, and postconstruction stormwater as required by Virginia DEQ and Department of Conservation and Recreation (DCR). All other methods of surface-disturbing exploration or site preparation for surface mineral extraction activity are defined as a "surface mineral mine" (§ 45.2-1101 of the Code of Virginia) and would not be exempt. The typical permitting and bonding requirements therefore apply for all other methods of surface-disturbing exploration. As noted in Chapter 1, current gold exploration activity in Virginia is being conducted by drilling, and no permits are required for these activities.

As discussed in Chapter 3, the hydrologic and geochemical conditions encountered by exploration drilling would determine whether surface water or groundwater systems might be affected, particularly if the drill holes are not plugged and appropriately sealed before being abandoned. Closing exploration drilling sites improperly could result in impacts to soil, vegetation, and habitat, while runoff and erosion from these areas could be harmful to surface water quality. The potential impacts from exploration drilling projects are likely to be limited in scale and much less significant than what could occur from gold mining and processing facilities (see Chapter 3). However, the current legal exemption of exploration drilling results in the potential for environmental damage and precludes regulators from requiring measures that could reduce or prevent impacts. In addition, exploration drilling activities do not require bonding.

Some states, including South Carolina and Idaho, also exempt exploration drilling projects from permitting and bonding requirements (see Table 5-11). In other states, including Montana, Nevada, and Colorado, and in certain counties in California, drilling is a permitted and bonded activity that requires plans for operations and reclamation (see Table 5-11). In Montana, exploration drilling requires an evaluation of potential environmental impacts prior to issuing the license or certificate of exploration (ARM 17.24.103; 75-1-201, MCA). Many states also have specific requirements for the reclamation of associated disturbance (roads, pads, and sumps) and for the construction of drill holes as monitoring wells, or plugging and abandoning the drill holes to limit potential environmental impacts to water resources (Montana, ARM 17.24.106; Nevada, NAC 534.420; Colorado, 2CCR407-5). In fact, Virginia law currently has provisions for the abandonment and plugging of private water wells (12VAC5-630-450), but a similar provision has not been promulgated for mineral exploration drilling.

Regarding the potential need for confidentiality during exploration to limit competition between companies, Colorado requires that exploration ("prospecting") applicants provide two forms with their Notice of Intent. One form contains both public and confidential information, which is used by the regulatory program for review. The second form contains only the information the applicant believes is public, redacting all confidential information. The public Notice of Intent is posted to the regulatory program's website within 5 days of submission, and public comments or requests for disclosure of confidential information must be received by the program within 10 working days (2CCR407-5.1.2).

Restricted Mining Permits for Small Mines

Under Virginia law, any mining operation that disturbs less than 1 acre of land and removes less than 500 tons of marketable minerals at any particular site is exempt from application fees, permit renewal fees, and bond requirements. However, the operator is still required to obtain a mine permit and safety license (§§ 45.2-1203 and 45.2-1200 of the Code of Virginia and 4VAC25-31-200). The mining operator must submit an application for a permit, a sketch of the mining site, and plans for operations and reclamation (§§ 45.2-1205 and 45.2-1206 of the Code of Virginia). The requirements for operations, drainage, and reclamation plans for Restricted Mining Permits are consistent with larger mines. This includes hydrologic studies and plans to minimize the adverse effects on water quantity and quality, if groundwater is encountered by the operation. Restricted Mining Permits are also subject to permit evaluations and inspections from the Mineral Mining Program, although no fees are paid to support the time and effort that regulators expend reviewing permits and carrying out other functions. Because these restricted permits are exempt from financial assurance (performance bond), the Commonwealth must pay the costs to conduct any necessary reclamation, closure, and long-term stewardship if an operator abandons the site (see section on "Financial Assurance").

Many of the known gold occurrences in Virginia are limited in size, and some may be small enough to qualify for a restricted mining permit. Under Virginia's current laws and regulations, mining activities as well as on-site processing could be included within the Restricted Mining Permit, but any structures, processing equipment, or waste disposal areas (fills/piles or impoundments) must fit within the 1-acre mining disturbance. Under current economic

TABLE 5-11 Exemptions for Small Mines and Exploration on State or Private Lands in Selected States

State	Public Notice for Exploration	Permit for Exploration	Exemptions for Small Mines
Virginia	No public notification or hearing for exploration drilling (exempt activity). The notification and hearing requirements for mining permits would apply to other methods of exploration (landowners within 1,000 feet, local government, utility services with 500 feet; hearing held if requested within 10 days of notice) (§ 45.2-1210 of the Code of Virginia, 4VAC25-31-170).	A mineral mining permit and financial assurance are required for searching, prospecting, exploring, or investigating for minerals through surface disturbance, but exploration drilling is exempt from these requirements (§§ 45.2-1101 and 45.2-1200 of the Code of Virginia, 4VAC25-31-70).	A Restricted Mining Permit applies if less than 1 acre of land is disturbed and less than 500 tons of minerals are removed at any site. Exempt from financial assurance (§ 45.2-1203 of the Code of Virginia and 4VAC25-31-200).
South Carolina	Public notice and public hearing requirements do not apply to exploration (S.C. Mining Act Section 48-20-50).	A certificate of exploration is required for exploration activities on 2 acres or less that involve the development of open pits, trenches, open cuts, or tunneling. A certificate of exploration is not required for drilling core holes, drilling bore holes, or conducting geophysical and geochemical sampling and analysis (S.C. Mining Act Section 48-20-50).	Disturbance of less than 5 acres to a depth of less than 20 feet with no processing facilities can be permitted under a General Mine Operating Permit (S.C. Mining Act Section 48-20-55).
Alaska	There is a 14-day agency notice with a notice to the public via the State Online Public Notice website.	Exploration operations on state lands that require permits include a facility that remains overnight; prospecting using hydraulic equipment methods; exploratory drilling over 300 feet deep; geophysical exploration for minerals subject to lease; or seismic surveys involving the use of explosives (Alaska DNR, 2022a).	Mined area less than 5 acres at one location in any year with cumulative unreclaimed mine area of less than 5 acres at one location, or where less than 5 acres and less than 50,000 cubic yards of gravel or other materials are disturbed or removed at one location in any year are exempt (AS 27.19.050).
Colorado	The Notice of Intent is provided in two forms by the applicant: one includes all information, while the other redacts confidential information. The redacted version is posted to the regulatory program's website, with a period of 10 working days for public comment after it is posted (2CCR407-1-5.1.2).	A Notice of Intent and financial assurance is required for "prospecting," which includes sinking shafts, tunneling, drilling core and bore holes, digging pits or cuts, and other associated disturbance works for the purpose of extracting samples prior to commencement of development or extraction operations (2CCR407-1.1(56)).	"Limited Impact Operations" include any operation that affects less than 5 acres or affects less than 10 acres and extracts less than 70,000 tons of minerals and overburden per year. A full mining permit is required for operations with metallurgical processing chemicals, or the exposure of toxic or acid-forming materials (CRS §§ 34-32-103 and 34-32-110).
Montana	No public notice at the time of application, but an EA or EIS document is developed. An EA may result in public notice and a comment period. An EIS requires notification, public meeting, and comment periods (MEPA, 75-1-102, MCA).	An exploration license and reclamation bond are required for all activities that result in disturbance of the surface. A bulk sample for metallurgical testing is limited to 10,000 tons (82-4-331 and 332, MCA; 82-4-303, MCA).	Exempt from permitting and limited bonding may apply if less than 5 acres are disturbed at one or two locations (82-4-303 and 305, MCA).

continued

TABLE 5-11 Continued

State	Public Notice for Exploration	Permit for Exploration	Exemptions for Small Mines
Idaho	No public notice required for exploration activities.	Exploration operations may require permitting if over 5 acres are disturbed for 12 consecutive months (§ 47-1503(7), Idaho Code). No application fee or financial assurance is required for exploration that is not a mining operation (IDAPA 20.03.02 060).	None. All surface mines operated after 1972 and all underground mines started after 2019 must have a reclamation plan and financial assurance (Eric Wilson, personal communication, 2022).
Nevada	No public notice unless exploration will disturb more than 5 acres (NAC 519A.410; NRS 519A.160).	A reclamation permit is required for exploration that will disturb more than 5 acres (NAC 519A.410; NRS 519A.160).	A reclamation permit is required for mining that will disturb more than 5 acres (NAC 519A.410; NRS 519A.160).
California	Exploratory activities could trigger California's Surface Mining and Reclamation Act (SMARA) depending on the nature and scope of the proposed exploratory project.	Local planning and environmental health departments often require permits for drilling and exploratory work. Exploratory activities that disturb more than one acre are subject to SMARA, which requires a permit, reclamation plan, and financial assurance.	SMARA applies to mining activities that disturb more than one acre or 1,000 cubic yards of material (Public Resource Code section 2714; California Code of Regulations Title 14 section 3505(a)).

NOTE: Several of these states have higher proportions of federal lands, which may result in projects triggering a NEPA process.

considerations, it seems unlikely that a small-scale mine under a Restricted Mining Permit would include complex processing facilities on-site at current gold prices. For example, a hypothetical example of a gold mine with grades of 0.29–1.55 ounces/ton (the range of historic gold mines in Virginia) and 500 tons of ore removed (without overburden or waste rock) would generate a total value of $290,000 to $1,550,000 at a gold price of $2,000 per ounce. This value would likely be insufficient to pay for the costs of a workforce, site exploration and development construction, mine production, and a significant level of processing. Thus, on-site processing within a Restricted Mining Permit may not be economically viable. It may be more likely for small operations to conduct partial steps toward gold separation and concentration, then transport that material to off-site locations for further processing and refining.

Given the small size of the envisioned operations, it is possible that potential environmental impacts from land disturbance would be limited if best management practices are followed for handling soil and rock materials, and for controlling runoff and erosion. Small operation footprints are not likely to result in development of large waste disposal sites or large impoundments for water, process solutions, or tailings. However, depending on the reactivity of the geologic materials, the methods that might be used for small-scale gold mining and processing, and the environmental setting of the mine site (e.g., proximity to streams), potential impacts could extend beyond the area disturbed directly by mining. Although potential failures of small impoundments would result in relatively small areas of direct inundation, the chemical impacts from metals, reagents, or other solutes could extend farther in the watershed; solutes from process solutions, reagents, or blasting by-products (e.g., nitrates) would likely be less persistent than any metals that may be deposited within relocated tailings or leached from on-site waste piles (see Chapter 3). Thorough site investigations, detailed designs for operations and reclamation, and detailed regulatory evaluations and oversight are essential for small-scale as well as larger-scale mining projects. As part of the permit, additional mitigation plans could be needed for the management of water, process solutions, facility air emissions, and/or waste materials.

Virginia's Restricted Mine Permit is similar to permits in other states that offer limited permitting requirements or full exemptions for certain operations based on the disturbance area, annual production volumes, and/or the commodity produced (see Table 5-11). These limited permitting requirements may be appropriate for certain mineral commodities and operations within nonreactive geologic settings and situations (e.g., sand and gravel, shallow rock quarries) where influence on water quantity or quality are very low, and within locations with low population density, where local-scale impacts are less likely to affect nearby residents. In Montana, because of the potential environmental impacts from insufficient project designs, operational, and/or reclamation practices, the exempt "Small Miner" operations may not utilize cyanide or other metal leaching agents without obtaining a full mine operating permit and providing a performance bond for the leaching facility portions of the site (ARM 17.24.185). In Colorado, permits with reduced requirements are available for "Limited Impact Operations," with different categories for activities with less than 5 or 10 acres of disturbance. However, these permits are not applicable if metallurgical processing chemicals are present on-site, toxic or acid-forming materials (i.e., sulfide minerals) may be exposed or disturbed, or there is potential for acid rock drainage to occur (CRS § 34-32-110). Those mining activities would be considered "Designated Mining Operations" and are required to obtain a full mining permit (CRS §§ 34-32-103 and 34-32-110).

Underground or Deep Mining

Virginia Energy reports there are currently two mineral mining permits that include (non-gold) underground operations. Both sites are in the process of closure, and there has not been a significant amount of underground mineral mining in the past 30 years (Michael Skiffington, personal communication, 2022). However, as discussed in Chapter 3, there may be potential for underground or "deep mining" to extract gold, whether through the development of new workings or the remining of historical mines.

Deep mining activity that has no significant effects on the surface is exempt from the definition of mineral mining and the applicability of codes and regulations. However, any surface facilities or associated surface disturbance in conjunction with underground mining would require a permit, and financial assurance for reclamation would be required if the area exceeds 1 acre. Additionally, given the climate and hydrology in Virginia, it is almost certain that underground mining would occur below the water table. This would require a hydrologic assessment and protection plan to minimize the adverse effects on water quantity and quality (4VAC25-31-130). The handling of groundwater would be addressed in the drainage plan; and the management, treatment, or discharge of water would be addressed by the protective methods established for the mining permit (4VAC25-31-130) and associated water protection permits administered by Virginia DEQ (e.g., VPDES).

Even without much surface disturbance, the operation of an underground gold mine can be highly complex. Best practices rely on a thorough site assessment that includes hydrologic, geochemical, and geotechnical characterizations; ground stability controls and safety measures; the management of water and waste materials; and quality assurance and monitoring programs. The exemption for deep mining makes it unclear what level of technical assessment and oversight is applicable for deep underground mines. The Division of Mineral Mines is allowed to evaluate operational plans and methods for underground mining to address the potential for significant surface effects (Michael Skiffington, personal communication, 2022; 4VAC25-31-130), and the operations, drainage, and reclamation plans for the permitted surface facilities must address mining methods (4VAC25-31-130).

The exemption of underground gold mining would also impact the calculation of financial assurance. The costs for the reclamation of an underground gold mine includes not only associated surface disturbance and facilities, but also the methods and costs for: implementing backfill or plugging methods to limit ground movement, groundwater flow, and/or chemical reactivity (sulfide oxidation) within the mine; plugging the access portals or ventilation shaft openings at the surface; and the management, treatment and discharge, and monitoring of water that may be required during reclamation and long-term stewardship. Under Virginia's current laws and regulations, some of these reclamation methods and costs would not be considered in the financial assurance (bond) for a permit that includes underground mining, because of the defined exemption and the bonding requirements that are based solely on acres of disturbance (see section on "Financial Assurance"). Other states do not exempt underground mining or differentiate underground activities from surface mining within the permitting requirements (Nevada, NAC 519A, NAC 445A; Montana, 82-4-335, MCA; 82-4-336, MCA; 82-4-338, MCA; ARM 17.24.116).

Processing Facilities

On-site processing facilities are included in the definition of a surface mineral mine and would be incorporated within a mine permit issued for gold mining (§ 45.2-1101 of the Code of Virginia). Virginia's laws and regulations do not limit the methods, reagents or process solutions, or equipment that might be utilized for gold processing in

Virginia. In contrast, processing facilities that are not located on-site with active mining or extraction ("toll mills") are not included in the definitions for a surface mineral mine or underground mineral mine (§ 45.2-1101 of the Code of Virginia). Therefore, Virginia's current laws and regulations do not require these facilities to obtain a permit from the Mineral Mining Program for the operation and reclamation of the site, although these facilities might need to obtain other permits. Based on previous descriptions of potential gold deposits and mining methods in Virginia (see Chapter 3), it is possible that small or Restricted Mining Permit mines would bring ore material or concentrates to a centralized facility for further processing.

The operations at toll mills may look very similar to processing facilities located at active mine sites, including multiple structures and types of equipment, storage and containment systems for process solutions (ponds, tanks, pumping systems), and disposal areas for tailings or other waste (impoundments or fills). The ore or concentrated material may come from many different sources, so the resulting waste material at toll mills may contain a wider range of contaminants than what may be found at a single mine and processing facility. Many of the waste materials generated at either toll mills or permitted on-site facilities would be exempt from regulation as hazardous waste under RCRA Subtitle C (Bevill exemption). Although a mining permit is not required, off-site processing facilities are subject to local zoning ordinances and Virginia DEQ's permitting requirements for protecting air quality and water quality (e.g., emission limits, runoff controls, discharge permits). Permits would be in place to limit emissions and some environmental impacts. Nevertheless, toll mills may not be regulated as stringently as processing facilities at permitted mine operations that use essentially the same techniques and may not include plans and financial assurances for facility and equipment demolition, reclamation, closure, and any necessary management of water, process solutions, and/or waste materials.

This legal gap creates a situation in which toll mills could substantially impact public health and the environment. Because they are outside of the regulatory framework, toll mills are more likely to have insufficient or incomplete site characterizations and project designs, and may not always implement best practices for operations, reclamation, and long-term stewardship of the facilities. In contrast to this legal gap in Virginia's regulatory framework, Montana regulations require that off-site mills or processing facilities obtain a full operating permit and provide a performance bond for reclamation, closure, and long-term stewardship (ARM 17.24.166). Specifications are provided about the terms of operation and reclamation of mills or processing facilities (ARM 17.24.165 through 171) and additional permits from Montana DEQ would be required to protect water quality and air quality.

Mine Permit Application

Prior to submitting a mine permit application to the Mineral Mining Program, the applicant must receive approval from local administrative officials with regard to zoning and land use requirements. The "heavy industrial" land use category is typically applied to mining operations. Most localities require some form of Conditional Use or Special Exception within areas of acceptable zoning, or a zoning change if unacceptable zoning currently exists (Michael Skiffington, personal communication, 2022). The localities often put additional conditions on the operations beyond the requirements in mineral mining codes and regulations (see Box 5-1; § 45.2-1227 of the Code of Virginia).

Following zoning and land use approvals, mining operations in Virginia must obtain a permit from the Mineral Mining Program. Application for a mineral mining permit is submitted with an initial permit fee and financial assurance (4VAC25-31-110), except for Restricted Mining Permits, which do not require a fee and financial assurance (§ 45.2-1203 of the Code of Virginia and 4VAC25-31-200).

Table 5-12 shows all of the components that are required for the application package (permit application checklist Form DMM-148; Virginia DMME, 2011). The general information requirements for the application forms are similar to those in other states, including the proposed mine location and adequate maps, name and contact information for the applicant, and their legal right to enter and mine the proposed property (Montana, 82-4-335, MCA; Nevada, NAC 445A.394; South Carolina, SC § 48-20-70). Much of this information may be addressed by the applicant with brief responses and the details are relatively easy for regulators to verify for accuracy (Form DMM-148 checklist). More complex narratives are required to address the necessary details for the operations, drainage, and reclamation plans and any associated technical studies.

TABLE 5-12 Required Components of Mineral Mining Plan in Virginia

Administrative Information	Permit/license application
	Public notification
	Relinquishment/succession
	Permit fees and bond fees
Operations Plan	Methods for mining and processing
	Topsoil handling and storage plan
	Spoil, overburden, and waste rock handling and disposal plan
	Plan for stockpiles, equipment storage, and maintenance areas
	Cut and fill slopes plan
	A copy of the Virginia Department of Transportation land use permit for roadways
	Plan for storage and disposal of scrap materials, service products, and solid/hazardous wastes
	Impoundments plan
Drainage Plan	Narrative of drainage system to be constructed before, during, and after mining
	A map or overlay showing the natural drainage system
	Design, maintenance, and abandonment plan for all sediment and drainage control structures
Reclamation Plan	Postmining land use plan
	Backfilling and regrading plan
	Revegetation plan
	Plans for closing or securing all entrances and reclaiming the surface areas of underground mines
Maps and Figures	Maps, cross-sections, and construction specifications of mine
	Map of all properties, and their owners, within 1,000 feet of the permit boundary
	Map of sensitive features within 500 feet of permit boundary
	Map of wetlands and riparian buffers that have been previously delineated
Technical Studies	Hydrologic studies and a plan to minimize adverse effects on water quality of quantity
	Preblast survey
	Wetland investigations

SOURCE: Permit Application Checklist, Form DMM-148.

Each application for a permit must be accompanied by a Mineral Mining Plan, which consists of separate documents for operations, drainage, and reclamation plans, along with supporting studies, maps, and figures. The primary components of these plans are summarized in Table 5-12. The Mineral Mining Plan is developed to "minimize adverse effects on the environment and facilitate integration of reclamation with mining operations" (4VAC25-31-360). It must describe the specifications for surface grading and restoration for postmining land use (§ 45.2-1206 of the Code of Virginia) and include a provision for the simultaneous reclamation of all affected land where practical (§ 45.2-1206 of the Code of Virginia; 4VAC25-31-130). A permit cannot be issued until at least 15 days after the application is submitted, except if everyone required to receive notice has issued a statement of no objection (4VAC25-31-170). Typically, the review process takes around 6 months for simple operations, or 1 year or two for larger and more complex operations (Michael Skiffington, personal communication, 2021). If the permit is not approved, the applicant would receive written objections and required modifications. The Director of Virginia Energy may reject the permit application if the operations would "constitute a hazard to the public safety or welfare," or if "a reasonable degree of reclamation or proper drainage control is not feasible." Modifications to the original plan must be submitted for review in the same manner as an original plan (§ 45.2-1205 of the Code of Virginia).

During the application review, the Mineral Mining Program reviews the adequacy of project plans and the supporting technical information. This means that program staff must have sufficient expertise, appropriate reference documents, and familiarity with current best practices to thoroughly review the permit application, identify potential flaws with the proposed plans for all stages of the project life cycle, and assess the adequacy of baseline information to support the plans. The Mineral Mining Program provides the primary adequacy review for the mining application, but the expertise of other government agencies and organizations, or their private contractors and consultants, may be necessary. Therefore, the Mineral Mining Program would benefit from a comprehensive understanding of potential environmental concerns and ability to identify them when outside assistance or expertise is needed, along with the authority and resources to hire expert consultants when necessary. Given the current lack of permitting for gold or mineral mining at Virginia Energy, this expertise may not be readily accessible within the agency. Local governments may require additional studies (e.g., surface water and groundwater, blasting, traffic and access, archaeological and historical resources) and impose additional requirements for operations to reduce impacts for public safety and potential nuisance (e.g., lights, noise, hours of operation). However, expertise and familiarity with potential mining impacts are likely not consistent across all county or community governments, so many environmental considerations may be overlooked or applied inconsistently among different jurisdictions. These shortcomings were expressed as a source of concern among some citizens living near exploration sites during the public listening sessions for this study.

An assessment of hydrologic baseline conditions is required for mining below the water table (4VAC-31-130) and the Mineral Mine Operator's Manual notes that a groundwater protection plan is needed to address the "potential for accidental releases of pollutants" (Virginia DMME, 2011) and to minimize the adverse effects to water quality and quantity (4VAC25-31-130). The Operator's Manual speaks generally about sources of water pollutants, but few specific details are provided about characterizing the primary geochemical factors that might degrade water quality (e.g., ore zones, host rock, or waste materials) and the protective mitigations or controls that could be implemented. A basic discussion of acid generation is provided in the context of testing soil or rec-lamation cover material, but additional guidance for the methods of geochemical characterization and predicting potential water quality impacts would be useful. The Operator's Manual notes generally that "mining operations that produce metals either as mine product, by-product, or waste, should complete a full assessment of the potential impacts of the operation on ground water quality," which might include contaminant transport computer models (Virginia DMME, 2011). The adequacy of such geochemical and hydrologic assessments is heavily dependent on the expertise and discretion of the applicants and regulators. Predicting water quality impacts from mineral mines has been an area of weakness in many states and federal jurisdictions, particularly for EIS documents developed from the 1980s to early 2000s (Kuipers and Maest, 2006). Best practices for predicting water quality impacts have continued to improve since that time, by including more detailed site characterization, waste characterization, and modeling of hydrologic and geochemical conditions.

Virginia's laws, codes, regulations, and Mineral Mine Operator's Manual do not reflect the importance of col-lecting a wide range of baseline information prior to mining. This baseline information is essential to evaluating a potential mine site and the best methods to extract gold and mitigate environmental impacts. Other states require these data to support the mine plans that accompany permit applications and to inform evaluations of environmental impacts. Baseline data may include geologic and geotechnical characterizations of site (overburden, waste rock, and ore), soils, vegetation, wildlife, surface water and groundwater hydrology and geochemistry, air quality, meteo-rology, aquatic biology, land use and ownership, recreation, cultural and historic resources, noise, transportation, and aesthetics (Colorado, 2CCR407-1-1.4; Montana, ARM 17.24.165; Nevada, NAC 445A.396). In California, the existing physical environmental conditions must be described from both local and regional perspectives, with special emphasis on local rare or unique environmental resources (California Code of Regulations 14 § 15125).

In order to follow a life-cycle approach (see Figure 3-1), project proponents would need to start collecting key data at the early stages to enable them, the regulators, and other stakeholders to make informed decisions about the design, operation, and closure of the project. Some data, such as stream flows, meteorology, geochemical weathering, and aquatic life surveys, may need to be collected over multiple years and over all seasons to reli-ably establish the environmental baselines needed to forecast and assess project impacts (see discussion of best practices in Chapter 3). Best practices by agencies are the sharing of data acquisition guidelines that applicants and stakeholders can see in advance. As the data are acquired, reported, and analyzed, there can be an ongoing dialogue about the sufficiency of those data so as to avoid any last-minute surprises about data needs.

Operations and Drainage Plans

In addition to the administrative information, maps, and technical studies described above, an application for a permit requires plans for both operations and drainage. Table 5-13 indicates the performance standards that would apply to all gold mining operations and drainage plans. Many of the performance standards adequately consider environmental protections and are similar to the general requirements in other states for aspects of mining operations, like site boundaries, barriers, and signage; soil salvage and stockpiles; road maintenance and dust control; avoidance of protected or sensitive features; and controls for runoff and erosion. In some cases, the standards in Virginia are prescriptive and quantify specific aspects of designs, like the runoff capacity for diversion structures and storage basins or the allowable slope angles for rock or fill structures, often based on material strength properties. In other cases, the codes and regulations are based on outcomes and do not provide specific guidance to achieve the standard, as in the stipulation that "Mining activities shall be conducted so that the impact on water quality and quantity are minimized" (4VAC25-31-360). In Virginia, the Mineral Mine Operator's Manual (Virginia DMME, 2011) provides guidance for many aspects of the operation, drainage, and reclamation plans, but these best practices are provided as recommendations and are not enforceable unless incorporated into a permit. This performance-based approach provides flexibility for the designs contained in the applicant's plans, but the codes and regulations provide little guidance for operators to achieve the objectives and few metrics for regulators to evaluate during the review of the application. The Mineral Mine Operator's Manual helps fill this gap for some aspects of the operation and drainage plans, but the manual does not address all factors that should be considered for gold mining activity. For example, the manual includes details about using geotextiles for temporary erosion control, drainage systems, and stabilizing roadways. However, there is no discussion of using durable geomembrane liner systems to contain water and waste materials during operations, or using such liners within capping systems to limit the potential for infiltration into reactive materials. Additionally, unless an operator incorporates the guidance details as specific conditions of their permit application, then these designs and methods are not enforceable.

Some states, including Arizona and New Mexico, have developed prescriptive descriptions of engineering designs and best practices covering topics such as designs for heaps or dumps, process solution ponds, geomembrane liner systems, leak detection and recovery systems, pipelines and tanks, and the construction and implementation of monitoring wells (New Mexico, 20.6.7.1 NMAC; ADEQ, 2004). Colorado has enacted requirements for phased construction, where inspections must verify acceptable progress before subsequent construction phases may continue, and prohibits the installation of liner systems where climatic conditions are not within design recommendations (2CCR407-1-7.3). The sections below expand on some of the more important performance standards in Virginia and compare them with those in other gold-producing states.

Water Withdrawal

Groundwater withdrawals in Virginia are not regulated west of I-95, outside of the Eastern Virginia Groundwater Management Area (9VAC25-600-20). Given the location of gold deposits described in Chapter 2, this means that groundwater withdrawal would not be directly regulated in the gold-producing region of Virginia. However, mineral mining permits that intercept groundwater are required to develop plans to minimize adverse effects on water quality or quantity (4VAC25-31-360), which might involve stipulations and mitigations to offset the effects of water withdrawal. Because almost all gold mining operations are expected to result in water withdrawal, inadequate implementation and oversight of water withdrawal plans could have significant repercussions for users of local groundwater.

Surface water withdrawals are regulated by the Water Board and the Virginia DEQ in places where the demand for surface water exceeds threshold limits (§ 62.1-242 et seq. of the Code of Virginia). Additionally, any permit for a major surface water withdrawal (more than 90 million gallons/month) and other impactful projects must provide a narrative description of the project as well as demonstrate that the project has avoided and minimized impacts to the aquatic environment (9VAC25-210-80, -90, and -110).

Process Solutions and Chemical Reagents

In Virginia, there are no specific regulations or restrictions on the nature of process solutions or chemical reagents that can be used within a gold mining operation. Nevertheless, compliance with applicable water quality

TABLE 5-13 Performance Standards for Operations Plan and Drainage Plan in Virginia

Air Quality	"Sources of dust shall be wetted down unless controlled by dry collection measures" (4VAC25-40-740). Control measures may be required for airborne contaminants, with regard to occupational health and safety (4VAC25-40-720).
Barriers and Screening	"Screening shall be provided for sound absorption and to improve the appearance of the mining site from public roads, public buildings, recreation areas, and occupied dwellings." Methods and specifications are determined by topography, berm or structure construction, vegetation types, and distance from permit boundary (4VAC25-31-420).
Boundaries and Signs	"A permanent sign shall be installed on the mining site adjacent to the principal access road and shall be visible and legible to access road traffic. The name of the permittee and the permit number shall be on the marker." (4VAC25-31-340). "The permit boundary of the mine shall be clearly marked with identifiable markings" or coincide with readily identifiable permanent features (e.g., streams, roads), when mine-related disturbance is within 100 feet of the permit boundary (4VAC25-31-140).
Drainage and Diversions (Runoff)	If necessary to cross or fill a drainageway, "properly engineered structures shall be provided to allow free-flowing drainage and minimize erosion. Where necessary, water-retarding structures shall be placed in drainageways" (4VAC25-31-470). "Surface water diversions shall be installed . . . where run-off has the potential for damaging property, causing erosion, contributing to water pollution, flooding or interfering with the establishment of vegetation." Temporary diversions (18 months or less) "shall convey the peak runoff of a 1-year, 24-hour storm," while diversions that "function more than 18 months shall be able to convey the peak run-off of a 10-year, 24-hour storm" (4VAC25-31-480).
Drainage and Diversions (Streams)	"All intermittent or perennial streams shall be protected from spoil by natural or constructed barriers. Stream channel diversions shall safely pass the peak run-off from a 10-year, 24-hour storm . . . the capacity shall be at least equal to the unmodified stream channel immediately upstream and downstream of the diversion" (4VAC25-31-460).
Impoundments (for Water, Liquids, or Tailings)	There are three subcategories of impoundments defined in codes and regulations. Specific requirements are provided for the design, construction, inspection, and closure of impoundments, based on the size and configuration of the feature (§ 45.2-1300 et seq. of the Code of Virginia; 4VAC-25-31-180, and 4VAC25-31-500).
Inactive Sites	A mining operation is complete and total reclamation shall begin when "no substantial mine-related activity has been conducted for a period of 12 consecutive months. . . . An operation may remain under permit for an indefinite period during which no mineral or overburden is removed if the following conditions are met: 1. All disturbed areas are reclaimed or adequately stabilized, or all erosion and sediment control systems are maintained in accordance with mining plans and proper engineering practices. 2. All drainage structures are constructed and maintained in accordance with mining plans and proper engineering practices. 3. All vegetation is maintained, including reseeding if necessary. 4. All improvements on site, including machinery and equipment, are maintained in a state of good repair and condition" (4VAC25-31-430).
Overburden, Refuse, Spoil and Waste Fills (NOT for Water, Liquids, or Tailings)	The plans and specifications "shall use current, prudent engineering practices." An engineering design report must include "calculations, drawings, and specifications" that account for the size and hazard potential of the fill, including: location and configuration, associated access, surface and subsurface drainage systems, and sediment control structures; cross-sections and profiles showing the original ground, fill profile, terraces, and constructed slopes; evaluation and preparation of the site and foundation, materials handling and placement, and sequencing of construction; slopes no steeper than 2H:1V for predominantly clay soils and no steeper than 3H:1V for predominantly sandy soils, or must exhibit a static safety factor of 1.5 for other steeper slopes. A closure and final reclamation plan for the fill and associated structures is required. "Fills shall be constructed, maintained and inspected to ensure protection of adjacent properties, preservation of public safety, and to provide prompt notice of any potentially hazardous or emergency situation." "On-site generated mine waste shall not be disposed of within the permitted mine area without prior approval." On-site generated mine waste may be approved as fill on the site, if adequately covered and vegetated (per reclamation plan). Inert waste generated from off-site "shall not be brought or disposed of in the mine permit area without prior approval" (4VAC25-31-400; 4VAC25-31-405).

TABLE 15-13 Continued

Processing Methods, Solutions, and Reagents	Codes and regulations do not limit the processing methods, process solutions, or chemical reagents that may be used in gold mining operations.
Protected Structures and Sensitive Features	"Mining activities shall be conducted in a manner that protects state waters, cemeteries, oil and gas wells, underground mines, public utilities and utility lines, buildings, roads, schools, churches, and occupied dwellings" (4VAC25-31-330).
Revegetation	"Disturbed land shall be stabilized as quickly as possible after it has been disturbed with a permanent protective vegetative cover. . . . Exposed areas subject to erosion on an active mining site shall be protected by a vegetative cover or by other approved methods. Simultaneous revegetation shall be incorporated into the mineral mining plan. Reclamation shall be completed on areas where mining has ceased" (4VAC25-31-520).
Roads	"Internal service roads and principal access roads shall be planned to minimize the impact of traffic, dust, and vehicle noise on developed areas outside the mining site." Methods must be employed to maintain the integrity of drainageways and limit damage to adjoining landowners and stream channels. Designs and specifications for ditches and culverts are provided in 4VAC25-31-350. Roads shall be surfaced with non-acid producing material and maintained to prevent the depositing of mud or debris on public loads, or introduce suspended solids into surface drainage. "Maintenance is required to ensure the proper functioning of the road and drainage system," and "dust from roads shall be adequately controlled" (4VAC25-31-350).
Sediment Control	Drainage from disturbed areas shall be directed into a sediment control structure before it is discharged from the permitted area. "Structures shall be located as close to the disturbed area as possible," but not located in perennial streams. Sediment control shall be installed prior to land disturbing activities within the drainage area, each primary sediment basin "shall provide at least 0.125 acre-feet of storage capacity for each acre of disturbed land draining to it. Storage basins shall be cleaned as necessary to ensure proper functioning before . . . reaching 60% capacity. Alternate sediment control measures that are as effective as sediment basins may be approved" (e.g., reduced basin storage capacity for small short-term disturbances, sediment channels, check dams, or mining methods that incorporate sediment control) (4VAC25-31-450).
Soil Stockpiles	A minimum quantity of soil shall be retained to cover and reclaim all disturbed areas "with six inches of soil or as specified in an approved operations plan." Soil shall be stored in a manner that remains available for reclamation use, with a maximum slope of 2H:1V, and it shall not be removed from the permitted area unless authorized. The stockpiled soil "shall be seeded with quick growing grasses or legumes for stabilization until used in final reclamation" (4VAC25-31-410).
Water Quality	All water discharge resulting from the mining of minerals "shall be between pH 6.0 and pH 9.0 unless otherwise approved by the director [of Virginia Energy]" (4VAC25-31-490). Discharges shall also be in compliance with standards established by the DEQ (9VAC25-260-20). Mining activities "shall be conducted so that the impact on water quality and quantity are minimized" (4VAC25-31-360). Mining below the water table "shall be done in accordance with the mining plan" (4VAC25-31-130).

NOTE: Some of these standards overlap with the required objectives for permits administered by other agencies (like Virginia DEQ), for the management of stormwater, the discharge of treated water, and the protection of surface water and groundwater.

standards is required for any water to be discharged from the facility, as described earlier in this chapter. In addition, according to Virginia Energy, site-specific requirements could be incorporated into the terms of a mining permit in order to protect water quality, based on the proposed mining and processing methods (Michael Skiffington, personal communication, 2021).

Water quality standards in Virginia are applicable to free cyanide (HCN and CN⁻), with criteria established for the protection of aquatic life in freshwater (acute = 0.022 mg/L, chronic 0.0052 mg/L) and in saltwater (acute and chronic = 0.001 mg/L). The standards for protecting human health address public water supplies derived from surface water (0.004 mg/L, 9VAC25-260-140), other surface waters related to fish consumption (0.4 mg/L, 9VAC25-260-140), and groundwater sources (0.005 mg/L, 9VAC25-280-40).

The aquatic life criteria in Virginia are consistent with the levels established by EPA for free cyanide (EPA, 2022j) and have been adopted by many states (e.g., Alaska,[13] Idaho, Montana, Nevada, and South Carolina). EPA guidance states that the analytical methods for total cyanide are allowed for screening, which would determine free cyanide, weak acid dissociable (WAD) metal cyanide complexes, and strong metal cyanide complexes (EPA, 2020b). Screening for free cyanide using an analytical method for total cyanide is not required, and laboratories or public water systems may choose to determine free cyanide without prior determination of total cyanide. However, if the total cyanide concentration exceeds 0.2 mg/L, then a measurement of free cyanide must be made using an approved free cyanide method to determine compliance (EPA, 2020b).

The federal level for public water supplies (0.2 mg/L free cyanide, 40 CFR 141.62(b)) is higher than the concentration allowed in Virginia. Other states have adopted the drinking water standard of 0.2 mg/L, although in some cases this may apply to groundwater and not surface water, and the form of cyanide listed in the respective regulations is variable. For example, Alaska and Colorado list free cyanide, Idaho lists WAD cyanide, and Montana, Nevada, and South Carolina list total cyanide. The standards adopted by other states to protect human health through fish consumption are also variable, although these specific criteria have not been established in every state. For example, Alaska has adopted standards for water and organisms (0.7 mg/L) and organisms only (220 mg/L), Idaho has adopted standards for water and organisms, or organisms only (0.140 mg/L), and South Carolina and Idaho have adopted standards for water and organisms, or organisms only (0.140 mg/L).

Some other states have specific regulation or restrictions applicable to cyanidation plants. Montana banned the use of cyanide for heap leaching or vat leaching for open pit ores, following a citizen's initiative in 1998. A common misconception is that the use of cyanide was banned entirely, but the language enacted in § 82-4-390, MCA, is specific to ores from open pits. This means that heap or vat leaching may still be permitted for ore obtained by underground mining or for legacy material produced from open pits prior to the conditional ban taking effect. In Montana, "Small Miner" sites (less than 5 acres) are exempt from most requirements for permitting and bonding (§ 82-4-305, MCA); however, these operations may not utilize cyanide or other metal-leaching agents without obtaining a full mine operating permit and providing a performance bond for the leaching facility portions of the site (Administrative Rules of Montana, ARM 17.24.185). Additional regulations apply to the use of cyanide or other metal-leaching agents regarding baseline information, operating plans, reclamation plans, performance standards, and bonding requirements (ARM 17.24.185 through 189). Arizona has included specific guidance about designing and operating leaching systems that includes monitoring for cyanide and related species (ADEQ, 2004). In Nevada, no facility may degrade state waters to the extent that the concentration of WAD cyanide exceeds 0.2 mg/L (NAC 445A.424). Idaho recently updated the rules for cyanidation (IDAPA 58.01.13) which provide some prescriptive details for the construction, operation, and closure of facilities that utilize cyanide as a primary leaching agent in order to ensure that pollutants associated with cyanidation are safely controlled and do not affect human or ecological health. Idaho requires that tailings impoundments contain no more than 50 mg/L WAD cyanide in the liquid fraction of the facility, and measures are required to prevent wildlife contact with any process water exceeding 50 mg/L WAD cyanide (IDAPA 58.01.13). Such examples, in combination with international guidance about current best practices and independent audits (see Chapter 3; International Cyanide Management Code, 2022), reinforce the concept that cyanidation may be a viable method for modern gold processing, but it requires specific regulatory considerations, due diligence and careful attention, and robust plans to address safety and management.

Impoundments

The Mineral Mining Program conducts permitting actions and regulatory oversight for impoundments at mine sites. The regulatory authority for impoundments is transferred from the Mineral Mining Program to the Dam Safety Program when the mine permit is terminated (Michael Skiffington, personal communication, 2022). Virginia regulations provide a list of technical documents that may be used as acceptable references for

---

[13] Alaska notes that the aquatic life criteria for free cyanide "shall be measured as weak acid dissociable (WAD) cyanide or equivalent approved EPA methods" (Alaska DEC, 2008).

impoundment designs and plans (4VAC25-31-500). These include specific publications (FEMA, 2013a,b), as well as others produced by USACE, the Natural Resources Conservation Service, the Bureau of Reclamation, the National Weather Service, and the U.S. Federal Energy Regulation Commission. Virginia regulations also provide specific designs and construction requirements for dams or mine refuse piles that impound liquids or semi-liquids (4VAC25-31-400). As summarized in Table 5-14, Virginia regulations provide technical specifications for three subcategories of water-retaining or silt-retaining impoundments, based on their size and characteristics (4VAC25-31-500). It is conceivable that gold mining impoundments in Virginia might fall into any of these size categories. The largest facilities (category A) pose more potential environmental risks than the two other categories (B and C), and therefore have more prescriptive requirements within codes and regulations for design standards, documentation, stability criteria, storm event management, inspections, closure plans, and emergency action planning. These prescriptive requirements include minimum static stability and seismic safety factors (1.5 and 1.2, respectively) for impoundments in category A, whereas impoundments in category B and C only have maximum allowances for slope steepness of 2 horizontal:1 vertical (2H:1V) in predominantly clay soils or 3H:1V in predominantly sandy soils.

The spillway design for Category A impoundments is dependent on their hazard classification, which ranges from high[14] to significant[15] to low hazard[16] (see Table 5-15; 4VAC50-20-40). Based on this classification, the spillways must be built to handle a design flood, which is defined as the probable maximum flood (PMF), half of the PMF (0.5PMF), or a 100-year storm, respectively (4VAC25-31-500). PMF is calculated from the probable maximum precipitation (see Figure 5-2), "the theoretically greatest depth of precipitation for a given duration that is meteorologically possible over a given size storm area at a particular geographical location at a particular time of year with no allowance made for future long-term climatic trends" (4VAC50-20-50). As noted in Table 5-15, the design flood can be reduced to the minimums identified in the table if an incremental hazard assessment is performed and shows it appropriate. An incremental hazard assessment is a comparative study of two floods of differing magnitude to identify the flood level above which there is no additional impact on downstream proper-ties. The "no additional impact" criterion commonly is defined as no more than a 2-foot increase of water level at impacted properties, although this definition is somewhat ambiguous as occasionally a 2-foot increase of water level changes the flood severity (which is defined as flood depth times velocity). While this approach is not used in many states, it is used in the design of dam structures that are regulated by the Federal Energy Regulatory Commission.

In contrast to the hazard dependent criteria for Category A impoundments, the design for storm events is much simpler for smaller Category B impoundments. Temporary Category B impoundments must only safely pass runoff from a 50-year storm, whereas permanent Category B impoundments must safely pass runoff from a 100-year storm.

Virginia's requirements and guidance for the construction of impoundments are less conservative than in some states and not consistent with best industry practices. For example, Nevada recommends that diversions around tailings storage facilities be designed and constructed to withstand 500-year storm events (NAC 519A.345; NDEP, 2016). Montana requires that spillways or other devices must protect against washouts during a 100-year flood (ARM 17.24.115), while a new tailings storage facility must store the PMF event, plus maximum operating water volume, plus sufficient freeboard for wave action, or a flood event design criterion less than the PMF but greater than the 500-year, 24-hour event if site-specific conditions determine that the PMF design standard is unnecessary (82-4-376(2)(cc), MCA). With regard to potential changes in the frequency and intensity of storm events, Montana requires that the design storm event for a tailings storage facility must include "evidence that the dynamic nature of climatology was considered" (82-4-376(2)(bb), MCA).

The prescriptive requirements provided for the maximum slope angles and minimum safety factors (static stability safety factor >1.5 and seismic safety factors >1.2) for the construction of impoundments in Virginia also fail to meet best industry practices. Draft International Committee on Large Dams guidelines for tailings

---

[14] High hazard indicates that failure would result in probable loss of life or serious economic damage (4VAC50-20-40).

[15] Significant hazard is when failure may cause loss of life or appreciable economic damage (4VAC50-20-40).

[16] Low hazard defined as failure that would result in no expected loss of life with minimal economic damage (4VAC50-20-40).

TABLE 5-14 Summary of Impoundment Requirements, Where Categories A, B, and C Are Determined Based on the Height Above Ground Level or the Volume of Material Impounded

	Impoundment Subcategory A	Impoundment Subcategory B	Impoundment Subcategory C
Description	"Structures that impound water or sediment to a height of 5 feet or more above the lowest natural ground area within the impoundment and have a storage volume of 50 acre-feet or more, or impound water or sediment to a height of 20 feet or more, regardless of storage volume."	Impoundments above the natural ground surface that do not meet or exceed the size criteria of Subcategory A.	Impoundments with impounding capability created solely by excavation (all contained below natural ground surface).
Design Standards	"Impounding structures shall be constructed, operated, and maintained such that they perform in accordance with their design and purpose throughout their life". They "shall be designed and constructed by, or under the direction of, a qualified professional engineer licensed in Virginia and experienced in the design and construction of impoundments. The designs shall meet the requirements of [4VAC25-31-500] and use current prudent engineering practices."	Impoundment "shall be designed and constructed using current, prudent engineering practice to safely perform the intended function."	Impoundment "shall be designed and constructed using prudent engineering practice to safely perform the intended function."
Document Requirements	"Plans and specifications shall consist of a detailed engineering design report that includes drawings and specifications," meeting the requirements found in 4VAC25-31-500.	NA	NA
Embankment Stability	"Impoundments meeting the size requirements and hazard potential of high, significant, or low shall have a minimum static safety factor of 1.5 for a normal pool with steady seepage saturation conditions and a seismic safety factor of 1.2."	Slopes shall be no steeper than 2H:1V in predominantly clay soils or 3H:1V in predominantly sandy soils.	Slopes shall be no steeper than 2H:1V in predominantly clay soils or 3H:1V in predominantly sandy soils.
Storm Events and Outlets	The design shall utilize a Spillway Design Flood event and Threshold for Incremental Damage Analysis, based on the classification of hazard potential (events ranging from 50-year storm to probable maximum flood). All structures shall allow draining within a reasonable period, a minimum of lowering the pool level by 6 inches per day, as determined by the engineer.	"Safely pass the runoff from a 50-year storm event for temporary (life of mine) structures and a 100-year storm event for permanent structures (to remain after mining is completed)."	"Be designed and constructed with outlet facilities capable of: protecting public safety, maintaining water levels to meet the intended use, being compatible with regional hydrologic practices."

TABLE 5-14 Continued

	Impoundment Subcategory A	Impoundment Subcategory B	Impoundment Subcategory C
Closure	"Closed and abandoned in a manner that ensures continued stability and compatibility with the postmining land use."	"Closed and abandoned to ensure continued stability and compatibility with the postmining use."	"Closed and abandoned to ensure continued stability and compatibility with the postmining use."
Inspections	Inspected and maintained to ensure that all structures function to design specifications. "Impoundments shall be inspected at least daily by a qualified person . . . who can provide prompt notice of any potentially hazardous or emergency situation as required under § 45.2-1302 of the Code of Virginia."	"Inspected and maintained to ensure proper functioning."	"Inspected and maintained to ensure proper functioning."
Protections	"Ensure protection of adjacent properties and preservation of public safety and . . . meet proper design and engineering standards" under § 45.2-1300 et seq. of the Code of Virginia and 4VAC25-31-500.	"Provide adequate protection for adjacent property owners and ensure public safety."	"Provide adequate protection for adjacent property owners and ensure public safety."
Emergency Action Plan	An Emergency Action Plan (EAP) is required to include: assigning responsibilities for decision making, implementing the EAP, and notifying all persons or organizations; procedures for timely and reliable detection, evacuation, and classification of emergency situations; actions and procedures to be followed before and during the development of emergency conditions; dam break inundation maps and appendix reports to support the development, training, and exercising the EAP; establishing time periods to review or revise the EAP.	NA	NA

SOURCE: The Mineral Mine Operator's Manual; 4VAC25-31-500A, B, and C.

TABLE 5-15 Design Criteria and Classification of Impoundments Under Subcategory A (Table 5-14)

Class of Impoundment	Spillway Design Flood	Minimum Threshold for Incremental Damage Analysis[a]
High Hazard	PMF	0.50 PMF
Significant Hazard	0.5 PMF	100-year storm
Low Hazard	100-year storm	50-year storm

NOTE: PMF = probable maximum flood.

[a] The proposed potential hazard classification and the proposed spillway design flood for an impounding structure may be lowered according to the results of an incremental damage analysis, but not below the minimum threshold values as indicated in the table (4VAC50-20-52).
SOURCE: The Mineral Mine Operator's Manual; 4VAC25-31-500.

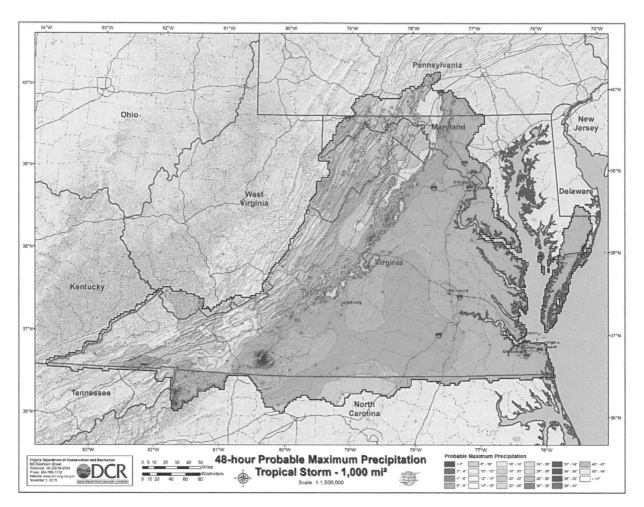

FIGURE 5-2 Statewide map of the 48-hour, 1,000-square mile probable maximum precipitation from tropical storms. SOURCE: Virginia DCR (2022a).

dam safety suggest that two principal stability conditions must be considered: static and post-liquefaction (ICOLD, 2020). The "static" condition in these draft guidelines differs from traditional "static" conditions in that the former considers the use of undrained shear strengths in geomaterials (tailings, perimeter dikes, foundation soils) that are contractive and at least partially saturated. In contrast, the traditional "static" conditions involve the use of drained (effective stress-based) shear strength parameters which commonly yield much larger factors of safety in soft, loose tailings materials. The post-liquefaction condition considers the stability of the structure after liquefaction has been triggered, regardless of the initiation mechanism. In contrast, the traditional "seismic" condition does not consider liquefaction and the potentially large-strength loss associated with soils (or tailings) liquefying. For these two conditions, static and post-liquefaction, the evolving best practice uses minimum recommended limit equilibrium factors of safety of 1.5 and 1.1, respectively. Furthermore, Virginia impoundment regulations provide simplified maximum slope angle requirements for various impoundment categories that are inappropriate. Specifically, a maximum slope angle of 2H:1V for clay soils is inappropriate and potentially unsafe, as the drained (effective-stress) fully softened friction angle of many clays ranges from 20° to 30° (Terzaghi et al., 1996). Standard recommended slopes angles for clay soils are not steeper than 3H:1V, regardless of the degree of soil compaction.

Finally, Virginia requirements for seismic design also are ambiguous and incompatible with best practices. Similar to flood and rainfall design levels, where design is based on probabilistic evaluations (e.g., rainfall from a 50-year storm), seismic design in other civil engineering applications is commonly based on a probabilistic framework. This framework provides a probability of exceedance of a particular ground motion parameter (e.g., 2 percent probability of exceedance of a particular ground motion parameter in 50 years, or a 2,475-year return period).

Montana has recently updated its requirements to provide a more robust inspections and monitoring program, which could provide useful guidance to Virginia. In 2015, Senate Bill 409 was enacted to update tailings storage facility (TSF) requirements (Section 82-4-301, MCA), based in part on the findings of the Mount Polley Independent Expert Investigation and Review Report (Morgenstern et al., 2015). The updates to the Metal Mine Reclamation Act (§ 82-4-300 et seq., MCA) are not prescriptive in detail, but ensure that TSFs allow for "adaptive management using evolving best engineering practices based on the recommendations of qualified, experienced engineers". The Montana statutes were strengthened with requirements for the qualifications and responsibilities of the Engineer of Record (§§ 82-4-303 and 82-4-375, MCA); multidisciplinary guidance and criteria for baselines studies, design documents, and plans for operations, maintenance, and surveillance throughout the facility life cycle (§§ 82-4-376 and 82-4-379, MCA); quality assurance monitoring and reporting during facility construction (§ 82-4-378, MCA); and additional technical oversight from an independent technical review board (three members) designated for each TSF (§ 82-4-377, MCA).

Reclamation Plan

Reclamation is defined as the "restoration or conversion of disturbed land to a stable condition that minimizes or prevents adverse disruption . . . and presents an opportunity for further productive use if such use is reasonable" (§ 45.2-1200 of the Code of Virginia). Postmining land use must be compatible with surrounding land use and Virginia Energy encourages productive uses of land (e.g., pasture, agricultural purposes, recreational areas). All permits and approvals for postmining land use must be obtained prior to implementation (4VAC25-31-360), and the reclamation plan submitted with the permit application must include

- A statement of the planned land use following reclamation, the proposed methods to assure concurrent reclamation, and a time schedule (§ 45.2-1206 of the Code of Virginia and 4VAC25-31-130);
- A description of the methods for grading, plans for removal of material (e.g., processing equipment, buildings, and other equipment), and revegetation of the disturbed area (4VAC25-31-130); and
- A description of the plans for closing or securing all surface entrances to underground workings (4VAC25-31-360) (this does not address the closure or reclamation of deeper mine tunnels, stopes, or related underground features).

Like the details within operations and drainage plans, site-specific conditions must be considered in the development and implementation of plans for reclamation, closure, and long-term stewardship. According to Kuipers (2000), the general principles for these plans should include "topsoil salvage and replacement; recontouring; revegetation; slope stability; stream protection; air and water resources protection; geochemical and acid rock drainage considerations; public health and safety; wildlife habitat restoration; and aesthetic impacts, including visual impacts." In addition to the general requirements provided above, Table 5-16 presents the performance standards for reclamation of mineral mines in Virginia. In many cases, these standards are brief and written with the intent of a particular outcome—for example: "Disturbed land shall be stabilized as quickly as possible after it has been disturbed with a permanent protective vegetative cover" (4VAC25-31-520). The Operator's Manual (Virginia DMME, 2011) provides considerable information and guidance about many aspects of reclamation (e.g., slope stability, revegetation and seed mixes, soil testing, designs for runoff control). Guidance is provided for seed purity and germination requirements (§ 3.2-4000 et seq. of the Code of Virginia) and the methods which might be used for native vegetation species. The exclusive use of native species during reclamation is not required, although using native species is an encouraged practice. Additionally, relatively few guidance details are provided about the

TABLE 5-16 Performance Standards for Reclamation Plan, Vegetation, and Bond Release Criteria for Concluded Mining Activities in Virginia

Acidic Material	"All acid material encountered during the mining operation shall be properly controlled to prevent adverse impacts on surface or groundwater quality. Upon completion of mining, acid materials shall be covered with a material capable of shielding them and supporting plant cover in accordance with the approved reclamation plan. Unless otherwise specified by the Director [of Virginia Energy], the minimum cover shall be four feet in depth" (4VAC25-31-380).
Erosion Control	"Riprap shall be used for the control of erosion on those areas where it is impractical to establish vegetation or other means of erosion control or in any areas where rock riprap is an appropriate means of reclamation. Placing of rock riprap shall be in accordance with drainage standards and the approved mineral mining plan. Other methods of stabilization may include gabions, concrete, shotcrete, geotextiles, and other means acceptable for the mineral mining plan" (4VAC25-31-510). "When a road is abandoned, steps shall be taken to minimize erosion and establish the post-mining use in accordance with the reclamation plan" (4VAC25-31-350).
Impoundments (for Water, Liquids, or Tailings)	For the largest category of impoundments, the designs, construction specifications, and other related data, including final closure and abandonment plans, shall be approved and certified by a qualified engineer and the licensed operator. For all size categories, the impoundments shall be closed and abandoned in a manner that ensures continued stability and compatibility with the post-mining land use. Every impoundment "shall be examined daily for visible structural weakness, volume overload, and other hazards by a qualified person designated by the licensed operator." "Inspections shall be performed more frequently when water and silt reaches 80 percent of the design capacity and during periods of rainfall that could create flooding conditions" (§§ 45.2-1301 and 45.2-1302 of the Code of Virginia, 4VAC-25-31-180, and 4VAC25-31-500).
Mineral Stockpiles	"Stockpiles of minerals shall be removed to ground level and the area shall be scarified and planted in accordance with the approved mineral mining plan. The Director [of Virginia Energy] shall allow a reasonable time for sale of stockpiles" (4VAC25-31-520 and 4VAC25-31-390).
Slopes and Stabilization	Spoil piles will be graded to minimize sediment run-off. "Slopes shall be graded in keeping with good conservation practices." "Long uninterrupted slopes shall be provided with drainage control structures, such as terraces, berms, and waterways," to accommodate surface water where necessary and to minimize erosion from runoff. "Slopes shall be stabilized, protected with a permanent vegetative or riprap covering" and not be in an eroded state at the time reclamation is complete. "Constructed cut or fill slopes shall not extend closer than 25 feet to any property boundary without the written permission of the adjoining property owner and the approval of the Director [of Virginia Energy]" (4VAC25-31-370, 4VAC25-31-390, and 4VAC25-31-530).
Underground Reclamation	"At the completion of mining, all entrances to underground mines shall be closed or secured and the surface area reclaimed in accordance with the mineral mining plan" (4VAC25-31-360).
Vegetation (Establishment)	"Critical areas" are defined as "problem areas such as those with steep slopes, easily erodible material, hostile growing conditions, concentration of drainage or other situations where revegetation or stabilization will be potentially difficult." "Crusted and hard soil surfaces shall be scarified prior to revegetation. Steep graded slopes shall be tracked [by] running a cleated crawler tractor or similar equipment up and down the slope. Application of lime and fertilizer shall be performed based on soil tests and the revegetation requirements in the approved reclamation plan. Vegetation shall be planted or seeded and mulched according to the mixtures and practices included in the approved reclamation plan. The seed used must meet the purity and germination requirements of the Virginia Department of Agriculture and Consumer Services. . . . Trees and shrubs shall be planted according to the specific post-mining land use, regional adaptability, and planting requirements included in the approved reclamation plan. . . . The use of grass, water bars, or diversion strips and natural vegetative drainage control may be required in the initial planting year" (4VAC25-31-10, 4VAC25-31-290, 4VAC25-31-530, and 4VAC25-31-540).

TABLE 15-16 Continued

Vegetation (Monitoring and Bond Release Criteria)	"The division's final inspection for bond release shall assess the adequacy of vegetation and shall be made no sooner than two growing seasons after the last seeding. . . . No noncritical areas larger than one-half acre shall be allowed to exist with less than 75% ground cover. Vegetation shall exhibit growth characteristics for long-term survival. Seeded portions of critical areas shall have adequate vegetative cover so the area is completely stabilized." Final inspection for bond release shall require the following vegetative cover, based on post-mining land use: • For intensive agriculture use, "planting and harvesting of a normal crop yield is required to meet the regulatory requirements for full or partial bond release. A normal yield for a particular crop is equal to the five-year average for the county." • "For forest and wildlife [use], at least 400 healthy plants per acre shall be established after two growing seasons." • For industrial, residential, or commercial use: "All areas not redisturbed by implementation of the post-mining use must be reclaimed and satisfactorily stabilized. All areas associated with construction of buildings or dwellings . . . [must be] covered by plans approved by the local governing body." "Areas not covered by such local government plans shall be reclaimed and stabilized" in accordance with the vegetation cover requirements above. • For other post-mining uses, all areas not directly used by the post-mining use should be stabilized in accordance with the vegetation cover requirements (4VAC25-31-290, 4VAC25-31-300, and 4VAC25-31-540).
Water Quality	All water discharge resulting from the mining of minerals "shall be between pH 6.0 and pH 9.0 unless otherwise approved by the Director [of Virginia Energy]." Discharges also need to be in compliance with applicable standards established by the DEQ. "Mining activities shall be conducted so that the impact on water quality and quantity are minimized. Mining below the water table shall be done in accordance with the mining plan under 4VAC25-31-130. Permanent lakes or ponds created by mining shall be equal to or greater than four feet deep, or otherwise constructed in a manner acceptable to the Director [of Virginia Energy]" (4VAC25-31-130, 4VAC25-31-360, 4VAC25-31-490, and 9VAC25-260-20).

complexities and potentially long-term necessity for managing reactive wastes or water treatment and discharge, which may be the most impactful and costly aspects for site reclamation and management. As discussed earlier, these guidance standards in the manual are not enforceable unless the details are specifically written into the permit.

Virginia's performance standards for reclamation (as detailed in Table 5-16) are similar to those found in other states, although guidance in each state is influenced by the specific ecosystems, land use, and climatic factors for their respective locations (Alaska-AS 29.19.020, 11 AAC 97.200-240; Montana-82-4-336, MCA and ARM 17.24.115; South Carolina-R.89-140 and R.89-330). Virginia has requirements for operators to provide an assessment of potential groundwater impacts and develop a protection plan (4VAC25-31-130), and set discharges that comply with applicable standards established by Virginia DEQ (9VAC25-260-20). Few requirements in Virginia address long-term stewardship situations in which managing water quantity and quality is required after mining and reclamation have ceased, a scenario that could develop for some gold mines. Specifically, few guidance details are provided about potential dewatering and other water quantity management systems, characterizing water quality and designing systems for water treatment and discharge (likely coordinated with VPDES permit), the challenges in modeling and managing pit lakes or saturated waste disposal areas (Flite, 2006; Nevada Department of Conservation and Natural Resources, 2021), or approaches to anticipating changes to water quality and quantity as a result of large storm events and climatic changes.

Pit lakes are one example of a feature that may represent a long-term source of water contamination persisting beyond mine closure. The required minimum depth of 4 feet for permanent lakes or ponds (4VAC25-31-130) may not be adequate to manage water quality conditions, but alternative methods are unclear and left to the discretion of the Division of Mineral Mining. Methods for mitigation should be described. These could include accelerated flooding, raising the flooded water level, and nutrient addition to facilitate bioremediation and stratification, as well as selective mining of problematic material from pit walls above the final lake level. But pumping and treatment should be regarded as the final option (INAP, 2014). In Colorado, legislative changes in 2019 (HB 1113) modified the requirements for reclamation plans, disallowing the option for perpetual water treatment. With a few exceptional circumstances, "a new or amended permit must demonstrate, by substantial evidence, a reasonably foreseeable end date for any water quality treatment necessary to ensure compliance with applicable water quality standards" (CRS § 34-32-116(7)).

Impoundments that store water and/or tailings are also examples of features that may necessitate long-term stewardship. The regulatory authority for impoundments is transferred from the Mineral Mining Program to the Dam Safety Program (within the Virginia DCR) when the mine permit is terminated (Michael Skiffington, personal communication, 2022). The conceptual phases of TSF reclamation and long-term stewardship are described by the Canadian Dam Association—these phases include active and passive care activities, which involve years of maintenance, monitoring, and evaluations to ensure that long-term stability is achieved. The principles for effective dam safety programs have been likened to a three-legged stool, with equally important legs consisting of (1) corporate responsibility by the facility owner and related stakeholders, (2) technical oversight and independent review, and (3) a strong, transparent regulatory environment (Morgenstern, 2011). The framework in Virginia for the operation and regulation of dams provides the minimum "stool legs," but updated requirements and guidance with specific focus on best practices for tailings management are needed.

Without more guidance regarding the complexities of TSF closure and maintenance, and approaches to perpetual water treatment from waste disposal areas, underground mines, and/or pit lakes, important details that may result in environmental impacts and affect long-term costs are left to the discretion and expertise of the permitting agency. Although performance-based regulations provide site-specific flexibility for the designs contained in the applicant's plans, Virginia's laws and regulations provide little guidance for operators to achieve the objectives and few metrics for regulators to evaluate during the review of the application. Sufficient guidance should be provided for planning and designing facilities, construction and quality assurance, operations and process optimization, monitoring and testing programs, methods for reclamation and revegetation, and any necessary long-term management and stewardship. In addition, the Operator's Manual, last updated in 2011, could be updated more frequently to incorporate legislative changes and administrative updates from the permitting program, data from case studies or research relevant to the environmental conditions in the Commonwealth, and current best practices and technical guidance from other states and national or international organizations.

Financial Assurance

Once the permit application is deemed complete, the applicant submits a financial assurance to Virginia Energy ("performance bond" or "bond"). The performance bond can be returned to the operator after all requirements in the approved operations plan and reclamation plan are met (§ 45.2-1208 of the Code of Virginia, 4VAC25-31-120, and 4VAC25-31-250). The bond liability is for the duration of the mining operation and for the period following reclamation until success of the final reclamation is demonstrated (4VAC25-31-230). The bond may be posted by an operator and a corporate surety, or the operator may submit cash, check, certificate of deposit, or irrevocable letter of credit in lieu of a bond (§ 45.2-1208 of the Code of Virginia and 4VAC25-31-260).

Virginia's procedures for performance bonding are consistent with the requirements in many other states. All states with gold mining operations approve of surety bonds and irrevocable letters of credit (typically issued by a bank) as acceptable forms of financial assurance, and most states accept trust funds, deeds, and various forms of cash or savings. These forms of performance bond are universally accepted because they are considered to be highly certain and relatively liquid (Kuipers, 2000). A corporate guarantee, also called self-bonding, is accepted in a few states (e.g., Nevada, Arizona, Idaho) and this bonding mechanism is based on evaluating an operator's ability to pay the cost of reclamation. Rather than providing a bond to the permitting agency, the permittee is required to demonstrate a specified ratio of assets to liability. The agency may require regular submittals of corporate financial statements and also require the permittee to establish a cash reserve to be used for reclamation. However, self-bonding does not insure the agencies and public against potential liability in the event of the company's financial failure; in the case of bankruptcy, the permitting agency is considered to be a creditor. Virginia and other states like Montana, Alaska, and South Carolina do not consider corporate guarantees to be acceptable, while New Mexico's statutes explicitly disallow them. Except where specifically allowed by some states, these bonding mechanisms are not employed by the Bureau of Land Management or the U.S. Forest Service. In 2019, Colorado enacted legislative changes through HB 1113 that repealed all self-bonding practices while retaining the other favorable bonding methods described previously (CRS § 34-32-117(3) and rules 2 CCR407-1-4.3.7 and 4.10).

Bond Amount Determination

The bond amount for a new mineral mine in Virginia is calculated at a flat rate of $3,000 per acre, based on the estimated acres of land to be affected by mining during the first year of operations (§ 45.2-1208 of the Code of Virginia). The minimum total bond for any mineral mining permit is $3,000, except for Restricted Mining Permits that are exempt from bonding, and Minerals Reclamation Fund participants (discussed below, 4VAC25-31-240). After mining operations commence, the bond is calculated annually at the time of permit renewal and covers the entire disturbed area, plus the estimated number of acres to be disturbed in the upcoming year (4VAC25-31-220). If additional areas are to be disturbed, the permittee must provide additional bond to cover the new acreage within 10 days of the anniversary of the permit (§ 45.2-1212 of the Code of Virginia; Figure 5-3). The bond must be posted prior to disturbing an area (4VAC25-31-220).

The financial assurance requirements in Virginia are not adequate to address the potential reclamation costs for gold mining. There may be examples of other mineral mining activities in Virginia which are adequately bonded (sand/gravel, stone quarries), but these may be exceptional cases, based on the recent reclamation costs

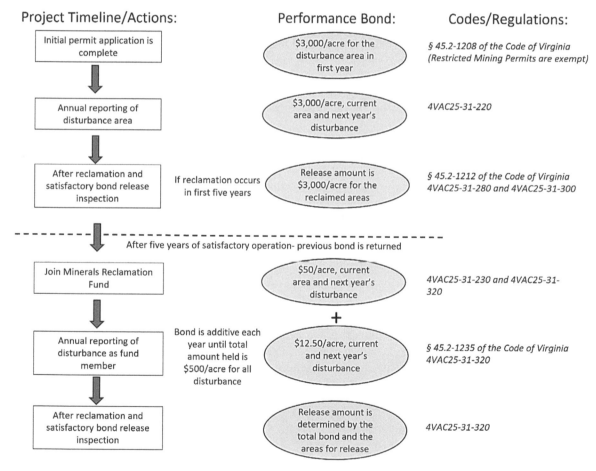

FIGURE 5-3 Current performance bond process for mining operations in Virginia, reflecting the general steps that would be involved in a performance bond during the lifespan of a hypothetical gold mining operation.

cited by Virginia Energy. Virginia's bonding might be appropriate for operations that have a low potential for extensive environmental impacts and thus for which land surface restoration is likely to be low cost (e.g., back-filling, grading, soil placement, revegetation) and the demolition and removal of minor facilities. However, it has been demonstrated that the required bond amount of $3,000/acre (or $500/acre for fund members) is not adequate for the reclamation costs for all current mineral mining operations. For example, Virginia Energy reports that two reclamation projects that were recently completed following bond forfeiture cost approximately $1,300/acre and $5,300/acre, and some mine operators have reported their own reclamation costs between $8,000/acre and $12,000/acre (Michael Skiffington, personal communication, 2022). Four companies, operating seven mine sites, have had their permits revoked and bonds forfeited in the past 10 years. All but one of those were the result of bankruptcies, with the other permit being revoked for failure to comply with a notice of noncompliance for required reclamation.

Given the potential environmental impacts and complexity of gold mining activities, the reclamation and long-term stewardship costs for gold mines could far exceed the currently established bonding rates due to the following factors, where applicable:

- The need for physical and chemical isolation of reactive overburden, other waste materials, or reactive mine surfaces (based on sulfide content or site-specific contaminants like mercury), likely including dumps and tailings impoundments;
- The need for water management within pits, underground mining, or impoundments, along with plans for monitoring and maintaining postclosure hydrologic balance;
- The need for water treatment for meteoric runoff, groundwater inflow, infiltration/seepage from reactive materials, or process solutions from the associated facilities, along with plans for postclosure treatment, water quality protection, and/or compliance monitoring;
- Processes for closure, demolition, and removal of mineral processing facilities and equipment and the management and disposal of reagents, process solutions, and/or other waste; and/or
- Postclosure monitoring of water quality and other reclamation requirements.

A reclamation bonding study by Kuipers (2000) documented approximately 150 metal mining operations in the western United States including gold mines and other metal mines. More than 20 years ago, the average bond level for these major mines was approximately $4,400 per acre, with costs ranging from less than $1,000/acre to greater than $50,000/acre, exceeding the amounts currently required for mining in Virginia. Federal agencies and multiple states have continued to update the requirements and guidance for determining financial insurance amounts since that time. Although an updated assessment of the current bonding costs at these 150 sites has not been conducted, the costs to perform mine reclamation have undoubtedly risen since this 2000 study. In Montana alone, the current bonding costs for gold mine permits vary from approximately $16,000 to more than $137,000,000 ($1,050 to $40,560 per disturbed acre respectively), depending on the disturbed surface area and level of activity at the site, the complexity of mining methods and reclamation plans (e.g., placer, open pit, or underground), and whether postclosure water management and site maintenance are required (Montana DEQ, 2022). There are many examples of gold mines which were not adequately bonded and developed significant postclosure water quality issues, resulting in very expensive long-term water treatment and site management being conducted by state and/or federal agencies following bond forfeiture (e.g., Zortman-Landusky Mines in Montana, Summitville Mine in Colorado, Brewer and Barite Hill Mines in South Carolina).

Rather than estimating the bond with a flat rate based on disturbed acreage, other state (see Table 5-17) and federal agencies require that financial assurances reflect the site-specific reclamation plan developed for the conditions of each proposed operation and these financial assurance amounts must be regularly reviewed and updated. In addition to initial closure and reclamation activities, these plans may include post-closure water management, water treatment, and other facility maintenance. It is not possible for this report to recommend or estimate the financial assurance amounts that might be necessary for potential gold mines in Virginia, in the absence of plans and conditions for specific projects. (Additional guidance documents for estimating costs and deriving bond calculations include BLM [2012], NDEP [2022], and USFS [2004].)

TABLE 5-17 Bond Calculations in Selected States

Alaska	The permitting commissioner "shall require an individual financial assurance in an amount not to exceed an amount reasonably necessary to ensure the faithful performance of the requirements of the approved reclamation plan." The maximum amount of $750/acre is applied for some mineral mines, but that bond limitation does not apply to a lode mine (i.e., lode mines are bonded at full, real cost estimates) (AK Stat § 27.19.040).
Colorado	Prior to enacting HB 1113 in 2019, only land reclamation costs were factored into the bond. Now the amount of the bond must be sufficient to ensure the completion of reclamation of affected lands if the agency has to complete the reclamation due to forfeiture, including all measures commenced or reasonably foreseen to ensure the protection of water resources including costs necessary to cover water quality protection, treatment, and monitoring as may be required by permit. An additional amount (5 percent of total bond) is required to address the agency's administrative costs while conducting reclamation (CRS § 34-32-117(4)).
Montana	All bonds must be based on "reasonably foreseeable activities that the applicant may conduct in order to comply with conditions of an operating permit or license" (§ 82-4-338(6), MCA). The bond "may not be less than the estimated cost to the state to ensure compliance with statutes, rules, and the permit, including the potential cost of department management, operation, and maintenance of the site upon temporary or permanent operator insolvency or abandonment" (§ 82-4-338(1), MCA).
South Carolina	The bond that is posted with the state agency covers the reclamation and closure of the site, but not necessarily the long-term stewardship related to protecting water quantity and quality. In the example of the Haile Gold Mine, a separate trust fund has been established through negotiations with conservation groups, which provides additional financial assurance beyond the money posted with the state agency.

Minerals Reclamation Fund (Bond Pool)

Each permittee with 5 years of satisfactory operation in the Commonwealth under the mineral mining codes and regulations is required to become a member of the Minerals Reclamation Fund (4VAC25-31-230 and 4VAC25-31-320). This fund is used solely for the Mineral Mining Program to conduct the reclamation of mining operations under the conditions of permit revocation and bond forfeiture (§ 45.2-1238 of the Code of Virginia).[17] All previously held performance bonds are released upon acceptance in the Minerals Reclamation Fund (§ 45.2-1236 of the Code of Virginia and 4VAC25-31-230). Eligible permittees enter the fund by making an initial payment of $50 for each disturbed acre and each acre to be affected during the next year. This is a significant decrease from the bond amount required prior to entering the fund ($3,000/acre). Thereafter, the member pays an annual fee of $12.50 per disturbed acre, plus each acre to be affected during the next year. These payments continue until the member has paid a total of $500 per disturbed acre, a cycle equivalent to 36 years of fund membership (§ 45.2-1235 of the Code of Virginia and 4VAC25-31-320) that no current operator has yet reached (Michael Skiffington, personal communication, 2022). Figure 5-3 depicts how the performance bond is modified through the lifespan of a hypothetical mining operation.

If the Minerals Reclamation Fund incurs expenditures from site reclamation following bond forfeiture, the money available in the Fund may be less than the total of all operator deposits (§ 45.2-1227 of the Code of Virginia). If the size of the Fund decreases to less than $2 million, the Director of Virginia Energy may suspend the return of payments and charge all members an equal amount for each affected acre, for a total amount sufficient to raise the fund to $2 million (§ 45.2-1237 of the Code of Virginia). When this happens, all members must post the required bond or other securities within 6 months or risk having their permit revoked (§ 45.2-1240 of the Code of Virginia). This situation has never occurred (Michael Skiffington, personal communication, 2022), likely due to the small scale and simplicity of current mining operations, which do not require detailed and lengthy strategies for the management of reactive waste materials or the management, treatment, and discharge of treated water. It is possible that complex reclamation activities could greatly decrease or potentially deplete the Minerals Reclamation Fund. It is likely that a mine operator with a revoked permit and forfeited bond has entered bankruptcy or similar financial hardship, which would further complicate or forestall repayment of the debt to the Commonwealth in such a case.

---

[17] Additional definitions and administrative components for the Minerals Reclamation Fund are provided in §§ 45.2-1233, 45.2-1234, 45.2-1242, and 45.2-1243 of the Code of Virginia.

Some states, including Arizona, California, Idaho, South Carolina, and Montana, do not accept or administer a bond pool as financial assurance for reclamation costs. Other states, including Nevada and Alaska, accept pool bonding as one type of bonding mechanism, but apply more prescriptive stipulations to pool members. For example, in Nevada, the state permitting agency determines the total bond amount, but the maximum bond coverage from the bond pool for the permittee is set at $3 million (NAC 519A.585). In Alaska, the permitting commissioner uses the projected cost of reclamation in relation to the size of the bonding pool to determine which mining operations are eligible to participate in the bonding pool. Alaska also excludes operations employing certain processes from the bonding pool, stating, for example, that "a mining operation may not be allowed to participate in the bonding pool if the mining operation will chemically process ore or has the potential to generate acid" (AK Stat § 27.19.040).

Bond Review and Audit

Virginia regulations stipulate that the bond amount be updated annually, based on the anniversary date for the permit and the extent of disturbance anticipated in the coming year. Other states require bond amounts to be updated regularly to account for any increase in disturbance area or any modifications to the operations and/or reclamation plans. In some states, including South Carolina and Montana, the amount of financial assurance is revaluated as part of modifying the permit, independent of the annual reporting or permit renewal cycle. Montana also requires annual site inspections and bond reviews, which might result in a bond recalculation, while a comprehensive recalculation must be performed at least every 5 years and following any major permit modifications (§ 82-4-338(3), MCA). The adjustments are not based solely on the increasing disturbance footprint, but they account for the actual costs to complete reclamation based on economic conditions and the complexity of the modified operation and reclamation plans. In California and Idaho, bond amounts are reviewed annually and adjusted if needed. Alaska requires financial assurances to be updated in tandem with the review of major permits, generally every 5 years, and the adequacy of the bonding is reviewed by an independent environmental consulting firm that audits the performance of both the operator and regulatory agencies (Kyle Moselle, personal communication, 2022). Adequate bond reviews also necessitate that regulatory agencies conduct site inspections regularly and information is utilized from site inspections and operator reports summarizing the progress of operations, reclamation, and monitoring.

Bond Release

At the time of annual permit renewal, a previously posted bond (or other security) may be released for each area disturbed in the past 12 months if reclamation work has been completed, or it may be transferred to additional acres to be disturbed in the upcoming year (§ 45.2-1212 of the Code of Virginia). Release is contingent on whether reclamation has been accomplished in accordance with the codes, regulations, and approved permit, including completion of the reclamation plan that supports the approved postmining land use (4VAC25-31-280). Virginia Energy's final inspection for bond release may be made no sooner than "two growing seasons after the last seeding" and the criteria for bond release are summarized in Figure 5-3 (all from 4VAC25-31-300). Bonds may be released incrementally, providing a financial incentive for operators to perform concurrent reclamation during mine operations.

Payments made to the Minerals Reclamation Fund may be repaid after reclamation is complete (§ 45.2-1212 of the Code of Virginia). Minerals Reclamation Fund deposits are held or retained according to the following formulas:

- "If the permit's fund balance divided by the number of acres remaining under bond is equal to or greater than $500, fund deposits for the permit will be released so that the remaining deposits equal $500 per acre for the acres remaining under bond" (4VAC25-31-320).
- "If the permit's fund balance divided by the number of acres remaining under bond is less than $500, the bond release amount will be determined by dividing the deposit amount by the number of bonded acres including the acres to be released, then multiplying by the number of acres to be released" (4VAC25-31-320).

Virginia's processes for the operator to request bond release and for the Director of Virginia Energy to confirm site conditions are comparable to the processes used in other states. The bond release criteria within the regulations provide guidance for determining the adequacy of reclamation methods and revegetation after at least 2 years of establishment. However, guidance is less clear about bond release criteria for postclosure water management and mitigations, which might include active or passive treatment systems. In the case of groundwater recovery in underground mines or pits, and the potential reliance on pit lake stratification to isolate potential contaminants, the requirements for water quality protections at closure may require a longer period of monitoring and verification. It is important that bonds are not released prematurely before the mitigation, management, or treatment methods have proven to be successful for achieving the long-term environmental requirements.

Bond Forfeiture

If a permit is revoked from a mine operator with less than 5 years of operation (not a Minerals Reclamation Fund member), then the available bond amount is forfeited to the Special Reclamation Fund (§ 45.2-1207 of the Code of Virginia). If a permit issued to a Minerals Reclamation Fund member is revoked, then the payments that the member has made to the Fund are forfeited to the Minerals Reclamation Fund (§ 45.2-1213 of the Code of Virginia). In either case, Virginia Energy must then use the forfeited payments to complete the reclamation plan for the permitted mining operation (§ 45.2-1238 of the Code of Virginia, 4VAC25-31-310). Bond liability extends to the entire permit area under conditions of forfeiture. After the completion of reclamation and payment of all fees, any remaining forfeited bond must be returned to the operator (§ 45.2-1213 of the Code of Virginia).

If the cost of reclamation exceeds the amount of the forfeited payments into the fund for a particular permit, the Director of Virginia Energy must draw on the rest of the Minerals Reclamation Fund for the cost of reclamation (§ 45.2-1238 of the Code of Virginia), and the amount by which the cost of reclamation exceeds the amount of the member's forfeited bond payments becomes a debt to the Commonwealth on the part of the permittee. The Director of Virginia Energy is authorized to collect such debts, and the money collected through appropriate legal action, minus the costs of legal action, is deposited in the fund (§ 45.2-1239 of the Code of Virginia).

Annual Reporting and Monitoring

Mineral mining permits need to be renewed each year within 10 days of the anniversary date in order to continue operations. The annual renewal of the permit must indicate the identity of the licensed operator, any agent, and their officers; the amount of minerals mined; and any changes to the information provided in the license application (§ 45.2-1129 of the Code of Virginia, 4VAC25-31-100, and 4VAC25-31-210). The mine operator is also required to annually update and extend the required site maps, which show the progress of the operations and mine workings, property lines, sensitive features, and other information provided in the initial application maps. If the time requirements are not met, the permit expires 10 days after the anniversary date (4VAC25-31-210).

The general administrative information required for mining permit renewals in Virginia is similar to the requirements in other states, but the details about the operator and about the status of land disturbance, completed reclamation, and mineral production are somewhat limited. Virginia's regulations also include a broad stipulation for "any other information, not of a private nature, that from time to time is required by [Virginia Energy]" (§ 45.2-1129 of the Code of Virginia). These general reporting requirements may not be adequate for the Mineral Mining Program to fully assess environmental compliance and identify any potential risks that arise during the course of mining operations. In particular, there are no prescriptive reporting requirements in Virginia for geochemical or geotechnical monitoring, water quantity or quality monitoring for surface water or groundwater, controls for dust or emissions, or invasive weed control and revegetation success. While mineral mining permits are reviewed and renewed annually, additional environmental information and monitoring data are likely submitted as permitting requirements for other regulatory agencies, like Virginia DEQ. Without a centralized structure in Virginia for the consolidation of annual reporting and monitoring data and few prescriptive requirements about what data should be reported, effective communication and coordination between different permitting agencies is critical. The comprehensive consideration of these data is necessary to ensure the Mineral Mining Program's annual analysis of mining operations.

Other states impose additional monitoring and reporting for mining permit renewals beyond those imposed in Virginia. For example, in addition to administrative details and inventories for all disturbed and reclaimed surface areas, Montana requires that annual reports include, if applicable, an inventory of available soil and reclamation materials; water balance analysis for all operations that use cyanide or metal-leaching agents or have the potential to generate acid; a comprehensive evaluation of water monitoring reports submitted throughout the year, including trend analyses for key site-specific parameters; updated accounting for cultural resource mitigations or management; monitoring results, material balances, and other information pertaining to geologic conditions; and an evaluation of monitoring and testing data required in the permit for sites that use cyanide or metal-leaching agents, reagent neutralization, or develop acid rock drainage or similar occurrences (ARM 17.24.118).

Inspections

The duties of the Mineral Mining Program's inspectors and the priorities for compliance inspections reflect the dual nature of the codes and regulations that apply to mine permits and safety licenses. Many of the inspection requirements are related to conditions affecting occupational health and safety. For example, complete safety inspections are required at least every 180 days for underground mineral mines and at least once per year for surface mineral mines if they are not inspected by MSHA (i.e., abandoned or temporarily idle mines; MSHA, 2013). Additional inspections can be made when deemed appropriate based on potential risks or when requested by miners or mine operators (§ 45.2-1148 of the Code of Virginia). To examine where any danger to miners might exist in an operational mine, or to people who might work or travel near an inactive mine (§ 45.2-1155 of the Code of Virginia), such inspections may include examining blasting practices; air flows, oxygen deficiency, and gas levels; entrances to abandoned areas; and roof and rib conditions (all from § 45.2-1155 of the Code of Virginia). Additional duties for mine inspectors involve reporting accidents involving serious personal injury or death, and responding to mine fires or mine explosions and taking charge of mine rescue and recovery operations (§ 45.2-1147 of the Code of Virginia). Some of this regulatory language appears to be adopted from the safety requirements for coal mining, where geologic conditions are more conducive to mine fires and explosions than in mineral (including gold) mines.

Although the mine safety aspects are very important, the purpose and frequency of compliance inspections for gold mining should also focus on areas of potential environmental risk and the associated protection or mitigation practices within the permit. Prescriptive elements are not provided in the codes or regulation for other key aspects of inspecting mineral mines during operations, reclamation, and long-term stewardship for the permit. The requirements merely state that the Director of Virginia Energy will "make investigations and inspections to ensure compliance" with the mineral mining codes and regulations (§ 45.2-1202 of the Code of Virginia). The internal policies in the Mineral Mining Program's Enforcement Policy and Procedures Manual (Virginia Energy, 2015a) establish that "reclamation inspection" frequency is based on the level of activity at the individual mine site. The manual stipulates that active sites, defined as sites where development, mining, reclamation, or other related activities occur, will receive a minimum of two inspections per calendar year (divided into each half of the year); for intermittent sites with cyclic production activities or temporarily idle sites, one inspection is required per year. In comparison, he frequency of mine inspections in Montana is also based on the characteristics and potential risks of the operations in accordance with the stipulation that "The department shall conduct an inspection at least once per calendar year for each permitted operation, and at least three times per year for each active operation that uses cyanide or other metal leaching solvents or reagents, has a permit requirement to monitor for potential acid rock drainage, or exceeds 1000 acres in permit area" (ARM 17.24.128).

For gold mining, the frequent and coordinated inspections of construction, operation, and reclamation activities by regulatory agencies are necessary to provide compliance oversight and guidance to ensure that operators implement best practices and function within the terms of the mining permit. Site inspections may be more effective when coordinated and performed jointly with regulatory agencies that administer the other required permits, such as those required to ensure water or air quality protections. This approach provides a mutual understanding of agency objectives and a more complete review of the operator's compliance, with the primary goal of assisting the operator in protecting the environment and attaining the intended postmining land use. Inspection reports and key findings should be shared among regulatory agencies and be made available for public review.

Regulatory oversight is recommended during facility construction to confirm that methods and materials are prepared and constructed in accordance with the approved designs and plans. Failure to maintain high-quality work during construction and installation may lead to future problems, like leaks or malfunctions in liner systems, or the failure of fill slopes or impoundments (Porter, 1997). Colorado has enacted requirements for phased construction, where inspections must verify acceptable progress before the operator proceeds with subsequent construction phases (2CCR407-1-7.3). Montana has enacted requirements for quality assurance during construction of tailings storage facilities, with certified monitoring and engineering reports to be submitted to the regulatory agency (§ 82-4-378, MCA).

Impoundment Inspections

The Mineral Mining Program inspects impoundments during the reclamation compliance inspections, which are conducted twice per year for each permit (Virginia Energy, 2015a). In addition to these inspections, Virginia regulations require that impoundment monitoring and daily inspections are performed by the mine operator's "qualified person," defined as a person "who is suited by training or experience for a given purpose or task" (4VAC25-31-10). The registered professional engineer who designed or oversaw the designs for an impoundment (the "Engineer of Record") may be available for consultation about the facility, as a resource for the operator and regulatory program (Michael Skiffington, personal communication, 2022). However, there are no specific requirements in codes or regulations for the Engineer of Record to conduct inspections of the impoundment during construction, operations, reclamation, or long-term stewardship. This lack of required involvement by the Engineer of Record is a shortcoming of the current regulations in Virginia, as it would be beneficial for the engineer to be involved with inspections during construction, operations, reclamation, and long-term stewardship.

In contrast to Virginia, Montana laws require that the Engineer of Record conduct annual impoundment inspections in addition to the more frequent inspections conducted by the operator, as specified in the state's Tailings Operation, Maintenance, and Surveillance Manual (§§ 82-4-379 and 82-4-381(1), MCA). The regulatory agency is also required to "conduct inspections, review records, and take other actions necessary to determine if the tailings storage facility is being operated in a manner consistent with the approved design document and the tailings operation, maintenance, and surveillance manual" (§ 82-4-381(4), MCA). Additionally, the designated independent review panel of engineers must conduct an impoundment inspection and comprehensive periodic review of associated designs, reports, models, and pertinent records at least every 5 years during active operations (§ 82-4-380, MCA).

Noncompliance, Suspension, and Revocation

The permittee and its employees and contractors must comply fully with the requirements of applicable codes and regulations (4VAC25-31-30). Any violation of the provisions in Chapter 12 of Title 45.2 of the Code of Virginia (Permits for Certain Mining Operations; Reclamation of Land) or of any order from the Director of Virginia Energy is a misdemeanor punishable by a maximum fine of $1,000, a maximum of 1 year in jail, or both (§ 45.2-1223 of the Code of Virginia). However, this penalty structure is not easily implemented by the Mineral Mining Program. In conjunction with the Virginia Office of the Attorney General, the Director of Virginia Energy may pursue charges for the violation of Chapter 12 provisions or any order from the Director of Virginia Energy. This statute defines the type of charge and sets the parameters for fines and punishment upon conviction. However, no fines can be assessed without adjudication from the appropriate court; therefore, the Director of Virginia Energy cannot directly issue fines for noncompliance (Michael Skiffington, personal communication, June 2022). Other states allow the calculation and direct issuance of fines or penalties based on the nature, extent, and impacts resulting from the violation, along with consideration for repeated offenses and the duration of the violation (South Carolina-S.C. Code 89-250, R.48-20-220; Montana-82-4-361, MCA).

The Mineral Mining Program's Enforcement Policy and Procedures Manual (Virginia Energy, 2015c) states that mine inspectors will initially notify operators of their noncompliance through a Special Order document, which must include the location, a description of the violation, and the remedial action required to resolve the violation. If the operator does not comply with the terms of the Special Order within the specified timeframe, then a Notice of Noncompliance is issued. The notice must specify how the operator has failed to obey the order and establishes a reasonable time frame within which the operator is required to comply with the order (§ 45.2-1213 of the Code of Virginia). The consequence for not complying with the terms of the Notice of Noncompliance is permit revocation and bond forfeiture.

Unlike the regulations for surface coal mines (4VAC25-130-842.12), there is not a specific provision in Virginia's codes or regulations for citizens to request an inspection to occur at a mineral mine when they have reason to believe that a violation or unlawful condition or practice has occurred. However, the Enforcement Policy and Procedures Manual (Virginia Energy, 2015b) establishes a policy to document and investigate citizen complaints regarding safety, health, or reclamation at mineral mines. Complaints that are specific to blasting at mineral mines are included within regulations (4VAC25-40-931) and the Enforcement Policy and Procedures Manual provides further details about how the mine inspector should review blasting records and seismic monitoring of air overpressure and ground vibration. The inspector may require the operator to perform additional seismic monitoring to conduct the complaint investigation. The mine inspector determines if the mine is being operated in accordance with codes and regulations and issues a Special Order if the inspector has reasonable cause to believe that a violation has occurred. If the complainant's contact information is available, the mine inspector also must contact the complainant after the investigation has been completed to notify them of the outcome, and share a copy of the complaint investigation report.

Virginia's procedures for issuing orders and Notices of Noncompliance are similar to those in other states, including South Carolina and Montana (South Carolina, R.48-20-160; Montana, 82-4-361, MCA). For example, Montana has a series of regulations that pertain to citizen complaints about blasting and the subsequent investigation that must be conducted by the permitting agency (82-4-356, MCA; ARM 17.24.157 through 159). To ensure compliance with requirements from the Office of Surface Mining Reclamation and Enforcement regarding the allowable frequency and decibel level of air blasts and the peak particle velocity for ground vibrations (30 CFR § 816.67(d)(4)), Montana also has a statutory provision that addresses citizen complaints about the loss of water quantity or quality, which requires an investigation from the permitting agency. If the inspection finds a preponderance of evidence in support of the complaint, this provision stipulates that the mine operator may be required to provide a replacement water supply to the complainant or risk having their permit suspended.

Conditions for Permit Suspension or Revocation

If the operator does not comply with the requirements set forth in the Notice of Noncompliance within the established time limits, the mine permit can be revoked and the bond forfeited (§ 45.2-1213; procedures found in 4VAC25-31-310) (see the section "Bond Forfeiture"). If the operator fails to comply with the terms of the permit, fails to renew a permit within the annual deadline, or defaults on the bond conditions, the permit *must* be revoked and the bond forfeited (4VAC25-31-310; Michael Skiffington, personal communication, 2022). The permit may also be revoked if the conditions are not met for stabilizing and maintaining mine permit areas that are temporarily inactive, as described in the next section (4VAC25-31-430).

If adverse environmental disruptions "seriously threaten or endanger the health, safety, welfare, or property rights of citizens of the Commonwealth," an injunction to prohibit further operations may be granted by the appropriate circuit court. An injunction does not relieve the operator of the duty to reclaim lands previously affected (all from § 45.2-1225 of the Code of Virginia). An appeal of an order from Virginia Energy must be submitted through certified mail and will suspend the permit revocation and bond forfeiture until the appeal is complete (§ 45.2-1226; § 2.2-4018 et seq.; 4VAC25-31-310).

Virginia's criteria for permit revocation and bond forfeiture are similar to those in other states, particularly when the corrective actions from a Notice of Noncompliance are not implemented or there is a site condition "which poses an immediate threat to public health, safety, or the environment" (South Carolina, R.48-20-160; Montana, 82-4-338 and 82-4-362, MCA). For example, in South Carolina and Montana, an overdue annual permit renewal, failure to provide additional bond at the appropriate milestones, or failure to comply with conditions of the permit *could* result in the suspension of the permit, which does not allow mining activity to occur until the noncompliance is resolved (South Carolina, R.48-20-160; Montana, 82-4-362, MCA). If corrective actions for these Montana and South Carolina examples are not resolved within the timeframe established through the suspension notice, *then* the suspension would be elevated to permit revocation and bond forfeiture.

Inactive or Abandoned Operations

Virginia regulations require the permittee to send notice of intent to stop the working of an underground mine for a period of 30 days, or a surface mine for a period of 60 days. This notice must occur at least 10 days prior to the intended discontinuation, or whenever the mine becomes inactive (§ 45.2-1130 of the Code of Virginia). A similar 10-day prior notification is required upon resumption of the work. Except for a surface mineral mine that is inspected by MSHA, the mining cannot resume until an inspector has inspected the mine and approved its use (§ 45.2-1130 of the Code of Virginia). A mine is determined to be complete and the permit can be revoked when no mine-related activity has been conducted for 12 consecutive months (4VAC25-31-430). However, a mine may remain inactive under a permit for an indefinite period if all disturbed areas are adequately stabilized, or all erosion and sediment control systems are maintained, and if drainage structures, vegetation, and machinery and equipment are well maintained (4VAC25-31-430).

According to Virginia Energy, some permit revocations and bond forfeitures have occurred involving sites that had become inactive, perhaps due to bankruptcy or other financial stress, and were not adequately reclaimed (Michael Skiffington, personal communication, 2022). Some mines may cease operations over a gradual timeframe, with a period of temporary inactivity. Virginia requires that reclamation commence after 12 months of inactivity, but there are also conditions under which a site may be allowed to remain inactive indefinitely. This could potentially lead to a scenario where the costs for the permittee to complete the final reclamation activities far exceed the costs to remain under this "care and maintenance" status, and this difference would increase over time due to deteriorating site conditions in the absence of adequate maintenance, and economic factors like inflation and market fluctuations for fuel, equipment, and personnel. If such a permit were eventually revoked and the bond forfeited before final reclamation occurred, the Commonwealth would be liable for the increased costs of completing final reclamation, which are unlikely to be sufficiently covered by the available bond amount.

It is common for state regulations to consider the potential closure and abandonment of a mine site, and allow a period of temporary cessation for a number of reasons. However, by establishing limits for the allowable length of temporary cessation and periodically revising the financial assurance calculation, states can reduce the risk of forfeited bonds being insufficient to cover final reclamation costs. Other states have variability in the timeframes that are allowed for temporary cessation and the stipulations or conditions that apply to the permittee while the site remains inactive. Examples include Arizona (allowing inactivity for up to 15 years; Code 27-926 and R11-2-207), Colorado (up to 10 years; 2CCR407-1-1.13), Montana (2 years; ARM 17.24.150 and 170), and South Carolina (2 years; S.C. Code Regs. § 89-270).

Banned Parties ("Bad Actors")

As part of the permit application in Virginia, applicants must specify whether the applicant or affiliates have ever had a mining permit revoked (in any state) or had a bond or security forfeited (§ 45.2-1205 of the Code of Virginia). If so, Virginia Energy must not issue a mineral mining permit, except when an operator who forfeited a bond pays the cost of reclamation in excess of the amount of the forfeited bond within 30 days of notice, or if their forfeited bond is equal to or greater than the cost of reclamation, in which cases the operator is then eligible for another permit (§ 45.2-1209 of the Code of Virginia).

The approach to deny a permit to certain parties, based on the applicant's record of having a revoked permit and forfeited bond, is common in other states to prevent "repeat offenders." Examples include South Carolina (S.C. Code 48-20-70 and 48-20-160) and Montana (§ 82-4-360, MCA). One aspect of Virginia's requirements that is less common is the consideration of revoked permits and forfeited bonds from other states. Although this may be beneficial by providing another layer of review and protection against potential bad actors, it may be difficult for Virginia regulators to adequately track, investigate, and enforce this aspect of the law, which relies on sufficient record keeping within other states and coordination between Virginia and other states.

Public Engagement

Public participation in the regulatory process for gold mining can occur during exploration, environmental review, permitting, operations, reclamation, and after closure. In the public listening sessions hosted by the committee, a recurring theme expressed by many community members was frustration and anger over the lack of communication and information provided by local and state officials regarding exploration activities in Buckingham County. Community members highlighted a variety of obstacles that hindered their ability to participate meaningfully and provide input into decisions regarding possible gold mining near their community. This concern highlights the importance of initiating community participation at the earliest stages of gold exploration and throughout the life cycle of a mine (see Chapter 3). Community members emphasized the challenges they faced in understanding the potential impacts of gold mining, especially because Buckingham County has no local radio station, television station, or newspaper, and an estimated one-third of community residents have no access to the internet. As a result of the lack of communication and available information, community members expressed a lack of trust in institutions that will be making decisions about the future of gold mining.

Another recurring theme in the listening sessions was concern regarding the disproportionate impact that mining could have on those experiencing existing environmental injustice and health disparities. In Virginia, racism has been recognized as a public health crisis (House Joint Resolution 537). Additionally, several actions by the Virginia legislature and executive branch have sought to address environmental justice. In 2019, the Commonwealth of Virginia commissioned a study about how to incorporate environmental justice into regulatory decision making. Among other outcomes, this study led to the passage of the Virginia Environmental Justice Act in 2020. In April 2021, an office of environmental justice was created within Virginia DEQ to seek meaningful engagement and change that would advance environmental justice. Its stated goal was to move beyond simple "check the box" exercises and establish substantive provisions that can build trust, understanding, and values alignment among interested stakeholders, governmental agencies, and proponents of potential gold mining projects (Virginia Natural and Historic Resources, 2022). Environmental justice issues will likely recur throughout Virginia in the absence of a regulatory framework that requires rigorous community engagement and protects the rights of local communities, especially those predominantly composed of underserved and marginalized groups.

The below sections discuss public participation opportunities and requirements related to various aspects of mining activities in Virginia and other states.

Tribal Consultation

There are 7 federally recognized tribes in Virginia (87 FR 4636) and 11 that are recognized by the Commonwealth (Secretary of the Commonwealth, 2022). These include the Mattaponi, Pamunkey, Chickahominy, Eastern Chickahominy, Rappahannock, Upper Mattaponi Tribe, Nansemond, Monacan, Cheroenhaka Nottoway, Nottoway, and Patawomeck Nations. There are no federal tribal reservations in Virginia and although some tribes have reacquired lands to support communal activities, there are only two state-recognized reservations (Pamunkey and Mattaponi), which are in King William County (Mattaponi Indian Reservation, 2022; Pamunkey Indian Reservation, 2021; Virginia Places, 2022). There are currently no tribal reservations in the areas of Virginia that are most favorable for gold mineralization, but traditional territories cover the state (see Figure 5-4). It is noteworthy that Rassawek, the historical capital of the Monacan Indian Nation, and other culturally important regions are located in the gold-producing areas of Virginia.

Under section 106 of the National Historic Preservation Act (NHPA; 54 USC 306108 and 36 CFR Part 800), tribal consultation is required when a federal agency action occurs on tribal lands, or a site that has religious or cultural significance to a tribe (54 USC 302706). The Regulations for Implementing the Procedural Provisions of NEPA (40 CFR Parts 1500-1508) encourage integration of the NEPA process with other planning and environmental reviews, such as section 106 of NHPA. Under NEPA, federal agencies are encouraged to consult with tribes and to invite tribes to be cooperating agencies in preparation of an EIS when potential impacts may affect tribal interests. EPA must consult with federally recognized tribal governments when issuing major air, waste, and water discharge permits pursuant to the Clean Air Act and the Clean Water Act (Executive Order 82). Finally, Executive Order 13175 encourages federal agencies to implement "meaningful and timely" consultation with tribes. As noted elsewhere, there are limited federal lands in Virginia, so there may be few federal actions that require consultation with tribes when permitting a gold mine in Virginia.

At the state level, Executive Order 82 (signed in November 2021) mandated that tribes must have input before the state approves certain development projects or permits. It mandated that the Department of Environmental

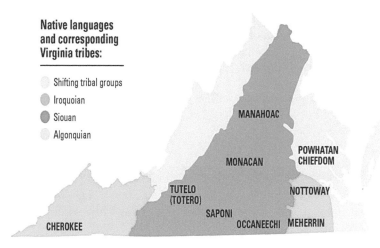

FIGURE 5-4 Traditional tribal territories and languages in Virginia.
SOURCE: Image from UVA (2022).

Quality, the Department of Conservation and Recreation, the Department of Historic Resources, and the Virginia Marine Resources set up a process to formally consult with the tribes when considering permit applications that could affect environmental, cultural, and historic resources. Relevant actions for this consultation included environmental impact reports (VA EIRs) for major state projects, burial permits for relocation of human remains, groundwater withdrawal permits for withdrawals greater than one million gallons per day, local government notifications for new and existing impoundment structures or dams, and construction or alteration of Virginia Regulated Impounding Structures Permits. However, VA EIR documents are not required for mining permits on private land, and the other conditions that would require tribal consultation may not directly apply to all potential gold mining permits. Furthermore, Senate Bill (SB) 482, which would have codified the Executive Order, was deferred to 2023 after passing the House. In 2022, a commission was established through HB 1136 to "[perform] a comprehensive review of Virginia law to assess ways in which it must be revised to reflect the government-to-government relationship the Commonwealth should maintain, by treaty and applicable federal law, with the sovereign, self-governing, federally recognized Tribal Nations located within the present-day external boundaries of the Commonwealth."

For comparison, other gold-producing states have passed legislation mandating tribal consultation. For example, New Mexico passed SB 196 in 2009, which enhanced government-to-government communication and collaboration between the state and tribal governments; Nevada passed AB 264 in 2019, which promotes collaboration between a state agency and tribes; California passed AB 52, which added provisions for a consultation process during evaluation and consideration of projects that are evaluated under the California Environmental Policy Act; and Montana requires consultation with tribes during the early development of EIS documents prepared under the Montana Environmental Policy Act (ARM 17.4.615).

Free, prior, and informed consent (FPIC) is the practice of ensuring the rights of indigenous people to consent or withhold consent to actions such as mining on their lands or territories. The United Nations Declaration on the Rights of Indigenous Peoples was adopted by the United States as a "call for a process of meaningful consultation with tribal leaders, but not necessarily the agreement of those leaders, before the actions addressed in those consultations are taken" (U.S. Department of State, 2011). The World Bank requires clients to secure FPIC (The World Bank, 2015), and the International Council for Metals and Mining (ICMM, 2022) and the Initiative for Responsible Mining Assurance have accepted the concept as best practice (IRMA, 2018). Recently, an interagency work group was established to reform federal mining laws, regulations, and permitting policies in the United States. This work group will "make recommendations for improvements necessary to ensure that new production meets strong environmental and community and Tribal engagement standards during all stages of mine development, from initial exploration through reclamation" (DOI, 2022).

In summary, federal, state, and industry entities have all indicated support for ongoing and meaningful tribal consultation. In practice, this includes an opportunity for review and comment on mining projects at the earliest possible stage, invitations for tribes to be cooperating agencies in preparation of environmental reviews, read receipts for emails sent to notify indigenous leaders, groups, and nongovernmental organizations of exploration and mining activity, as well as a variety of other forms of communication. It should not be assumed that only one tribe has historic ties to an area, as a legal precedent may affect them all (Luke Tyree, personal communication, 2022).

Exploration

As described above, exploration drilling does not require a permit in Virginia and therefore does not require public notice. Other types of surface-disturbing exploration would require a permit, but the public notification requirements are very limited (see the section "Mine Permit Application"). This is similar to practices in South Carolina, but different from the practices in Montana and California, where public notification and comment periods may be required depending on the assessment of impacts (see Table 5-11).

The current permitting exemption for exploration drilling creates a barrier for public communication that is unlike other mineral mining activities in Virginia (see Table 5-11), and results in a lack of shared information, transparency, and engagement among regulators, nearby residents, communities, and other stakeholders. As noted above, a lack of robust public notice for exploration activities has created distrust among potentially impacted communities. The lack of permitting requirements for exploration drilling is not commensurate with the level of public interest, concerns, and uncertainty related to such activities, as shown in the example in Buckingham County (see Box 5-4). There may be other opportunities for public information and engagement for exploration drilling projects, if required by local governance or associated permits from other regulatory programs, but a consistent statewide approach is lacking in the codes and regulations for mineral mining.

Environmental Assessment

One of the most common types of public participation occurs during environmental review. The federal NEPA process has formal requirements for sharing of data and information and includes provisions to make data, proposals, and a wide range of project-related documents widely available to the public. The lead federal agency hosts

---

BOX 5-4
Case Study of Exploration Drilling in Buckingham County

In the example of exploration drilling conducted by Aston Bay in Buckingham County, the permitting exemption meant there were no requirements for notification, or allowing objections or requests for a hearing, for the neighboring landowners, local government, or utility companies. Based on public feedback provided to the committee (open meetings on December 15, 2021, and January 24, May 25, and May 26, 2022), some of the concerns raised about this exploration drilling activity include

- Impacts to water quantity or quality with the potential to affect nearby water users.
- A general lack of notification, involvement, and consideration for the local government, communities, and other stakeholders.

Transparency and effective communication are critical in all stages of a project's life cycle. Given the absence of required public notification and open exchange of information in this example, uncertainty and concerns about exploration drilling and feelings of mistrust against exploration or mining companies and governmental bodies are not surprising. Stakeholders are left to reach their own conclusions about project details, potential risks, and future outcomes, but they may lack technical expertise encompassing the complexity and variability of gold deposits and feasibility studies related to exploration data.

a website that describes and explains the proposed project and houses project-related information and documents that the public can view and download. There are also prescribed public notice and comment opportunities at multiple times during the NEPA process. For example, NEPA requires a 45-day public comment period after the issuance of a draft EIS and the required review of the draft EIS by EPA is made publicly available. However, a full NEPA process is unlikely to occur at a potential gold mining site in Virginia, except where a CWA 404 permit is needed. Outside of the context of a major federal action, the only other review of environmental impacts applying to proposed gold mining operations would be a Virginia EIR for mining occurring on state lands, which is likely to be uncommon. Even when a Virginia EIR is developed, public notification and comment requirements for the environmental review of gold mining projects are minimal in comparison to an EIS under NEPA (see Table 5-2).

Permitting

Depending on the particulars of the project, there may be several opportunities for public engagement during the process for issuing a permit for a gold mine in Virginia. For example, the Water Board is required to issue notice of a draft VPDES permit in a local newspaper (Troutman Sanders LLP, 2008); a public comment period may be offered for NSR permits (9VAC5-80-1170); and the issuance of a Virginia Water Protection Permit for water withdrawal requires notification and an opportunity for a public hearing (§ 62.1-44.15:20 of the Code of Virginia). Of particular importance in almost any mining operation is the requirement that each permit applicant to the Virginia Mineral Mining Program must notify certain members of the public about a new permit application via certified mail (§ 45.2-1210 of the Code of Virginia and 4VAC25-31-170; Virginia Energy, 2022f). These members of the public include "property owners within 1,000 feet of the permit boundary, the Chief Administrative Official of the local political subdivision [county or city] where the prospective mining operation would take place, and all public utilities on or within 500 feet of permit boundary." Notably, this proximity criterion could omit nearby communities or other stakeholders that do not reside in the immediate area. In addition, no new notices are required for a permit renewal or for the expansion of a permit (§ 45.2-1210 of the Code of Virginia), yet these permitting and renewal of permitting actions are critical milestones for mining operations and warrant meaningful engagement with landowners, communities, and other stakeholders.

The details that are contained in the notifications (see Box 5-5) demonstrate that current Virginia laws do not provide enough information to support meaningful dialogue about the proposed mining operations and inform substantive comments and questions from the public. Property owners only have 10 days from receipt of the permit notification to submit written objections or request a hearing. This public hearing serves as an informal "information-gathering" forum that provides additional information to the public and where objections or concerns

---

BOX 5-5
Requirements for Public Notification

The following information must be included in the public notifications (§ 45.2-1210 of the Code of Virginia and 4VAC25-31-170):

- Date of notification and the name and address of the permit applicant issuing the notice;
- Name and address of the notification recipient;
- Location of the proposed mine (city or county), the distance of the proposed mine site to the nearest town or easily identified landmark, and the tax map identification number of the parcels of the prosed mine site;
- Statement of the applicant's intent to seek a mining and reclamation permit from Virginia Energy, noting that the mining permit must address requirements for regrading, revegetation, and erosion controls of mineral mine sites; and
- Notification that property owners have 10 days from receipt of the permit notification to submit written objections or request a hearing.

about the operation can be publicly addressed. A hearing may be the first opportunity for the public and other stakeholders to learn about details of the project and formulate their questions and concerns, but this may not occur if the hearing is not requested within the 10-day timeframe. The inadequate review timeline puts the onus on neighboring property owners to have a prior understanding of the proposed operation and familiarity with potential environmental impacts from mining in order to submit substantive and meaningful objections to the permit application. This limitation makes the current process insufficient to provide for a meaningful exchange of comments between project proponents, regulators, and the public.

In some states, including South Carolina and Montana, the state permitting program is required to provide public notice at a local and regional scale. Notifications are issued when new mining permit applications are received, as well as when an operator submits an application for major modifications to an existing permit (S.C. Code Regs. § 89-100; 82-4-353, MCA; and ARM 17.24.119). In Colorado, applicants for new permits or modifications to an existing permit must publish a public notice in a local newspaper (up to 4 consecutive weeks), issue notices to nearby owners of surface and mineral rights, and provide a copy of the application materials for public review at the county clerk or recorder's office (2CCR407-1-1.6).

In other states, including Alaska and Montana, the state agency with the primary permitting responsibilities hosts a proposed project website that contains a wide variety of project-related documents. These documents include baseline data; project plans and plan revisions; permitting timelines; a schedule of public meeting, notice, and comment opportunities; preliminary lists of permits and explanations of agency roles and responsibilities; public comments and agency responses to comments; and official correspondence. The website functions as a portal for the public to learn about the proposed project and helps to make the permitting process more intelligible to the public. It can also facilitate a dialogue between the public, project proponents, and regulators that promotes mutually acceptable resolutions of issues of concern. The chronology of the webpage becomes part of the administrative record and can be of particular importance if there are legal challenges to permitting decisions.

Closure and Bond Release

Because Virginia has a flat bonding rate, there is no public notice or opportunity for public input during the calculation of initial bonds or bond adjustments that may occur over the life of a mineral mining operation in Virginia. For comparison, in Montana, a comprehensive bond review is conducted at least every 5 years (§ 82-4-338(3), MCA), which includes a public notice issued by the permitting agency and a 30-day period for public comment. The permitting agency may not release or decrease the bond amount without providing local and statewide notice of the opportunity to request a hearing (§ 82-4-338(5), MCA). Similarly, Alaska has robust public notice and meeting requirements tied to the reissuance of permits and the associated recalculation of financial assurances (see the example of the Fort Knox Mine; Alaska DNR [2022b]). As with public engagement for other aspects of permitting actions, it has been recognized that "full and unrestricted public participation should be provided in the process of establishing reclamation and closure plans and bond amounts, and as a part of bond release" (Kuipers, 2000).

CONCLUSIONS AND RECOMMENDATION

Gold mining has a long history in Virginia dating from the 1800s (see Chapter 1). At present, however, there are few metal mining activities in the state and no active commercial gold mines (Virginia Energy, 2022d). Instead, most mineral mining in Virginia is focused on the extraction of sand, clay, limestone, granite, slate, mineral sands, and kyanite (Virginia Energy, 2022a). Given these current mineral extraction activities, it is not surprising that the present regulatory framework is geared toward projects such as sand and gravel mining, which currently make up approximately half of the active mineral mining permits issued by the state (Virginia Energy, 2022d). Although most of Virginia's mineral mining laws and regulations seem suitable for the types of mines now operating in the State, gold mining raises a host of environmental and public health issues that merit additional attention and suggest a need for changes in law and regulation. Virginia's present regulatory structure is not adequate to protect against potential land and water quality degradations that could accompany gold mining activities. Specific conclusions and a recommendation to improve the existing regulatory framework are summarized below.

Review of Impacts

Virginia's current regulatory system lacks an effective and consistent process for review of environmental impacts from potential gold mining projects. As a result, it is unlikely that a robust collection, evaluation, and review of site-specific data regarding potential impacts of gold mining activities and their impact on the public health and welfare of surrounding communities will take place. NEPA requires federal agencies to consider the potential environmental effects on natural resources, as well as social, cultural, and economic resources, before permitting. Virginia law does not require a NEPA-like review of environmental impacts for private lands, where gold mining is most likely to occur. Additionally, while baseline studies in Virginia appear to be recommended, they are not required. This means that in the absence of a major federal action that triggers the federal NEPA process, there may be limited collection of baseline information and no formal documentation of the regulatory program's analysis, disclosure of impacts, or decision making for a range of environmental resources or factors. Some states have a state-specific NEPA-like process that allows for a consistent approach to collecting and considering baseline information and other material relevant to environmental impacts (e.g., Montana and California). Other states have regulation, code, and guidance documents that emphasize the importance of baseline studies (e.g., Colorado, Nevada, Montana, California). The protection of air and water quality would be strengthened if Virginia adopted laws and promulgated regulations that required up-front, robust data collection and a NEPA-like analysis that discusses and evaluates reasonable alternatives.

Exemptions

Virginia provides exemptions from regulatory oversight for off-site processing and exploratory drilling which are not commensurate with the potential impacts from those operations.

- **Off-site processing:** Gold processing facilities in Virginia that are not located on site with active mining or extraction ("toll mills") would not require a permit from the Mineral Mining Program for the operation and reclamation of the site. Toll mills may look very similar to permitted on-site processing facilities and similar environmental impacts may result from toll mills. In fact, the waste materials at toll mills may contain a broader range of potential contaminants if the source materials come from different locations. While toll mills may be required to obtain permits from other agencies to protect air quality and water quality, the lack of regulatory oversight by the Mineral Mining Program means that site characterization, project plans and designs, and the implementation of best practices for operations, reclamation, and long-term stewardship may not be adequately addressed.
- **Exploratory drilling:** Virginia's current laws and regulations exempt exploratory drilling for mineral resources. Impacts on the environment during initial exploration are generally minor, localized, and easily reclaimed. However, advanced exploration methods may be associated with greater impacts (see Chapter 3). While surface impacts including erosion and runoff may be regulated by Virginia DEQ and DCR, there are currently no mineral mining regulations for exploration in Virginia that mandate the plugging of drill holes or the covering of drill cuttings from the hole. If best practices are not utilized for these closure activities, pollution of the local groundwater and surface water could occur. This exemption for exploratory drilling also means that public notice to citizens and local communities is not required. Greater oversight of exploration drilling would ensure community participation starting at the earliest appropriate stage and continuing throughout the life cycle of a potential gold mine, and would lessen the likelihood of these localized impacts, especially in regard to more advanced and intensive drilling programs. This oversight could include requirements to file plans for drilling, closure, and reclamation, and a requirement to provide notice to those around the exploration site.

Underground gold mining without significant surface effects is also currently exempt from regulations under Virginia's mineral mining codes and regulations. While significant surface effects related to disturbances and facilities would require a permit, the exemption for underground gold mining could cause important aspects of underground mines to be excluded from operations and closure plans of the surface permit. Additionally, the level of technical assessment and oversight for underground gold mines by Virginia Energy is not clear.

Financial Assurance

Virginia's bonding requirements are insufficient to cover the costs of reclamation and long-term stewardship of gold mining and processing operations, which poses a fiscal and environmental risk to the Commonwealth in the case of the bankruptcy of mining enterprises or abandonment of their mining sites.

- **Bonding rates:** Virginia's bonding rates are based solely on disturbed acreage. This type of bond calculation often leads to undercollection of bonds for gold mining and processing operations because it focuses only on aspects of land reclamation and does not account for additional costs like postclosure water management. Additionally, Virginia offers a bond pool, called the Minerals Reclamation Fund, with even lower per-acre rates and pooled risk. The complex reclamation and long-term stewardship activities that might be necessary for some gold mining projects could greatly deplete or potentially exhaust the Minerals Reclamation Fund used by the Commonwealth to guarantee reclamation. The regular recalculation of potential costs using verifiable engineering estimates would constitute an improved model for determining bonding rates. This model would estimate the costs for reclamation and long-term stewardship for all aspects of the operation over the project's life, including any postclosure water management, treatment, and monitoring that may be required to achieve long-term hydrologic, physical, and chemical stability. The integrity of the Minerals Reclamation Fund could be maintained using a similar bond calculation model, or by establishing membership criteria that are based on the operation's characteristics and its potential impacts.

- **Exemptions from bonding:** Virginia's exemptions from bonding for underground gold mining, small-scale gold mining, and toll mills do not reflect the costs necessary to conduct reclamation and long-term stewardship at those operations. No financial assurance is provided to the Commonwealth for these exempt operations, which poses a fiscal and environmental risk to the Commonwealth and its citizens.

- **Bond release:** Virginia does not have clear guidance regarding the criteria for bond release for projects that require complex closure and reclamation. To ensure successful mine reclamation, bonds should only be released following the demonstration that performance standards for reclamation have been achieved over a sufficient period of time. These performance standards may include requirements for slope stability, vegetation establishment, water quality, and hydrologic balance. Incremental bond release for areas at which successful reclamation has been demonstrated can encourage the timely completion of reclamation.

Performance Standards and Their Enforcement

To incorporate best practices, build a mutual understanding among permittees and regulators, and better support protection of human health and the environment, Virginia agencies will need to review the regulatory performance standards pertinent to gold mining and update guidance documents. Virginia's performance-based laws and regulations provide flexibility for the site-specific designs of each project, but do not provide sufficient guidance for operators to achieve objectives and do not offer sufficient metrics for regulators to evaluate during the review of applications and inspection of activities. Fiscal and environmental risks to the Commonwealth would be reduced with improved guidance and performance standards on best practices for the collection of baseline information, geochemical characterization, water management, waste rock management, tailings management, and impoundment design. Specifically, performance standards for impoundment designs could recommend a probabilistic framework for designing for seismic events and a consideration of the predicted increased frequency and magnitude of major storm events due to climate change. Performance standards would also be improved with conservative recommendations for slope angles and safety factors that reflect best geotechnical practices and incorporate the potential for undrained loading and liquefaction in saturated tailings. Additionally, decision makers may want to reconsider the current practice of using incremental damage assessments to calculate design floods requirements for impoundments.

The capacity to regulate is as important as a strong regulatory framework and is a concern for Virginia given the limited experience with the regulation of metal and gold mining. The capacity to regulate requires robust funding of the regulatory entities, as well as diverse and appropriate technical expertise of the regulators, supplemented by periodic reviews of evolving best practices. In addition, effective coordination between multiple regulatory entities is critical for protecting air quality and water quality, particularly when evaluating, permitting, and monitoring compliance for stormwater and process water management, treatment technologies, and methods for discharge. Given the lack of experience of Virginia regulatory entities in regulating metal and gold mining, regulators' current expertise and familiarity with best practices may be limited. There are also key gaps in Virginia's capacity to implement and enforce some of its laws and regulations, such as the inability to directly issue penalties or fines for noncompliance without lengthy adjudication, and the lack of requirements for impoundment inspections by the associated Engineer of Record. Higher-level technical reviews, third-party reviews, or audits would enhance the evaluations of Mineral Mining Plans and inspections of individual permits.

Public Engagement and Environmental Justice

The current requirements for public engagement in Virginia are inadequate and compare unfavorably with other states, the federal government, and modern best practices because they require the provision of limited information, place the burden of public notification on the mine permit applicant, and apply only to a limited scope of recipients. Industry best practices are adopting a greater emphasis on public engagement, consultation, and partnership with communities before and after mining activities are initiated, as well as free, prior, informed consent to govern interactions with tribes. In Virginia, there is a scarcity of project details in the new permit notifications, a short deadline provided for filing objections or a request for hearing, and a limited number of area residents that are required to be notified, with no specific inclusion of tribal communities. In addition, Virginia Energy does not make technical reports, designs, and other components of the permit application package readily available for public review. Finally, there are no requirements in Virginia for public notice or opportunity for public input for exploratory drilling or when an application is renewed, a permitted project is expanded, or a bond is released. These permitting actions are critical milestones for the mining operation, and they warrant meaningful engagement with nearby landowners, communities, and other stakeholders. Without changes in Virginia's regulatory requirements to provide for expanded public outreach, additional informational meetings, and longer review timelines, Virginia fails to meet the current best practices of public engagement and lags behind other states.

Current Virginia regulations that are applicable to mineral mining will need to be amended to reach the goals set out in the Environmental Justice Act. In 2020, the Virginia legislature passed the Virginia Environmental Justice Act to better incorporate environmental justice into regulatory decision making in the Commonwealth. In the context of potential gold mining projects, an emphasis on environmental justice requires a regulatory structure that recognizes existing environmental injustice, population vulnerabilities, and economic and health disparities, and aims to reduce existing disparities and prevent future disparate impacts. This regulatory structure should ensure that those experiencing existing environmental injustice and health disparities are notified in a timely fashion about potential gold mining projects, are able to consult meaningfully with potential gold mining project proponents, and can contribute to decision making.

***

As detailed above, Virginia's present regulatory structure is not adequate to protect against the potential environmental degradations that could accompany gold mining activities. Stronger requirements for bonding, public engagement, and the review of environmental impacts are necessary; as well as updated regulatory capabilities, exemptions, performance standards, and guidance documents in order to protect human health and the environment.

RECOMMENDATION: To protect against the potential impacts of gold mining, the General Assembly and state agencies should update Virginia's laws and its regulatory framework.

References

911 Metallurgist. 2016. Pulp formulas & metallurgical formulas. https://www.911metallurgist.com/blog/basic-metallurgical-formulas (accessed July 12, 2022).

AAP (American Academy of Pediatrics). 2021. Lead exposure in children. https://www.aap.org/en/patient-care/lead-exposure/lead-exposure-in-children (accessed July 5, 2022).

ADEQ (Arizona Department of Environmental Qualiy). 2004. Arizona mining guidance manual BADCT. Arizona Department of Environmental Quality.

Adler, R., and J. Rascher. 2007. A strategy for the management of acid mine drainage from gold mines in Gauteng. Vol. CSIR/NRE/PW/ER/2007/0053/C: Contract Report for Thutuka (Pty) Ltd., submitted by the Water Resource Governance Systems Research Group.

AIME (American Institute of Mining Engineers). 1917. Transactions of the American Institute of Mining Engineers. Vol. 49: Princeton University.

Alaska DEC (Department of Environmental Conservation). 2008. Alaska water quality criteria manual for toxic and other deleterious organic and inorganic substances: Alaska Department of Environmental Conservation.

Alaska DNR (Department of Natural Resources). 2022a. Application for permits to mine in Alaska (APMA). https://dnr.alaska.gov/mlw/mining/apma (accessed July 5, 2022).

Alaska DNR. 2022b. Fort Knox mine. https://dnr.alaska.gov/mlw/mining/large-mines/fort-knox (accessed July 7, 2022).

Alpers, C. N. 2017. Arsenic and mercury contamination related to historical gold mining in the Sierra Nevada, California. Geochemistry: *Exploration, Environment, Analysis* 17(2):92–100.

AOML (Atlantic Oceanographic and Meteorological Laboratory). 2022a. Continental United States hurricane impacts/landfalls 1851–2021. https://www.aoml.noaa.gov/hrd/hurdat/All_U.S._Hurricanes.html (accessed June 17, 2022).

AOML. 2022b. Continental United States tropical storms impacts/landfalls 1851–1970, 1983–2021. https://www.aoml.noaa.gov/hrd/hurdat/uststorms.html (accessed June 17, 2022).

Arndt, N. 2011. Greenschist facies. In Encyclopedia of astrobiology, edited by R. Amils, M. Gargaud, J. Cernicharo Quintanilla, H. J. Cleaves, W. M. Irvine, D. Pinti, and M. Viso. Springer: Berlin, Heidelberg.

Ashley, R. P. 2002. Geoenvironmental model for low-sulfide gold-quartz vein deposits. Open-File Report 02-195, edited by R. R. I. Seal and N. K. Foley. U.S. Geological Survey. Pp. 176–195.

Ashley, R. P., and K. Savage. 2001. Analytical data for waters of the Harvard Open Pit, Jamestown Mine, Tuolumne County, California, March 1998–September 1999. U.S. Department of the Interior, U.S. Geological Survey.

Ashworth, K. K. 1983. Genesis of gold deposits at the Little Squaw Mines, Chandalar Mining District, Alaska. Western Washington University.

Assi, M. A., M. N. M. Hezmee, M. Y. M. Sabri, and M. A. Rajion. 2016. The detrimental effects of lead on human and animal health. *Veterinary World* 9(6):660.

Associated Press. 1990. Biologists search for dead animals after cyanide spill into river. https://apnews.com/article/59ec9558 4076b6a9b2da47c706dcfc87 (accessed September 27, 2022).

Aston Bay Holdings Ltd. 2022a. Buckingham gold property. https://astonbayholdings.com/projects/virginia-usa/buckingham-gold-property (accessed June 21, 2022).

Aston Bay Holdings Ltd. 2022b. Virginia properties. https://astonbayholdings.com/news/150-m-extension-to-gold-bearing-buckingham-vein-revealed (accessed June 21, 2022).

Atlas Copco (Epiroc operated as Atlas Copco prior to 2018). 2007. Mining methods in underground mining. Second Edition. Örebro, Sweden: Atlas Copco Rock Drills AB.

ATSDR (Agency for Toxic Substances and Disease Registry). 2007. Toxicological profile for arsenic. ATSDR, U.S. Department of Health and Human Services.

ATSDR. 2012. Toxicological profile for cadmium. ATSDR, U.S. Department of Health and Human Services.

ATSDR. 2019. Toxicological profile for antimony and compounds. ATSDR, U.S. Department of Health and Human Services.

ATSDR. 2020. Toxicological profile for lead. ATSDR, U.S. Department of Health and Human Services.

ATSDR. 2022. Toxicological profile for mercury. ATSDR, U.S. Department of Health and Human Services.

Aydin, C. I., B. Ozkaynak, B. Rodríguez-Labajos, and T. Yenilmez. 2017. Network effects in environmental justice struggles: An investigation of conflicts between mining companies and civil society organizations from a network perspective. *PLoS One* 12(7):e0180494.

Aylmore, M. 2016. Alternative lixiviants to cyanide for leaching gold ores. In Gold ore processing, edited by M. D. Adams. Amsterdam: Elsevier. Pp. 447–484.

Azam, S., and Q. Li. 2010. Tailings dam failures: A review of the last one hundred years. *Geotechnical News* 28(4):50–54.

Babedi, L., B. von der Heyden, M. Tadie, and M. Mayne. 2022. Trace elements in pyrite from five different gold ore deposit classes: A review and meta-analysis. Geological Society, London, Special Publications 516.

Bailey, B. L., L. J. Smith, D. W. Blowes, C. J. Ptacek, L. Smith, and D. C. Sego. 2013. The Diavik Waste Rock Project: Persistence of contaminants from blasting agents in waste rock effluent. *Applied Geochemistry* 36:256–270.

Barcelos, D. A., F. V. Pontes, F. A. da Silva, D. C. Castro, N. O. Dos Anjos, and Z. C. Castilhos. 2020. Gold mining tailing: Environmental availability of metals and human health risk assessment. *Journal of Hazardous Materials* 397:122721.

Basoy, V. 2015. The world's highest grade gold mines. https://www.mining.com/the-worlds-highest-grade-gold-mines (accessed August 26, 2022).

Bass, C. E. 1940. The Vaucluse gold mine, Orange County, Virginia. *Economic Geology* 35(1):79–91.

Benchmark Resources. 2022. McLaughlin Mine Project. http://benchmarkresources.com/projects/mclaughlin-mine-project (accessed June 17, 2022).

Biagioni, C., M. D'Orazio, G. O. Lepore, F. d'Acapito, and S. Vezzoni. 2017. Thallium-rich rust scales in drinkable water distribution systems: A case study from northern Tuscany, Italy. *Science of the Total Environment* 587:491–501.

Bleiwas, D. I. 2012. Estimated water requirements for gold heap-leach operations. Open-File Report 2012-1085. Reston, VA: U.S. Geological Survey.

BLM (Bureau of Land Management). 2012. Surface management handbook. Vol. H-3809-1. Washington, DC: U.S. Department of the Interior.

Bobb, J. F., L. Valeri, B. Claus Henn, D. C. Christiani, R. O. Wright, M. Mazumdar, J. J. Godleski, and B. A. Coull. 2015. Bayesian kernel machine regression for estimating the health effects of multi-pollutant mixtures. *Biostatistics* 16(3):493–508.

Bojakowska, I., and A. Paulo. 2013. Thallium in mineral resources extracted in Poland. Paper read at E3S Web of Conferences.

Borowitz, J. L., G. E. Isom, and D. V. Nakles. 2005. Human toxicology of cyanide. In Cyanide in water and soil: Chemistry, risk, and management, edited by D. A. Dzombak, R. S. Ghosh, and G. M. Wong-Chong. Boca Raton, FL: CRC Press. Pp. 249–262.

Bräuner, E. V., R. B. Nordsborg, Z. J. Andersen, A. Tjønneland, S. Loft, and O. Raaschou-Nielsen. 2014. Long-term exposure to low-level arsenic in drinking water and diabetes incidence: A prospective study of the diet, cancer and health cohort. *Environmental Health Perspectives* 122(10):1059–1065.

Brierley, C. L. 2016. Biological processing: Biological processing of sulfidic ores and concentrates—integrating innovations. In Innovative process development in metallurgical industry, edited by V. I. Lakshmanan. Berlin: Springer. Pp. 109–135.

Brochu, S. 2010. Assessment of ANFO on the environment. Quebec: Defence Research and Development Canada Valcartier.

Brodowsky, K. 2019. Industrielastwagen cat 775g. https://commons.wikimedia.org/wiki/File:Industrielastwagen_CAT_7 75G_019300_20190716_173633_02.jpg (accessed June 17, 2022).

Brown, W. R., 1969. Geology of the Dillwyn quadrangle, Virginia. Report of Investigations 10. Charlottesville, Virginia: Virginia Division of Mineral Resources. 77 pp.

Butler-Dawson, J., K. A. James, L. Krisher, D. Jaramillo, M. Dally, N. Neumann, D. Pilloni, A. Cruz, C. Asensio, and R. J. Johnson. 2022. Environmental metal exposures and kidney function of Guatemalan sugarcane workers. *Journal of Exposure Science & Environmental Epidemiology* 32(3):461–471.

Butterman, W. C., and E. B. Amey. 2005. Mineral commodity profiles, gold. Open-File Report 02-303. Reston, VA: U.S. Geological Survey.

Callahan, M. A., and K. Sexton. 2007. If cumulative risk assessment is the answer, what is the question? *Environmental Health Perspectives* 115(5):799–806.

Camargo, J. A., A. Alonso, and A. Salamanca. 2005. Nitrate toxicity to aquatic animals: A review with new data for freshwater invertebrates. *Chemosphere* 58(9):1255–1267.

Campanella, B., L. Colombaioni, E. Benedetti, A. Di Ciaula, L. Ghezzi, M. Onor, M. D'Orazio, R. Giannecchini, R. Petrini, and E. Bramanti. 2019. Toxicity of thallium at low doses: A review. *International Journal of Environmental Research and Public Health* 16(23):4732.

Canadian Dam Association. 2021. Tailings dam breach analysis. Canadian Dam Association.

The Canadian Press. 2016. Mining company faces new charges in worker's cyanide poisoning death in NE Ontario. Global News.

Cañedo-Argüelles, M., B. J. Kefford, C. Piscart, N. Prat, R. B. Schäfer, and C.-J. Schulz. 2013. Salinisation of rivers: An urgent ecological issue. *Environmental Pollution* 173:157–167.

CDC (U.S. Centers for Disease Control and Prevention). 2016. Lead poisoning investigation in northern Nigeria. https://www.cdc.gov/onehealth/in-action/lead-poisoning.html (accessed June 21, 2022).

CEQ (Council on Environmental Quality). 2020. States and local jurisdictions with NEPA-like environmental planning requirements. https://ceq.doe.gov/laws-regulations/states.html (accessed June 21, 2022).

CEQ. 2021. A citizen's guide to NEPA: Having your voice heard. Washington, DC: Executive Office of the President of the United States.

Chaffee, M. A., and S. Sutley. 1994. Analytical results, mineralogical data, and distributions of anomalies for elements and minerals in three mother lode-type gold deposits, Hodson Mining District, Calaveras County, California. Reston, VA: U.S. Geological Survey.

Chambers, D. M., and K. Zamzow. 2019. Documentation of acidic mining exploration drill cuttings at the pebble copper–gold mineral prospect, southwest Alaska. *Environments* 6(7):78.

Chesapeake Bay Program. 2022. 2025 Watershed Implementation Plan (WIPs). https://www.chesapeakeprogress.com/clean-water/watershed-implementation-plans (accessed September 22, 2022).

Chinnasamy, S. S., P. Hazarika, D. Pal, R. Sen, and G. Govindaraj. 2021. Pyrite textures and trace element compositions from the granodiorite-hosted gold deposit at Jonnagiri, eastern Dharwar craton, India: Implications for gold mineralization processes. *Economic Geology* 116(3):559–579.

Cision PR Newswire. 2020. OceanaGold announces results of updated Haile technical report.

Clark, D., Jr., and R. Hothem. 1991. Mammal mortality at Arizona, California, and Nevada gold mines using cyanide extraction. *California Fish and Game* 77(2):61–69.

Clausen, R., T. Montoya, K. Chief, S. Chischilly, J. Yazzie, J. Turner, L. M. Jacobs, and A. Merchant. 2020. Gold metal waters: The Animas River and the Gold King Mine spill, edited by B. Clark and P. McCormick. Louisville: University Press of Colorado.

Clements, W. H., D. B. Herbst, M. I. Hornberger, C. A. Mebane, and T. M. Short. 2021. Long-term monitoring reveals convergent patterns of recovery from mining contamination across 4 western US watersheds. *Freshwater Science* 40(2):407–426.

Cleven, R. F. M. J., and M. Van Bruggen. 2000. The cyanide accident in Barskoon (Kyrgyzstan). *RIVM* 609026 001.

Cohen, R. M., R. F. Charles, and D. C. Skipp. 2007. Evaluating ground water supplies in fractured metamorphic rock of the Blue Ridge province in northern Virginia.

Costello, M. 2016. Electrowinning. In Gold ore processing, edited by M. D. Adams. Amsterdam: Elsevier.

Cox, D. P., and W. C. Bagbey. 1992. Model 22b: Descriptive model of Au-Ag-Te veins. In Mineral deposit models, U.S. Geological survey bulletin. Vol. 1693, edited by D. P. Cox and D. A. Singer. P. 124.

CRS (Congressional Research Service). 2020. Federal land ownership: Overview and data. Vol. R42346. Washington, DC: CRS.

CRS. 2022. Supreme Court revisits scope of waters of the United States (WOTUS) under the Clean Water Act. Washington, DC: CRS.

Cust, J., and S. Poelhekke. 2015. The local economic impacts of natural resource extraction. *Annual Review of Resource Economics* 7(1):251–268.

Daniel, D., and Y. Wu. 1994. Compacted clay liners and covers for arid sites. Discussion and closure. *Journal of Geotechnical Engineering* 120(8).

Davis, G. H., S. J. Reynolds, and C. F. Kluth. 2011. Structural geology of rocks and regions. New York: John Wiley & Sons.

Davis, R., and D. M. Franks. 2011. The costs of conflict with local communities in the extractive industry. Paper read at Proceedings of the First International Seminar on Social Responsibility in Mining, Santiago, Chile.

Degnan, J. R., J. Bohlke, K. Pelham, D. M. Langlais, and G. J. Walsh. 2016. Identification of groundwater nitrate contamination from explosives used in road construction: Isotopic, chemical, and hydrologic evidence. *Environmental Science & Technology* 50(2):593–603.

Dentith, M., and S. T. Mudge. 2014. Geophysics for the mineral exploration geoscientist. Cambridge, UK: Cambridge University Press.

Diez Roux, A. V., and C. Mair. 2010. Neighborhoods and health. *Annals of the New York Academy of Sciences* 1186(1):125–145.

The Diggings. 2022. https://thediggings.com (accessed June 20, 2022).

DOI (U.S. Department of the Interior). 1991. South Carolina gold mining operations suspended. In Minerals Today. Washington, DC: Bureau of Mines.

DOI. 2022. Interior department launches interagency working group on mining reform. https://www.doi.gov/pressreleases/interior-department-launches-interagency-working-group-mining-reform (accessed October 26, 2022).

DOL (U.S. Department of Labor). 2022. MNM safety alert—explosive and blasting safety. https://www.msha.gov/news-media/alerts-hazards/mnm-safety-alert-explosive-and-blasting-safety (accessed June 17, 2022).

Domingo-Relloso, A., M. Grau-Perez, I. Galan-Chilet, M. J. Garrido-Martinez, C. Tormos, A. Navas-Acien, J. L. Gomez-Ariza, L. Monzo-Beltran, G. Saez-Tormo, and T. Garcia-Barrera. 2019. Urinary metals and metal mixtures and oxidative stress biomarkers in an adult population from Spain: The Hortega Study. *Environment International* 123:171–180.

Donato, D., O. Nichols, H. Possingham, M. Moore, P. Ricci, and B. Noller. 2007. A critical review of the effects of gold cyanide-bearing tailings solutions on wildlife. *Environment International* 33(7):974–984.

Donato, D., D. Madden-Hallett, G. Smith, and W. Gursansky. 2017. Heap leach cyanide irrigation and risk to wildlife: Ramifications for the International Cyanide Management Code. *Ecotoxicology and Environmental Safety* 140:271–278.

Dong, K., F. Xie, W. Wang, Y. Chang, D. Lu, X. Gu, and C. Chen. 2021. The detoxification and utilization of cyanide tailings: A critical review. *Journal of Cleaner Production* 302:126946.

Donkor, A. K., J.-C. J. Bonzongo, V. K. Nartey, and D. K. Adotey. 2005. Heavy metals in sediments of the gold mining impacted Pra River basin, Ghana, West Africa. *Soil & Sediment Contamination* 14(6):479–503.

Duckett, R., O. Flite, and M. O'Kane. 2012. ARD management Rio Tinto Ridgeway gold mine. Paper presented at 19th Annual British Columbia-MEND ML/ARD Workshop.

Dunbar, W. S., J. Fraser, A. Reynolds, and N. C. Kunz. 2020. Mining needs new business models. *The Extractive Industries and Society* 7(2):263–266.

Dunne, R. 2016. Flotation of gold and gold-bearing ores. In Gold ore processing, edited by M. D. Adams. Amsterdam: Elsevier. Pp. 315–338.

Durand, J. 2012. The impact of gold mining on the Witwatersrand on the rivers and karst system of Gauteng and North West Province, South Africa. *Journal of African Earth Sciences* 68:24–43.

Dyno Nobel. 2022. Product hub, packaged explosives technical information and safety datasheets. https://www.dynonobel.com/resource-hub/products/metal/packaged-explosives (accessed September 1, 2022).

Dzombak, D. A., R. S. Ghosh, and T. C. Young. 2005. Physical–chemical properties and reactivity of cyanide in water and soil. In Cyanide in water and soil: Chemistry, risk, and management, edited by D. A. Dzombak, R. S. Ghosh, and G. M. Wong-Chong. Boca Raton, FL: CRC Press. Pp. 69–104.

Eagles-Smith, C. A., E. K. Silbergeld, N. Basu, P. Bustamante, F. Diaz-Barriga, W. A. Hopkins, K. A. Kidd, and J. F. Nyland. 2018. Modulators of mercury risk to wildlife and humans in the context of rapid global change. *Ambio* 47(2):170–197.

Ebbs, S. D., G. M. Wong-Chong, B. S. Bond, J. T. Bushey, and E. F. Neuhauser. 2005. Biological transformation of cyanide in water and soil. In Cyanide in water and soil: Chemistry, risk, and management, edited by D. A. Dzombak, R. S. Ghosh, and G. M. Wong-Chong. Boca Raton, FL: CRC Press. Pp. 105–134.

Eisler, R. 1985. Cadmium hazards to fish, wildlife, and invertebrates: A synoptic review. Washington, DC: U.S. Fish & Wildlife Service, U.S. Department of the Interior.

Eisler, R. 1991. Cyanide hazards to fish, wildlife, and invertebrates: A synoptic review. Washington, DC: U.S. Fish & Wildlife Service, U.S. Department of the Interior.

Eisler, R. 1993. Zinc hazards to fish, wildlife, and invertebrates: A synoptic review. Washington, DC: U.S. Fish & Wildlife Service, U.S. Department of the Interior.

Eisler, R. 2004. Arsenic hazards to humans, plants, and animals from gold mining. *Reviews of Environmental Contamination and Toxicology* 133–165.

Eisler, R., D. R. Clark, S. N. Wiemeyer, and C. J. Henny. 1999. Sodium cyanide hazards to fish and other wildlife from gold mining operations. In Environmental impacts of mining activities. Berlin: Springer. Pp. 55–67.

Elliott, E. D., and D. C. Esty. 2021. The end environmental externalities manifesto: A rights-based foundation for environmental law. *NYU Environmental Law Journal* 29:505.

Emmett, E. A. 2021. Asbestos in high-risk communities: Public health implications. *International Journal of Environmental Research and Public Health* 18(4):1579.

Encyclopedia Britannica. 2022a. Mining. https://www.britannica.com/technology/mining (accessed July 12, 2022).

Encyclopedia Britannica. 2022b. https://www.britannica.com (accessed July 12, 2022).

Engels, J. 2021. Conventional impoundment storage—the current techniques. https://www.tailings.info/disposal/conventional.htm (accessed June 17, 2022).

Entwistle, J. A., A. S. Hursthouse, P. A. Marinho Reis, and A. G. Stewart. 2019. Metalliferous mine dust: Human health impacts and the potential determinants of disease in mining communities. *Current Pollution Reports* 5(3):67–83.

EPA (U.S. Environmental Protection Agency). 1991. Mine site visit: Brewer Gold Company. Washington, DC: EPA.

EPA. 1994a. Environmental justice 1994 annual report: Focusing on environmental protection for all people. Vol. 3103. Washington, DC: EPA.

EPA. 1994b. Technical resource document: Extraction and beneficiation of ores and minerals. Volume 2. Gold. Washington, DC: EPA.

EPA. 1994c. Treatment of cyanide heap leaches and tailings. Washington, DC: EPA.

EPA. 1998a. Guidelines for ecological risk assessment. Washington, DC: EPA.

EPA. 1998b. Technical background document: Identification and description of mineral processing sectors and waste streams. Washington, DC: EPA.

EPA. 1999. The class V underground injection control study: Mining, sand, or other backfill wells. Vol. 10. Washington, DC: EPA.

EPA. 2003. Framework for cumulative risk assessment. Washington, DC: EPA.

EPA. 2005. Interim record of decision. Summary of remedial alternative selection. Brewer Gold Mine, Jefferson, Chesterfield County, South Carolina. Atlanta, GA: EPA Region 4.

EPA. 2011a. A field-based aquatic life benchmark for conductivity in central Appalachian streams. Washington, DC: National Center for Environmental Assssment.

EPA. 2011b. Ore mining and dressing preliminary study report. Washington, DC: EPA.

EPA. 2014a. Chapter 5: General policies. In Water quality standards handbook. Washington, DC: EPA.

EPA. 2014b. Record of decision. Brewer Gold Mine site, Chesterfield County, South Carolina. Atlanta, GA: EPA Region 4.

EPA. 2014c. Reference guide to treatment technologies for mining-influenced water. Washington, DC: EPA.

EPA. 2015. A handbook of constructed wetlands. Washington, DC: EPA.

EPA. 2020a. Superfund proposed plan for interim record of decision Barite Hill Superfund Site Operable Unit 1 (OU1). Washington, DC: EPA.

EPA. 2020b. Clarification of free and total cyanide analysis for safe drinking water act (SDWA) compliance. Cincinnati, OH: EPA.

EPA. 2021a. Air quality statistics report. https://www.epa.gov/outdoor-air-quality-data/air-quality-statistics-report (accessed June 21, 2022).

EPA. 2021b. Clean Water Act programs utilizing the definition of WOTUS. https://www.epa.gov/wotus/clean-water-act-programs-utilizing-definition-wotus (accessed June 21, 2022).

EPA. 2021c. National menu of best management practices (BMPs) for stormwater. https://www.epa.gov/npdes/national-menu-best-management-practices-bmps-stormwater (accessed June 17, 2022).

EPA. 2021d. Third five-year review report for Brewer Gold Mine. Washington, DC: EPA.

EPA. 2021e. Air Emissions Monitoring for Permits. https://www.epa.gov/air-emissions-monitoring-knowledge-base/air-emissions-monitoring-permits (accessed June 22, 2022).

EPA. 2022a. Brewer Gold Mine. https://response.epa.gov/site/site_profile.aspx?site_id=612 (accessed July 15, 2022).

EPA. 2022b. Chesapeake Bay total maximum daily load (TMDL). https://www.epa.gov/chesapeake-bay-tmdl (accessed June 22, 2022).

EPA. 2022c. Criteria air pollutants. https://www.epa.gov/criteria-air-pollutants (accessed July 12, 2022).

EPA. 2022d. Electronics donation and recycling. https://www.epa.gov/recycle/electronics-donation-and-recycling (accessed August 11, 2022).

EPA. 2022e. Estimated nitrate concentrations in groundwater used for drinking. https://www.epa.gov/nutrient-policy-data/estimated-nitrate-concentrations-groundwater-used-drinking (accessed August 26, 2022).

EPA. 2022f. Federal requirements for class V wells. https://www.epa.gov/uic/federal-requirements-class-v-wells (accessed August 22, 2022).

EPA. 2022g. Initial list of hazardous air pollutants with modifications. https://www.epa.gov/haps/initial-list-hazardous-air-pollutants-modifications (accessed June 22, 2022).

EPA. 2022h. Legislative and regulatory timeline for mining waste. https://www.epa.gov/hw/legislative-and-regulatory-timeline-mining-waste (accessed June 22, 2022).

EPA. 2022i. More information about class V well types. https://www.epa.gov/uic/more-information-about-class-v-well-types (accessed August 22, 2022).

EPA. 2022j. National recommended water quality criteria—aquatic life criteria table. https://www.epa.gov/wqc/national-recommended-water-quality-criteria-aquatic-life-criteria-table (accessed June 22, 2022).

EPA. 2022k. Potential well water contamination and their impacts. https://www.epa.gov/privatewells/potential-well-water-contaminants-and-their-impacts (accessed June 16, 2022).

EPA. 2022l. Regulations for emissions from nonroad vehicles and engines. https://www.epa.gov/regulations-emissions-vehicles-and-engines/regulations-emissions-nonroad-vehicles-and-engines (accessed August 15, 2022).

EPA. 2022m. Special wastes. https://www.epa.gov/hw/special-wastes (accessed June 22, 2022).

EPA. 2022n. Sulfate in drinking water. https://archive.epa.gov/water/archive/web/html/sulfate.html (accessed June 17, 2022).

EPA. 2022o. Summary of the Safe Drinking Water Act. https://www.epa.gov/laws-regulations/summary-safe-drinking-water-act (accessed August 22, 2022).

EPA. 2022p. Superfund site: Brewer Gold Mine, Jefferson, SC. https://cumulis.epa.gov/supercpad/SiteProfiles/index.cfm?fuseaction=second.cleanup&id=0405550 (accessed June 17, 2022).

EPA. 2022q. Underground injection control in EPA region 3 (DE, DC, MD, PA, VA, and WV). https://www.epa.gov/uic/underground-injection-control-epa-region-3-de-dc-md-pa-va-and-wv (accessed August 22, 2022).

EPA. 2022r. Who has to obtain a Title V permit. https://www.epa.gov/title-v-operating-permits/who-has-obtain-title-v-permit (accessed July 18, 2022).

EPA Region 10. 2003. EPA and hardrock mining: A source book for industry in the northwest and Alaska. Seattle, Washington, DC: EPA Region 10.

Ettinger, A. S., A. R. Zota, C. J. Amarasiriwardena, M. R. Hopkins, J. Schwartz, H. Hu, and R. O. Wright. 2009. Maternal arsenic exposure and impaired glucose tolerance during pregnancy. *Environmental Health Perspectives* 117(7):1059–1064.

Farjana, S. H., M. P. Mahmud, and N. Huda. 2021. Life cycle assessment for sustainable mining. Amsterdam: Elsevier.

Farzan, S. F., Y. Chen, J. R. Rees, M. S. Zens, and M. R. Karagas. 2015a. Risk of death from cardiovascular disease associated with low-level arsenic exposure among long-term smokers in a US population-based study. *Toxicology and Applied Pharmacology* 287(2):93–97.

Farzan, S. F., M. R. Karagas, J. Jiang, F. Wu, M. Liu, J. D. Newman, F. Jasmine, M. G. Kibriya, R. Paul-Brutus, and F. Parvez. 2015b. Gene–arsenic interaction in longitudinal changes of blood pressure: Findings from the health effects of arsenic longitudinal study (HEALS) in Bangladesh. *Toxicology and Applied Pharmacology* 288(1):95–105.

Fashola, M. O., V. M. Ngole-Jeme, and O. O. Babalola. 2016. Heavy metal pollution from gold mines: Environmental effects and bacterial strategies for resistance. *International Journal of Environmental Research and Public Health* 13(11):1047.

FEMA (Federal Emergency Management Agency). 2013a. Federal guidlines for dam safety: Emergency action planning for dams. Washington, DC: FEMA 64.

FEMA. 2013b. Selecting and accommodating inflow design floods for dams. Washington, DC: FEMA P-94.

Fendorf, S., and B. D. Kocar. 2009. Biogeochemical processes controlling the fate and transport of arsenic: Implications for south and southeast Asia. *Advances in Agronomy* 104:137–164.

Fletcher, D. E., W. A. Hopkins, T. Saldaña, J. A. Baionno, C. Arribas, M. M. Standora, and C. Fernández-Delgado. 2006. Geckos as indicators of mining pollution. *Environmental Toxicology and Chemistry: An International Journal* 25(9):2432–2445.

Flite, O. P., III. 2006. Onset and persistence of biogenic meromixis in a filling pit lake—a limnological perspective. Clemson University.

Foley, N. K., and R. A. Ayuso. 2012. Gold deposits of the Carolina slate belt, southeastern United States: Age and origin of the major gold producers. Open-File Report 2012-1179. Reston, VA: U.S. Geological Survey.

Foley, N., R. A. Ayuso, and R. Seal. 2001. Remnant colloform pyrite at the Haile gold deposit, South Carolina: A textural key to genesis. *Economic Geology* 96(4):891–902.

Frankel, A., A. McGarr, J. Bicknell, J. Mori, L. Seeber, and E. Cranswick. 1990. Attenuation of high-frequency shear waves in the crust: Measurements from New York state, South Africa, and Southern California. *Journal of Geophysical Research: Solid Earth* 95(B11):17441–17457.

French, M., N. Alem, S. J. Edwards, E. Blanco Coariti, H. Cauthin, K. A. Hudson-Edwards, K. Luyckx, J. Quintanilla, and O. Sánchez Miranda. 2017. Community exposure and vulnerability to water quality and availability: A case study in the mining-affected Pazña Municipality, Lake Poopó Basin, Bolivian Altiplano. *Environmental Management* 60(4):555–573.

Fromm, P. O. 1980. A review of some physiological and toxicological responses of freshwater fish to acid stress. *Environmental Biology of Fishes* 5(1):79–93.

Fuerstenau, M. C., B. Wakawa, R. Price, and R. Wellik. 1974. Toxicity of selected sulfhydryl collectors to rainbow trout. *Transactions of the Society of Mining Engineers, AIME* 256(4):337–341.

Fullam, M., B. Watson, A. Laplante, and S. Gray. 2016. Advances in gravity gold technology. In Gold ore processing, edited by M. D. Adams. Amsterdam: Elsevier. Pp. 301–314.

Gallagher, S. S., G. E. Rice, L. J. Scarano, L. K. Teuschler, G. Bollweg, and L. Martin. 2015. Cumulative risk assessment lessons learned: A review of case studies and issue papers. *Chemosphere* 120:697–705.

Gensemer, R. W., D. K. DeForest, R. D. Cardwell, D. Dzombak, R. Santore, and M. Stewart. 2006. Reassessment of cyanide ambient water quality criteria: An integrated approach to protection of the aquatic environment. *Proceedings of the Water Environment Federation* 2006(6):5709–5718.

Global Tailings Review. 2020. Global industry standards on tailings management. https://globaltailingsreview.org/wp-content/uploads/2020/08/global-industry-standard_EN.pdf.

Goldfarb, R. J., T. Baker, B. Dubé, D. I. Groves, C. J. Hart, and P. Gosselin. 2005. Distribution, character, and genesis of gold deposits in metamorphic terran. In One Hundredth Anniversary Volume, edited by J. W. Hedenquist, J. F. H. Thompson, R. J. Goldfarb, and J. P. Richards. Society of Economic Geologists.

GoldPrice.Org. 2022. Gold price. https://goldprice.org/gold-price-history.html (accessed June 16, 2022).

Good, R. S., O. M. Fordham, Jr., and C. R. Halladay. 1977. Geochemical reconnaissance for gold in the Caledonia and Pendleton Quadrangles in the Piedmont of central Virginia. *Virginia Minerals* 23(2):13–22.

Graton, L., and W. Lindgren. 1906. Reconnaissance of some gold and tin deposits of the southern Appalachians, with notes on the dahlonega mines. Vol. 293. Reston, VA: U.S. Geological Survey.

Green, A., A. D. Jones, K. Sun, and R. L. Neitzel. 2015. The association between noise, cortisol and heart rate in a small-scale gold mining community—a pilot study. *International Journal of Environmental Research and Public Health* 12(8):9952–9966.

Griffith, M. B. 2017. Toxicological perspective on the osmoregulation and ionoregulation physiology of major ions by freshwater animals: Teleost fish, crustacea, aquatic insects, and mollusca. *Environmental Toxicology and Chemistry* 36(3):576–600.

Griffiths, S. R., D. B. Donato, G. Coulson, and L. F. Lumsden. 2014. High levels of activity of bats at gold mining water bodies: Implications for compliance with the International Cyanide Management Code. *Environmental Science and Pollution Research* 21(12):7263–7275.

Grimalt, J. O., M. Ferrer, and E. Macpherson. 1999. The mine tailing accident in Aznalcollar. *Science of the Total Environment* 242(1–3):3–11.

Groves, D. I., R. J. Goldfarb, F. Robert, and C. J. Hart. 2003. Gold deposits in metamorphic belts: Overview of current understanding, outstanding problems, future research, and exploration significance. *Economic Geology* 98(1):1–29.

Groves, D. I., M. Santosh, and L. Zhang. 2020. A scale-integrated exploration model for orogenic gold deposits based on a mineral system approach. *Geoscience Frontiers* 11(3):719–738.

Gusek, J. J., and L. A. Figueroa. 2009. Mitigation of metal mining influenced water. Vol. 2. Littleton, Colorado: Society for Mining, Metallurgy, and Exploration, Inc.

Guthrie, C. 2020. Group 11 launches as in-situ gold miner. Mining Magazine. https://www.miningmagazine.com/sustainability/news/1394353/group-11-launches-as-in-situ-gold-miner (accessed September 27, 2022).

Hackley, P. C., J. D. Peper, W. C. Burton, and J. W. Horton. 2007. Northward extension of Carolina slate belt stratigraphy and structure, south-central Virginia: Results from geologic mapping. *American Journal of Science* 307(4):749–771.

Haines, T. A. 1981. Acidic precipitation and its consequences for aquatic ecosystems: A review. *Transactions of the American Fisheries Society* 110(6):669–707.

Hammarstrom, J. M., A. N. Johnson, R. R. Seal II, A. L. Meier, P. L. Briggs, and N. M. Piatak. 2006. Geochemical and mineralogical characterization of the abandoned Valzinco (lead-zinc) and Mitchell (gold) mine sites prior to reclamation, Spotsylvania County, Virginia. Reston, VA: U.S. Geological Survey.

Hammerschmidt, J., J. Güntner, B. Kerstiens, and A. Charitos. 2016. Roasting of gold ore in the circulating fluidized-bed technology. In Gold ore processing, edited by M. D. Adams. Amsterdam: Elsevier. Pp. 393–409.

Haptonstall, J. 2011. Shrinkage stoping. In SME mining engineering handbook. Vol. 1. Littleton, CO: Society for Mining, Metallurgy, and Exploration.

Hassan, I., S. R. Chowdhury, P. K. Prihartato, and S. A. Razzak. 2021. Wastewater treatment using constructed wetland: Current trends and future potential. *Processes* 9(11):1917.

Hassan, N. A., M. Sahani, R. Hod, and N. A. Yahya. 2015. A study on exposure to cyanide among a community living near a gold mine in Malaysia. *Journal of Environmental Health* 77(6):42–49.

Henny, C. J., R. J. Hallock, and E. F. Hill. 1994. Cyanide and migratory birds at gold mines in Nevada, USA. *Ecotoxicology* 3(1):45–58.

Herath, H. M. A. S., T. Kawakami, S. Nagasawa, Y. Serikawa, A. Motoyama, G. T. Chaminda, S. Weragoda, S. Yatigammana, and A. Amarasooriya. 2018. Arsenic, cadmium, lead, and chromium in well water, rice, and human urine in Sri Lanka in relation to chronic kidney disease of unknown etiology. *Journal of Water and Health* 16(2):212–222.

Herrera, R., K. Radon, O. S. von Ehrenstein, S. Cifuentes, D. M. Muñoz, and U. Berger. 2016. Proximity to mining industry and respiratory diseases in children in a community in northern Chile: A cross-sectional study. *Environmental Health* 15(1):1–10.

Herrera, R., U. Berger, O. S. Von Ehrenstein, I. Díaz, S. Huber, D. Moraga Muñoz, and K. Radon. 2018. Estimating the causal impact of proximity to gold and copper mines on respiratory diseases in Chilean children: An application of targeted maximum likelihood estimation. *International Journal of Environmental Research and Public Health* 15(1):39.

Heylmun, E. 2001. Gold in Virginia. *International California Mining Journal* 71(1):3.

Hibbard, J. P., J. S. Beard, W. S. Henika, and J. W. Horton, Jr. 2016. Geology of the western Piedmont in Virginia. In Geology of Virginia. Vol. 18, edited by C. Bailey, W. C. Sherwood, L. Eaton, and D. Powars. Martinsville, VA: Virginia Museum of Natural History Special Publication. Pp. 87–124.

Hill, E., and P. Henry. 1996. Cyanide. In Noninfectious diseases of wildlife, 2nd ed., edited by A. Fairbrother, L. N. Locke, G. L. Hoff. Ames, IA: Iowa State University Press. Pp. 99–107.

Hilmers, A., D. C. Hilmers, and J. Dave. 2012. Neighborhood disparities in access to healthy foods and their effects on environmental justice. *American Journal of Public Health* 102(9):1644–1654.

Hilson, G., and R. Maconachie. 2020. Artisanal and small-scale mining and the sustainable development goals: Opportunities and new directions for sub-Saharan Africa. *Geoforum* 111:125–141.

Holmes, D. 2022. Shocking news about historic gold mining contamination. Piedmont Environmental Council. https://www.pecva.org/region/orange/historic-gold-mining-contamination.

Horowitz, H. M., D. J. Jacob, Y. Zhang, T. S. Dibble, F. Slemr, H. M. Amos, J. A. Schmidt, E. S. Corbitt, E. A. Marais, and E. M. Sunderland. 2017. A new mechanism for atmospheric mercury redox chemistry: Implications for the global mercury budget. *Atmospheric Chemistry and Physics* 17(10):6353–6371.

Horton, J. W., Jr., B. E. Owens, P. C. Hackley, W. C. Burton, P. E. Sacks, and J. P. Hibbard. 2016. Geology of the eastern Piedmont in Virginia. In Geology of Virginia, edited by C. Bailey, W. C. Sherwood, L. Eaton, and D. Powars. Martinsville, VA: Virginia Museum of Natural History Special Publication. Pp. 125–158.

Hughes, K. S., J. Hibbard, R. T. Sauer, and W. C. Burton. 2014. Stitching the western Piedmont of Virginia: Early paleozoic tectonic history of the Ellisville Pluton and the Potomac and Chopawamsic Terranes. In 44th Annual Virginia Geological Field Conference. Vol. 9. Martinsville, VA: Virginia Museum of Natural History. P. 33.

Hynes, H., and R. Lopez. 2007. Cumulative risk and a call for action in environmental justice communities. *Journal of Health Disparities Research and Practice* 1(2):3.

Iakovides, M., G. Iakovides, and E. G. Stephanou. 2021. Atmospheric particle-bound polycyclic aromatic hydrocarbons, n-alkanes, hopanes, steranes and trace metals: PM2. 5 source identification, individual and cumulative multi-pathway lifetime cancer risk assessment in the urban environment. *Science of the Total Environment* 752:141834.

IARC (International Agency for Research on Cancer). 1993. Mercury and mercury compounds. In Beryllium, cadmium, mercury, and exposures in the glass manufacturing industry; IARC monographs on the evaluation of carcinogenic risks to humans. Vol. 58. Lyon, France: IARC Working Group on the Evaluation of Carcinogenic Risks to Humans.

IARC. 2006. Inorganic and organic lead compounds. Vol. 87. Lyon, France: IARC monographs on the evaluation of carcinogenic risks to humans.

IARC. 2012a. Arsenic and arsenic compounds. In A review of human carcinogens. Part C: Arsenic, metals, fibres, and dusts. Vol. 100c. Lyon, France: IARC monographs on the evaluation of carcinogenic risks to humans.

IARC. 2012b. Cadmium and cadmium compounds. In Beryllium, cadmium, mercury, and exposures in the glass manufacturing industry. Vol. 58. Lyon, France: IARC monographs on the evaluation of carcinogenic risks to humans.

IARC. 2014. Diesel and gasoline engine exhausts and some nitroarenes. Vol. 105. Lyon, France: IARC monographs on the evaluation of carcinogenic risks to humans.

IARC. 2022. Agents classified by the IARC monographs. https://monographs.iarc.who.int/agents-classified-by-the-iarc (accessed July 9, 2022).

ICMM (International Council on Mining and Metals). 2022. Respect indigenous peoples. https://www.icmm.com/en-gb/our-work/social-performance/indigenous-peoples-and-human-rights/respect-indigenous-peoples (accessed July 5, 2022).

ICOLD (International Commission on Large Dam). 2020. Tailings dam safety bulletin—preliminary draft for review.

Idaho Transportation Department. 2011. Geosynthetics. In Best management practices manual. https://apps.itd.idaho.gov/Apps/env/BMP/PDF%20Files%20for%20BMP/Chapter%205/PC-8%20%20Geosynthetics.pdf (accessed July 5, 2022).

IFC (International Finance Corporation). 2007. Environmental, health and safety general guidelines. https://www.ifc.org/wps/wcm/connect/topics_ext_content/ifc_external_corporate_site/sustainability-at-ifc/policies-standards/ehs-guidelines (accessed July 5, 2022).

INAP (International Network for Acid Prevention). 2014. The global acid rock drainage guide. http://www.gardguide.com/images/5/5f/TheGlobalAcidRockDrainageGuide.pdf (accessed June 17, 2022).

International Cyanide Management Code. 2021. Guidance for use of the mining operations verification protocol. Washington, DC: International Cyanide Management Institute.

International Cyanide Management Code. 2022. The cyanide code. https://cyanidecode.org (accessed June 17, 2022).

International Cyanide Management Institute. 2021. Results that matter everyday. https://cyanidecode.org/wp-content/uploads/2022/06/2021_Annual_Report_Secure.pdf.

IPCC (Intergovernmental Panel on Climate Change). 2021. Climate change 2021: The physical science basis. Contribution of Working Group I to the sixth assessment report of the Intergovernmental Panel on Climate Change, edited by V. Masson-Delmotte, P. Zhai, A. Pirani, S. L. Connors, C. Péan, S. Berger, N. Caud, Y. Chen, L. Goldfarb, M. I. Gomis, M. Huang, K. Leitzell, E. Lonnoy, J. B. R. Matthews, T. K. Maycock, T. Waterfield, O. Yelekçi, R. Yu, and B. Zhou. Cambridge, UK, and New York: Cambridge University Press. https://doi.org/10.1017/9781009157896.

IRMA (Initiative for Responsible Mining Assurance). 2018. Free, prior and informed consent (FPIC). https://responsiblemining.net/wp-content/uploads/2018/08/Chapter_2.2_FPIC.pdf (accessed July 6, 2022).

ISEE (International Society of Explosive Engineers). 2011. The blasters' handbook. Edited by J. J. Stiehr and I. S. Dean. 18th ed. Cleveland, OH: ISEE.

Ishihara, K. 1984. Post-earthquake failure of a tailings dam due to liquefaction of pond deposit. Paper read at International Conference on Case Histories in Geotechnical Engineering.

Ivanova, G., and J. Rolfe. 2011. Assessing development options in mining communities using stated preference techniques. *Resources Policy* 36(3):255–264.

Ivanova, G., J. Rolfe, S. Lockie, and V. Timmer. 2007. Assessing social and economic impacts associated with changes in the coal mining industry in the Bowen basin, Queensland, Australia. *Management of Environmental Quality: An International Journal* 18(2):211–228.

Jaacks, J. A., L. G. Closs, and J. A. Coope. 2011. Geochemical prospecting. In SME mining engineering handbook, edited by P. Darling. Littleton, CO: Society for Mining, Metallurgy, and Exploration.

Jamieson, J. W., M. D. Hannington, S. Petersen, and M. K. Tivey. 2016. Volcanogenic massive sulfides. In Encyclopedia of marine geosciences. Encyclopedia of earth sciences series, edited by J. Harff, M. Meschede, S. Petersen, and J. Thiede. Dordrecht: Springer.

Janz, D. M., D. K. DeForest, M. L. Brooks, P. M. Chapman, G. Gilron, D. Hoff, W. A. Hopkins, D. O. McIntyre, C. A. Mebane, and V. P. Palace. 2010. Selenium toxicity to aquatic organisms. In Ecological assessment of selenium in the aquatic environment, edited by P. M. Chapman, W. J. Adams, M. L. Brooks, C. G. Delos, S. N. Luoma, W. A. Maher, H. M. Ohlendorf, T. S. Presser, and D. P. Shaw. Pensacola, FL: Society of Environmental Toxicology and Chemistry. Pp. 141–231.

Johnson, N. E. 1983. A study of the vein copper mineralization of the Virgilina District, Virginia and North Carolina. Virginia Polytechnic Institute and State University.

Johnson, P., B. Zoheir, W. Ghebreab, R. Stern, C. Barrie, and R. Hamer. 2017. Gold-bearing volcanogenic massive sulfides and orogenic-gold deposits in the Nubian Shield. *South African Journal of Geology* 120(1):63–76.

Johnston, A., R. L. Runkel, A. Navarre-Sitchler, and K. Singha. 2017. Exploration of diffuse and discrete sources of acid mine drainage to a headwater mountain stream in Colorado, USA. *Mine Water and the Environment* 36(4):463–478.

Jones, S. P. 1909. Second report on the gold deposits of Georgia. Vol. 19. Geological Survey of Georgia.

Junior Mining Network. 2022. OceanaGold achieves full year 2021 guidance on record annual production from Haile. https://www.juniorminingnetwork.com/junior-miner-news/press-releases/712-tsx/ogc/114453-oceanagold-achieves-full-year-2021-guidance-on-record-annual-production-from-haile.html.

Kappes, R., C. Fortin, and R. Dunne. 2013. Current status of the chemistry of gold flotation in industry. *CIM Journal* 4(3).

Kaur, P., N. Gunawardena, and J. Kumaresan. 2020. A review of chronic kidney disease of unknown etiology in Sri Lanka, 2001–2015. *Indian Journal of Nephrology* 30(4):245.

KCA (Kappes, Cassiday & Associates). 2020. Mercury retorts. https://www.kcareno.com/mercury-retorts (accessed September 9, 2022).

Kish, S. A., and H. J. Stein. 1989. Post-Acadian metasomatic origin for copper-bearing vein deposits of the Virgilina District, North Carolina and Virginia. *Economic Geology* 84(7):1903–1920.

Kivinen, S., J. Kotilainen, and T. Kumpula. 2020. Mining conflicts in the European Union: Environmental and political perspectives. *Fennia* 198.

Knopf, A. 1929. The mother lode system of California. Vol. 157. Reston, VA: U.S. Geological Survey.

Koerner, R. 1994. Designing with geosynthetics, 3th ed. Englewood Cliffs, NJ: Prentice-Hall.

Koski, R. A., L. Munk, A. L. Foster, W. C. Shanks III, and L. L. Stillings. 2008. Sulfide oxidation and distribution of metals near abandoned copper mines in coastal environments, Prince William Sound, Alaska, USA. *Applied Geochemistry* 23(2):227–254.

Kotze, M., B. Green, J. Mackenzie, and M. Virnig. 2016. Resin-in-pulp and resin-in-solution. In Gold ore processing, edited by M. D. Adams: Elsevier. Pp. 561–583.

Kramer, A., S. Gaulocher, M. Martins, and L. d. S. Leal Filho. 2012. Surface tension measurement for optimization of flotation control. *Procedia Engineering* 46:111–118.

Kraus, U., and J. Wiegand. 2006. Long-term effects of the Aznalcóllar mine spill—heavy metal content and mobility in soils and sediments of the Guadiamar River valley (SW Spain). *Science of the Total Environment* 367(2–3):855–871.

Krauss, R. 2002. Environmental manager, McLaughlin Mine, an oral history conducted in 1994–2001 by Eleanor Swent in the Knoxville Mining District. https://archive.org/details/knoxvillemining08swenrich.

Kuipers, J. R. 2000. Hardrock reclamation bonding practices in the western United States. National Wildlife Federation.

Kuipers, J. R., and A. S. Maest. 2006. Comparison of predicted and actual water quality at hardrock mines. Kuipers & Associates, Buka Environmental.

Kyle, J., P. Breuer, K. Bunney, R. Pleysier, and P. May. 2011. Review of trace toxic elements (Pb, Cd, Hg, As, Sb, Bi, Se, Te) and their deportment in gold processing. Part 1: Mineralogy, aqueous chemistry and toxicity. *Hydrometallurgy* 107(3–4):91–100.

Kyle, J., P. Breuer, K. Bunney, and R. Pleysier. 2012. Review of trace toxic elements (Pb, Cd, Hg, As, Sb, Bi, Se, Te) and their deportment in gold processing. Part II: Deportment in gold ore processing by cyanidation. *Hydrometallurgy* 111:10–21.

Laney, F. B. 1917. The geology and ore deposits of the Virgilina District of Virginia and North Carolina. North Carolina Geological and Economic Survey.

Leduc, G. 1981. Ecotoxicology of cyanides in freshwater. In Cyanide in biology, edited by B. Vennesland, E. E. Conn, C. J. Knowles, J. Westley, and F. Wissing. New York: Academic Press. Pp. 487–494.

Leduc, G. 1984. Cyanides in water: Toxicological significance, edited by L. Weber. Vol. 2, Aquatic toxicology. New York: Raven Press.

Leduc, G., R. C. Pierce, and I. R. McCracken. 1982. The effects of cyanides on aquatic organisms with emphasis upon freshwater fishes. Ottawa, Ontario: National Research Council of Canada.

Lee, I., Y. J. Park, M. J. Kim, S. Kim, S. Choi, J. Park, Y. H. Cho, S. Hong, J. Yoo, and H. Park. 2021. Associations of urinary concentrations of phthalate metabolites, bisphenol A, and parabens with obesity and diabetes mellitus in a Korean adult population: Korean National Environmental Health Survey (KoNEHS) 2015–2017. *Environment International* 146:106227.

LeHuray, A. 1982. Lead isotopic patterns of galena in the Piedmont and Blue Ridge ore deposits, southern Appalachians. *Economic Geology* 77(2):335–351.

Lenhardt, W. 2009. The impact of earthquakes on mining operations. *BHM Berg- und Hüttenmännische Monatshefte* 154(6):249–254.

Lesniak, J. A., and S. M. Ruby. 1982. Histological and quantitative effects of sublethal cyanide exposure on oocyte development in rainbow trout. *Archives of Environmental Contamination and Toxicology* 11(3):343–352.

Lewis, J., J. Hoover, and D. MacKenzie. 2017. Mining and environmental health disparities in Native American communities. *Current Environmental Health Reports* 4(2):130–141.

Lide, D. R. 1999. Handbook of chemistry and physics. Boca Raton, FL: CRC Press.

Liévanos, R. S., P. Greenberg, and R. Wishart. 2018. In the shadow of production: Coal waste accumulation and environmental inequality formation in eastern Kentucky. *Social Science Research* 71:37–55.

Linden, M. A., J. R. Craig, and T. N. Solberg. 1985. Mineralogy and chemistry of gold in the Virgilina District, Halifax County, Virginia. Vol. 31. *Virginia Minerals* 31(2):17–22.

Linker, L., G. Shenk, and J. Martin. 2019. Final phase 3 WIP sediment targets and basin-to-basin exchanges. https://www.chesapeakebay.net/channel_files/29907/ii.a.b._sediment_targets__b2b_final_planning_targets_for_mb_10-17-19.pdf (accessed June 22, 2022).

Lo, Y.-C., C. A. Dooyema, A. Neri, J. Durant, T. Jefferies, A. Medina-Marino, L. de Ravello, D. Thoroughman, L. Davis, and R. S. Dankoli. 2012. Childhood lead poisoning associated with gold ore processing: A village-level investigation—Zamfara state, Nigeria, October–November 2010. *Environmental Health Perspectives* 120(10):1450–1455.

Lockie, S., M. Franettovich, V. Petkova-Timmer, J. Rolfe, and G. Ivanova. 2009. Coal mining and the resource community cycle: A longitudinal assessment of the social impacts of the Coppabella coal mine. *Environmental Impact Assessment Review* 29(5):330–339.

Lonsdale, J. T. 1927. The geology of the gold-pyrite belt of the northeastern Piedmont Virginia. Vol. 30. Virginia Geological Survey.

Lyu, Z., J. Chai, Z. Xu, Y. Qin, and J. Cao. 2019. A comprehensive review on reasons for tailings dam failures based on case history. *Advances in Civil Engineering* 2019:18.

MacDonald Gibson, J., and K. J. Pieper. 2017. Strategies to improve private-well water quality: A North Carolina perspective. *Environmental Health Perspectives* 125(7):076001.

Macrotrends. 2022. Gold prices—100 year historical chart. https://www.macrotrends.net/1333/historical-gold-prices-100-year-chart (accessed June 16, 2022).

Maest, A. S. 2022. The potential environmental and human health effects of gold mining in Virginia. Submitted to the Committee on Potential Impacts of Gold Mining in Virginia, National Academies of Sciences, Engineering, and Medicine, on behalf of the Southern Environmental Law Center.

Maest, A. S., J. R. Kuipers, C. L. Travers, and D. A. Atkins. 2005. Predicting water quality at hardrock mines. Methods and models, uncertainties and state-of-the-art. Kuipers & Associates and Buka Environmental.

Majumdar, S., S. Singh, P. R. Sahoo, and A. Venkatesh. 2019. Trace-element systematics of pyrite and its implications for refractory gold mineralisation within the carbonaceous metasedimentary units of Palaeoproterozoic South Purulia shear zone, eastern India. *Journal of Earth System Science* 128(8):1–25.

Mangan, M., J. Craig, and J. Rimstidt. 1984. Submarine exhalative gold mineralization at the London-Virginia mine, Buckingham County, Virginia. *Mineralium Deposita* 19(3):227–236.

Manning, T., and D. Kappes. 2016. Heap leaching of gold and silver ores. In Gold ore processing, edited by M. D. Adams. Amsterdam: Elsevier. Pp. 413–428.

Martel, R., L. Trépanier, B. Lévesque, G. Sanfaçon, P. Brousseau, M.-A. Lavigne, L.-C. Boutin, P. Auger, D. Gauvin, and L. Galarneau. 2004. Carbon monoxide poisoning associated with blasting operations close to underground enclosed spaces. Part 1. Co production and migration mechanisms. *Canadian Geotechnical Journal* 41(3):371–382.

Martinez-Pagan, P., A. Faz-Cano, and M. Rosales-Aranda. 2009. ERT studies on tailings ponds of the Sierra Minera Cartagena-La Union, SE Spain. Paper read at AGU Fall Meeting Abstracts.

Mashishi, D. T., C. Wolkersdorfer, and H. Coetzee. 2022. The Whitehill Formation as a natural geochemical analogue to the Witwatersrand basin's mine water issues, South Africa. *Environmental Science and Pollution Research* 29(18):27195–27208.

Mattaponi Indian Reservation. 2022. Official site of the Mattaponi Indian Tribe & Reservation. https://www.mattaponination.com (accessed August 26, 2022).

McCarthy, T. S. 2011. The impact of acid mine drainage in South Africa. *South African Journal of Science* 107(5):1–7.

McLemore, V. T. 2008. Basics of metal mining influenced water. Vol. 1. Littleton, CO: Society for Mining, Metallurgy, and Exploration.

Mercier-Langevin, P., B. Dubé, B. Lafrance, M. D. Hannington, A. Galley, J. Moorehead, and P. Gosselin. 2007. Metallogeny of the Doyon-Bousquet-Laronde mining camp. Abitibi greenstone belt, Quebec. In Mineral deposits of Canada: A synthesis of major deposit-types, district metallogeny, the evolution of geological provinces, and exploration methods, edited by W. D. Goodfellow. Vol. 5. Geological Association of Canada. Pp. 673–701.

Merriam-Webster. 2022. https://www.merriam-webster.com (accessed July 12, 2022).

Metso:Outotec. 2022. Biox® process. https://www.mogroup.com/portfolio/biox-process (accessed June 17, 2022).

Meza-Montenegro, M. M., A. J. Gandolfi, M. E. Santana-Alcántar, W. T. Klimecki, M. G. Aguilar-Apodaca, R. Del Río-Salas, M. De la O-Villanueva, A. Gómez-Alvarez, H. Mendivil-Quijada, and M. Valencia. 2012. Metals in residential soils and cumulative risk assessment in Yaqui and Mayo agricultural valleys, northern Mexico. *Science of the Total Environment* 433:472–481.

Mindat. 2022a. Definition of gangue. https://www.mindat.org/glossary/gangue (accessed July 12, 2022).

Mindat. 2022b. Meta clastic-sedimentary-rock. https://www.mindat.org/min-51362.html (accessed July 12, 2022).

Mindat. 2022c. Moss mine, Tabscott, gold-pyrite belt, Goochland Co., Virginia, USA. https://www.mindat.org/loc-16591.html (accessed July 11, 2022).

Minerals Council of Australia. 2014. Under the microscope: The true costs of gold production. https://www.coindesk.com/markets/2014/06/28/under-the-microscope-the-true-costs-of-gold-production (accessed June 17, 2022).

The Mining Association of Canada. 2017. A guide to the management of tailings facilities.

Mining People International. 2016. The worlds deepest, biggest and deadliest open pit mines. https://www.miningpeople.com.au/news/the-worlds-deepest-biggest-and-deadliest-open-pit-mines (accessed July 12, 2022).

Minnesota Department of Health. 2018. Nitrate and methemoglobinemia. https://www.health.state.mn.us/communities/environment/water/docs/contaminants/nitratmethemog.pdf (accessed July 12, 2022).

Mobley, R. M., G. M. Yogodzinski, R. A. Creaser, and J. M. Berry. 2014. Geologic history and timing of mineralization at the Haile gold mine, South Carolina. *Economic Geology* 109(7):1863–1881.

Montana DEQ (Department of Environmental Quality). 2022. Query from workflow database for the hard rock mining section.

Moran, R. 1998. Cyanide uncertainties: Observations on the chemistry, toxicity, and analysis of cyanide in mining-related waters. Washington, DC: Mineral Policy Center.

Moran, R. 1999. Cyanide in mining: Some observations on the chemistry, toxicity and analysis of mining-related waters. *Central Asia Ecology* 99.

Morel, F. M., A. M. Kraepiel, and M. Amyot. 1998. The chemical cycle and bioaccumulation of mercury. *Annual Review of Ecology and Systematics* 29:543–566.

Morgenstern, N. R. 2011. Petroleum history society oil sands oral history project transcript. The Oil Sands Oral History. https://glenbow.ucalgary.ca/wp-content/uploads/2019/06/Morgenstern_Norbert.pdf.

Morgenstern, N. R. 2018. Geotechnical risk, regulation, and public policy. *Soils and Rocks, São Paulo* 41(2):107–129.

Morgenstern, N. R., S. G. Vick, and D. Van Zyl. 2015. Independent expert engineering investigation and review panel. Report on Mount Polley tailings storage facility breach. Victoria, BC: Province of British Columbia.

Morin, K. A., and N. M. Hutt. 2009. Mine-water leaching of nitrogen species from explosive residues. *Proceedings of Geo-Halifax* 20–24.

Morrice, E., and R. Colagiuri. 2013. Coal mining, social injustice and health: A universal conflict of power and priorities. *Health & Place* 19:74–79.

Morton, J. 2020. Haile ramps up. Online. Engineering and Mining Journal. https://www.e-mj.com/features/haile-ramps-up (accessed September 26, 2022).

MSHA (Mine Safety and Health Administration). 2013. Metal and nonmetal general inspection procedures handbook. PH13-IV-1. U.S. Department of Labor, Mine Safety and Health Administration.

Multotec Group. 2019. Comminution. https://www.multotec.com/en/comminution (accessed July 12, 2022).

Naicker, K., E. Cukrowska, and T. McCarthy. 2003. Acid mine drainage arising from gold mining activity in Johannesburg, South Africa and environs. *Environmental Pollution* 122(1):29–40.

Natural Resources Canada. 2022. Gold facts. https://www.nrcan.gc.ca/our-natural-resources/minerals-mining/minerals-metals-facts/gold-facts/20514 (accessed June 16, 2022).

Navas-Acien, A., E. K. Silbergeld, R. Pastor-Barriuso, and E. Guallar. 2008. Arsenic exposure and prevalence of type 2 diabetes in US adults. *JAMA* 300(7):814–822.

NCEI (National Centers for Environmental Information). 2022. State climate summary: Virginia. https://statesummaries.ncics.org/chapter/va (accessed June 17, 2022).

NDEP (Nevada Division of Environmental Protection). 2016. Preparation requirements and guidelines for permanent closure plans and final closure reports. Bureau of Mining Regulation and Reclamation.

NDEP. 2022. Standardized Reclamation Cost Estimator (SRCE). https://ndep.nv.gov/land/mining/reclamation/reclamation-cost-estimator (accessed July 5, 2022).

Nevada Department of Conservation and Natural Resources. 2021. Pit lakes are tricky to manage: 4 tools we use to preserve water quality. http://dcnr.nv.gov/blogs/pit-lakes-are-tricky-to-manage-4-tools-we-use-to-preserve-water-quality (accessed July 5, 2022).

Nevada Division of Minerals. 2020. Major Mines of Nevada 2020: Mineral Industries in Nevada's Economy. Reno, NV: Nevada Bureau of Mines and Geology.

The Nile Machinery Co. 2022. Gold CIP production line gold mining machine gold processing plant. https://www.nile-mining.com/showroom/Gold-CIP-Production-Line-Gold-Mining-Machine-Gold-Processing-Plant.html (accessed June 17, 2022).

NIOSH (National Institute for Occupational Safety and Health). 2022. Pneumoconioses. https://www.cdc.gov/niosh/topics/pneumoconioses/default.html (accessed June 21, 2022).

Nishenko, S., and G. Bollinger. 1990. Forecasting damaging earthquakes in the central and eastern United States. *Science* 249(4975):1412–1416.

NWS (National Weather Service). 2016. The hurricane history of central and eastern Virginia. https://www.weather.gov/media/akq/miscNEWS/hurricanehistory.pdf (accessed June 17, 2022).

NWS. 2022. HDSC PMP documents. https://www.weather.gov/owp/hdsc_pmp (accessed June 17, 2022).

OceanaGold. 2022a. Haile Gold Mine. https://oceanagold.com/operation/haile (accessed June 20, 2022).

OceanaGold. 2022b. Modern mining, historic mine. https://www.hailegoldmine.com/mining (accessed June 21, 2022).

Olvera Alvarez, H. A., A. A. Appleton, C. H. Fuller, A. Belcourt, and L. D. Kubzansky. 2018. An integrated socio-environmental model of health and well-being: A conceptual framework exploring the joint contribution of environmental and social exposures to health and disease over the life span. *Current Environmental Health Reports* 5(2):233–243.

Owens, B. E., and B. J. Peters. 2018. Ferruginous quartzites in the Chopawamsic Terrane, Piedmont province, Virginia: Evidence for an ancient back-arc hydrothermal system. *Economic Geology* 113(2):421–438.

Owens, B. E., S. D. Samson, and S. E. King. 2013. Geochemistry of the Arvonia Formation, Chopawamsic Terrane, Virginia: Implications for source area weathering and provenance. *American Journal of Science* 313(3):242–266.

Pakalnis, R. T., and P. B. Hughes. 2011. Sublevel stoping. In SME mining engineering handbook. Vol. 1. Littleton, CO: Society for Mining, Metallurgy, and Exploration.

Paktunc, A. D. 1999. Characterization of mine wastes for prediction of acid mine drainage. In Environmental impacts of mining activities. Berlin: Springer. Pp. 19–40.

Pamunkey Indian Reservation 2021. Reservation. https://pamunkey.org/reservation (accessed August 26, 2022).

Pannier, B. 2020. Even two decades after massive cyanide spill, Kyrgyz poisoning victims get scant compensation. Radio Free Europe/Radio Liberty.

Pardee, J. T., and C. F. Park. 1948. Gold deposits of the southern Piedmont. Vol. 213. U.S. Government Printing Office.

Park, C. F. 1936. Preliminary report on gold deposits of the Virginia Piedmont. Vol. 44. Virginia Geological Survey.

Park, S. K., Z. Zhao, and B. Mukherjee. 2017. Construction of environmental risk score beyond standard linear models using machine learning methods: Application to metal mixtures, oxidative stress and cardiovascular disease in NHANES. *Environmental Health* 16(1):1–17.

Pascoe, C., R. Keim, and P. Haarala. 2022. Kensington gold operations, Alaska: Technical report summary. Coeur Mining, Inc.

Pavich, M. J., G. Leo, S. Obermeier, and J. R. Estabrook. 1989. Investigations of the characteristics, origin, and residence time of the upland residual mantle of the Piedmont of Fairfax County, Virginia. U.S. Geological Survey Proffesional Paper 1352.

Pavlides, L. 1981. The central Virginia volcanic-plutonic belt: An island arc of Cambrian (?) age. Reston, VA: U.S. Geological Survey.

Pavlides, L., J. E. Gair, and S. L. Cranford. 1982. Central Virginia volcanic-plutonic belt as a host for massive sulfide deposits. *Economic Geology* 77(2):233–272.

Peiyue, W. 2021. Gas poisoning kills 6 at illegal Shanxi Gold Mine. Sixth Tone. https://www.sixthtone.com/news/1009322/gas-poisoning-kills-6-at-illegal-shanxi-gold-mine (accessed September 27, 2022).

Perry, A., and R. L. Kleinmann. 1991. The use of constructed wetlands in the treatment of acid mine drainage. *Natural Resources Forum* 15(3):178–184.

Peter, A. J., and T. Viraraghavan. 2005. Thallium: A review of public health and environmental concerns. *Environment International* 31(4):493–501.

Peters, J. L., M. P. Fabian, and J. I. Levy. 2014. Combined impact of lead, cadmium, polychlorinated biphenyls and non-chemical risk factors on blood pressure in NHANES. *Environmental Research* 132:93–99.

Peters, S. C., and J. D. Blum. 2003. The source and transport of arsenic in a bedrock aquifer, New Hampshire, USA. *Applied Geochemistry* 18(11):1773–1787.

Petersen, M. D., A. M. Shumway, P. M. Powers, C. S. Mueller, M. P. Moschetti, A. D. Frankel, S. Rezaeian, D. E. McNamara, N. Luco, and O. S. Boyd. 2020. The 2018 update of the US National Seismic Hazard Model: Overview of model and implications. *Earthquake Spectra* 36(1):5–41.

Petkova, V., S. Lockie, J. Rolfe, and G. Ivanova. 2009. Mining developments and social impacts on communities: Bowen basin case studies. *Rural Society* 19(3):211–228.

PJSC (Public Joint Stock Company) Gaysky GOK. 2017. Loader. https://commons.wikimedia.org/wiki/File:Load_haul_dump_machine.jpg (accessed June 17, 2022).

Plumlee, G. S., K. Smith, M. Montour, W. Ficklin, and E. Mosier. 1999. Geologic controls on the composition of natural waters and mine waters draining diverse mineral-deposit types. *Reviews in Economic Geology: The Environmental Geochemistry of Mineral Deposits* 6A-6B:373–432.

Pollitz, F. F., and W. D. Mooney. 2015. Regional seismic-wave propagation from the M5.8 23 August 2011, Mineral, Virginia, earthquake. In The 2011 Mineral, Virginia, earthquake and its significance for seismic hazards in eastern North America, edited by J. Horton. Vol. 509. Geological Society of America Special Paper. Pp. 295–304.

Pommen, L. W. 1983. The effect of water quality of explosives use in surface mining, volume 1: Nitrogen sources, water quality, and prediction and management of impacts. MOE Technical Report 4. BC Ministry of the Environment.

Pond, G. J., M. E. Passmore, F. A. Borsuk, L. Reynolds, and C. J. Rose. 2008. Downstream effects of mountaintop coal mining: Comparing biological conditions using family- and genus-level macroinvertebrate bioassessment tools. *Journal of the North American Benthological Society* 27(3):717–737.

Porter, E. 1997. Examining state mining laws of Colorado and Montana. University of Montana.

Pourbaix, M. 1966. Atlas of electrochemical equilibria in aqueous solutions. London: Pergamon Press.

Prein, A. F., C. Liu, K. Ikeda, S. B. Trier, R. M. Rasmussen, G. J. Holland, and M. P. Clark. 2017. Increased rainfall volume from future convective storms in the US. *Nature Climate Change* 7(12):880–884.

Price, W. 2009. Prediction manual for drainage chemistry from suphidic geologic materials. Mine Environment Neutral Drainage (MEND) Report 1.20.1.

Pyle, G. G., R. D. Plomp, L. Zink, and J. L. Klemish. 2022. Invertebrate metal accumulation and toxicity from sediments affected by the Mount Polley mine disaster. *Environmental Science and Pollution Research* 1–16.

Que, S., K. Awuah-Offei, and V. Samaranayake. 2015. Classifying critical factors that influence community acceptance of mining projects for discrete choice experiments in the United States. *Journal of Cleaner Production* 87:489–500.

Ramani, R. V. 2012. Surface mining technology: Progress and prospects. *Procedia Engineering* 46:9–21.

Rehman, K., F. Fatima, I. Waheed, and M. S. H. Akash. 2018. Prevalence of exposure of heavy metals and their impact on health consequences. *Journal of Cellular Biochemistry* 119(1):157–184.

Rehman, M., L. Liu, Q. Wang, M. H. Saleem, S. Bashir, S. Ullah, and D. Peng. 2019. Copper environmental toxicology, recent advances, and future outlook: A review. *Environmental Science and Pollution Research* 26(18):18003–18016.

Reid, A. J., A. K. Carlson, I. F. Creed, E. J. Eliason, P. A. Gell, P. T. Johnson, K. A. Kidd, T. J. MacCormack, J. D. Olden, and S. J. Ormerod. 2019. Emerging threats and persistent conservation challenges for freshwater biodiversity. *Biological Reviews* 94(3):849–873.

Revey, G. 1996. Practical methods to control explosives losses and reduce ammonia and nitrate levels. *Mining Engineering* 61–64.

Rickwood, C., M. King, and P. Huntsman-Mapila. 2015. Assessing the fate and toxicity of thallium I and thallium III to three aquatic organisms. *Ecotoxicology and Environmental Safety* 115:300–308.

Rico, M., G. Benito, A. Salgueiro, A. Díez-Herrero, and H. Pereira. 2008. Reported tailings dam failures: A review of the European incidents in the worldwide context. *Journal of Hazardous Materials* 152(2):846–852.

Rimstidt, J. D., J. A. Chermak, and P. M. Gagen. 1994. Rates of reaction of galena, sphalerite, chalcopyrite, and arsenopyrite with Fe (III) in acidic solutions. ACS Publications.

Robertson, P. K., L. De Melo, D. Williams, and G. W. Wilson. 2019. Report of the expert panel on the technical causes of the failure of Feijão Dam I. http://www.b1technicalinvestigation.com/pt (accessed June 17, 2022).

Roble, S. 2022. Natural heritage resources of Virginia: Rare animals. Natural heritage rare species lists (2022-summer). Richmond, VA: Virginia Department of Conservation and Recreation, Division of Natural Heritage.

Rosseland, B. O., T. D. Eldhuset, and M. Staurnes. 1990. Environmental effects of aluminium. *Environmental Geochemistry and Health* 12(1):17–27.

Ruby, S. M., D. R. Idler, and Y. P. So. 1986. The effect of sublethal cyanide exposure on plasma viteliogenin levels in rainbow trout (Salmo gairdneri) during early vitellogenesis. *Archives of Environmental Contamination and Toxicology* 15(5):603–607.

Saiki, M. K., B. A. Martin, T. W. May, and C. N. Alpers. 2010. Mercury concentrations in fish from a Sierra Nevada foothill reservoir located downstream from historic gold-mining operations. *Environmental Monitoring and Assessment* 163(1):313–326.

Sanders, A. P., M. J. Mazzella, A. J. Malin, G. M. Hair, S. A. Busgang, J. M. Saland, and P. Curtin. 2019. Combined exposure to lead, cadmium, mercury, and arsenic and kidney health in adolescents age 12–19 in NHANES 2009–2014. *Environment International* 131:104993.

Sandhaus, D., and J. Craig. 1986. Gahnite in the metamorphosed stratiform massive sulfide deposits of the Mineral District, Virginia, USA. *Tschermaks mineralogische und petrographische Mitteilungen* 35(2):77–98.

Santore, R. C., D. M. Di Toro, P. R. Paquin, H. E. Allen, and J. S. Meyer. 2001. Biotic ligand model of the acute toxicity of metals. 2. Application to acute copper toxicity in freshwater fish and daphnia. *Environmental Toxicology and Chemistry: An International Journal* 20(10):2397–2402.

Satarug, S. 2018. Dietary cadmium intake and its effects on kidneys. *Toxics* 6(1):15.

Savage, K., R. Ashley, and D. Bird. 2009. Geology of the Jamestown mine, Mother Lode gold district, CA, and geochemistry of the Harvard mine pit lake. https://www.waterboards.ca.gov/academy/courses/ard/day2/day2_sec6d_harvardmine_ra.pdf (accessed July 5, 2022).

SCDHEC (South Carolina Department of Health and Environmental Control). 2022. Groundwater resources. https://scdhec.gov/BOW/groundwater-use-reporting/groundwater-resources (accessed July 18, 2022).

Schellenbach, W. L., and M. P. Krekeler. 2012. Mineralogical and geochemical investigations of pyrite-rich mine waste from a kyanite mine in central Virginia with comments on recycling. *Environmental Earth Sciences* 66(5):1295–1307.

Schoemaker, K. 2022. Doré pour. https://www.lbma.org.uk/wonders-of-gold/items/dor%C3%A9-pour (accessed June 17, 2022).

Schoenberger, E. 2016. Environmentally sustainable mining: The case of tailings storage facilities. *Resources Policy* 49:119–128.

Schwartz, M. O. 2000. Cadmium in zinc deposits: Economic geology of a polluting element. *International Geology Review* 42(5):445–469.

SCOTUS Blog. 2022. Sackett v. Environmental Protection Agency. https://www.scotusblog.com/case-files/cases/sackett-v-environmental-protection-agency (accessed July 5, 2022).

Seal, R. R., D. P. Haffner, and A. L. Meier. 1998. Environmental characteristics of the abandoned Greenwood Mine area, Prince William Forest Park, Virginia: Implications for mercury geochemistry, Open-File Report 98-326. Reston, VA: U.S. Geological Survey.

Seal, R. R., II, A. N. Johnson, J. M. Hammarstrom, and A. L. Meier. 2002. Geochemical characterization of drainage prior to reclamation at the abandoned Valzinco Mine, Spotsylvania County, Virginia. Report 02-360. Reston, VA: U.S. Geological Survey.

Seal, R. R. I., and J. M. Hammarstrom. 2002. Field guide to the environmental geochemistry of abandoned mines in the Virginia gold-pyrite belt. Geological Society of Washington.

Secretary of the Commonwealth. 2022. What is state recognition? https://www.commonwealth.virginia.gov/virginia-indians/state-recognized-tribes (accessed August 6, 2022).

Sexton, K. 2012. Cumulative risk assessment: An overview of methodological approaches for evaluating combined health effects from exposure to multiple environmental stressors. *International Journal of Environmental Research and Public Health* 9(2):370–390.

Shealy Environmental Services Inc. 1991. A macroinvertebrate assessment conducted for Brewer Gold Mine Company, Chesterfield County, South Carolina. West Columbia, SC.

Sillitoe, R. H. 2020. Gold deposit types: An overview. In Geology of the world's major gold deposits and provinces. edited by R. H. Sillitoe, R. J. Goldfarb, F. Robert, and S. F. Simmons. Vol. 23. Society of Economic Geologists. Pp. 1–28.

Sillman, S., J. A. Logan, and S. C. Wofsy. 1990. The sensitivity of ozone to nitrogen oxides and hydrocarbons in regional ozone episodes. *Journal of Geophysical Research: Atmospheres* 95(D2):1837–1851.

Simovic, L., and W. Snodgrass. 1985. Natural removal of cyanide in gold milling effluents-evaluation of removal kinetics. *Water Quality Research Journal* 20(2):120–135.

Sims, J. 2018. Fort Knox mine; Fairbanks North Star Borough, Alaska, USA. National Instrument 43-101 Technical Report.

Sincovich, A., T. Gregory, A. Wilson, and S. Brinkman. 2018. The social impacts of mining on local communities in Australia. *Rural Society* 27(1):18–34.

Singh, R., N. Gautam, A. Mishra, and R. Gupta. 2011. Heavy metals and living systems: An overview. *Indian Journal of Pharmacology* 43(3):246.

SMGB (State Mining and Geology Board). 2007. Report on backfilling of open-pit metallic mines in California. SMGB.

Solomonson, L. 1981. Cyanide as a metabolic inhibitor. In Cyanide in biology, edited by B. Vennesland, E. Conn, C. Knowles, J. Westley, and F. Wissing. New York: Academic Press. P. 11.

Sparling, D. W., and T. P. Lowe. 1996. Environmental hazards of aluminum to plants, invertebrates, fish, and wildlife. *Reviews of Environmental Contamination and Toxicology* 1–127.

Spears, D. B., and M. L. Upchurch. 1997. Metallic mines, prospects, and occurences in the gold-pyrite belt of Virginia. Vol. 147. Charlottesville, VA: Virginia Division of Mineral Resources.

SRK Consulting. 2020. NI 43-101 technical report, Haile Gold Mine, Lancaster County, South Carolina. Denver, CO.

The State. 2021. Big mine fined again for toxic discharges near small SC town. https://www.thestate.com/news/local/environment/article255642286.html (accessed September 27, 2022).

Staunton, W. 2016. Carbon-in-pulp. In Gold ore processing, edited by M. D. Adams. Amsterdam: Elsevier. Pp. 535–552.

Stephan, G. 2011. Cut-and-fill mining. In SME mining engineering handbook. Vol. 1. Littleton, CO: Society for Mining, Metallurgy, and Exploration.

Stephenson, H., and D. Castendyk. 2019. The reclamation of Canmore Creek—an example of a successful walk away pit lake closure. *Journal of Petroleum Technology*. https://jpt.spe.org/reclamation-canmore-creek-example-successful-walk-away-pit-lake-closure (accessed September 27, 2022).

Stillings, L. L. 2017. Selenium. In Critical mineral resources of the United States—economic and environmental geology and prospects for future supply. Vol. 1802. U.S. Geological Survey. Pp. Q1–Q55.

Stumm, W., and J. J. Morgan. 1996. Aquatic chemistry: chemical equilibria and rates in natural waters, 3rd ed. New York: John Wiley & Sons.

Sweet, P. C. 1971. Gold mines and prospects in Virginia. *Virginia Minerals* 17(3):25–36.

Sweet, P. C. 1980. Gold in Virginia. Vol. 19. Charlottesville, VA: Virginia Division of Mineral Resources.

Sweet, P. C. 1995. Update on unreported occurrences of gold-silver in Virginia. *Virginia minerals* 41(2):9–16.

Sweet, P. C. 2007. Gold: Virginia division of geology and mineral resources fact sheet. Charlottesville, VA: Virginia Department of Energy.

Sweet, P. C., and J. A. Lovett. 1985. Additional gold mines, prospects, and occurrences in Virginia. *Virginia Minerals* 31(4):41–52.

Sweet, P. C., and D. Trimble. 1982. Gold occurrences in Virginia, an update. *Virginia minerals* 28(4):33–41.

Sweet, P. C., W. L. Lassetter, and W. C. Sherwood. 2016. Non-fuel mineral resources in Virginia. In Geology of Virginia, edited by C. M. Bailey, W. C. Sherwood, L. S. Eaton, and D. S. Powars. Vol. 18. Martinsville, VA: Virginia Museum of Natural History. Pp. 407–441.

Taber, S. 1913. Geology of the gold belt in the James River basin, Virginia, University of Virginia.

Tarr, A. C., and R. L. Wheeler. 2006. Earthquakes in Virginia and vicinity, 1774–2004. Open-File Report 2006-1017. Reston, VA: U.S. Geological Survey.

Teague, A., C. Swaminathan, and J. Van Deventer. 1998. The behaviour of gold bearing minerals during froth flotation as determined by diagnostic leaching. *Minerals Engineering* 11(6):523–533.

Terzaghi, K., R. B. Peck, and G. Mesri. 1996. Soil mechanics in engineering practice, 3rd ed. New York: John Wiley and Sons.

Thomas, K., and M. Pearson. 2016. Pressure oxidation overview. In Gold ore processing, edited by M. D. Adams. Amsterdam: Elsevier. Pp. 341–358.

Tirima, S., C. Bartrem, I. von Lindern, M. von Braun, D. Lind, S. M. Anka, and A. Abdullahi. 2016. Environmental remediation to address childhood lead poisoning epidemic due to artisanal gold mining in Zamfara, Nigeria. *Environmental Health Perspectives* 124(9):1471–1478.

Tourigny, G., A. C. Brown, C. Hubert, and R. Crepeau. 1989. Synvolcanic and syntectonic gold mineralization at the Bousquet Mine, Abitibi greenstone belt, Quebec. *Economic Geology* 84(7):1875–1890.

Tourigny, G., D. Doucet, and A. Bourget. 1993. Geology of the Bousquet 2 mine; an example of a deformed, gold-bearing, polymetallic sulfide deposit. *Economic Geology* 88(6):1578–1597.

Trapp, H., Jr., and M. A. Horn. 1997. Ground water atlas of the United States: Segment 11, Delaware, Maryland, New Jersey, North Carolina, Pennsylvania, Virginia, West Virginia. Reston, VA: U.S. Geological Survey.

Troutman Sanders LLP. 2008. Virginia environmental law handbook, 4th ed. Lanham, Maryland: Government Institutes.

Tuttle, M., M. Carter, and J. Dunahue. 2015. Paleoliquefaction study of the earthquake potential of the central Virginia seismic zone. Paper read at Geological Society of America.

Tuttle, M. P., K. Dyer-Williams, M. W. Carter, S. L. Forman, K. Tucker, Z. Fuentes, C. Velez, and L. M. Bauer. 2021. The liquefaction record of past earthquakes in the central Virginia seismic zone, eastern United States. *Seismological Research Letters* 92(5):3126–3144.

Tutu, H., T. McCarthy, and E. Cukrowska. 2008. The chemical characteristics of acid mine drainage with particular reference to sources, distribution and remediation: The Witwatersrand basin, South Africa as a case study. *Applied Geochemistry* 23(12):3666–3684.

University of Wisconsin Population Health Institute. 2022. County Health Rankings and Roadmaps. https://www.county healthrankings.org (accessed June 21, 2022).

U.S. Census Bureau. 2020. Census redistricting data (Public Law 94-171) summary file.

U.S. Department of State. 2011. Announcement of U.S. Support for the United Nations declaration on the rights of Indigenous peoples: Initiatives to promote the government-to-government relationship & improve the lives of Indigenous peoples.

USACE (U.S. Army Corps of Engineers). 1993. Standard practice for shotcrete. EM 1110-2-2005.

USACE. 2022. Haile Gold Mine supplemental environmental impact statement.

USACHPPM (U.S. Army Center for Health Promotion and Preventive Medicine). 2007. Wildlife toxicity assessment for thallium. 37-EJ1138-01O. USACHPPM.

USFS (U.S. Forest Service). 1995. Anatomy of a mine from prospect to production. General Technical Report INT-GTR-35 Revised. Ogden, UT: U.S. Department of Agriculture.

USFS. 2004. Training guide for reclamation bond estimation and administration.

USGS (U.S. Geological Survey). 1995. Environmental considerations of active and abandoned mine lands: Lessons from Summitville, Colorado. Vol. 2220. Reston, VA: U.S. Geological Survey.

USGS. 2006. Recycled cell phones—a treasure trove of valuable metals.

USGS. 2007a. Divisions of geologic time—major chronostratigraphic and geochronologic units.

USGS. 2007b. Royal Mountain King Mine. https://mrdata.usgs.gov/mrds/show-mrds.php?dep_id=10310580 (accessed July 12, 2022).

USGS. 2017a. Antimony. In Critical mineral resources of the United States—economic and environmental geology and prospects for future supply. Vol. 1802-C. U.S. Geological Survey, U.S. Department of the Interior.

USGS. 2017b. Drilling deep for science and society. https://www.usgs.gov/news/featured-story/benefits-understanding-earth-its-core (accessed June 17, 2022).

USGS. 2017c. Kensington (Coeur-Alaska). https://mrdata.usgs.gov/ardf/show-ardf.php?ardf_num=JU261 (accessed July 1, 2022).

USGS. 2019. Arsenic and drinking water. https://www.usgs.gov/mission-areas/water-resources/science/arsenic-and-drinking-water (accessed June 20, 2022).

USGS. 2022a. Earthquake hazards program. https://earthquakes.usgs.gov (accessed June 17, 2022).

USGS. 2022b. Willis Mountain Mine. https://mrdata.usgs.gov/mrds/show-mrds.php?dep_id=10069811 (accessed July 12, 2022).

USGS. 2022c. Final list of critical minerals. 87 *Federal Register* 10381:23295–23296.

USGS. 2022d. Mineral commodity summaries: Gold. https://pubs.usgs.gov/periodicals/mcs2022/mcs2022-gold.pdf (accessed September 27, 2022)

UVA (University of Virginia). 2022. Indigenous/UVA relating. https://eocr.virginia.edu/monacan (accessed August 6, 2022).

Vandenberg, J., M. Schultze, C. D. McCullough, and D. Castendyk. 2022. The future direction of pit lakes: Part 2, corporate and regulatory closure needs to improve management. *Mine Water and the Environment* 41:544–556.

VanDerwerker, T., L. Zhang, E. Ling, B. Benham, and M. Schreiber. 2018. Evaluating geologic sources of arsenic in well water in Virginia (USA). *International Journal of Environmental Research and Public Health* 15(4):787.

Vargas, C., and G. P. Campomanes. 2022. Practical Experience of Filtered Tailings Technology in Chile and Peru: An Environmentally Friendly Solution. *Minerals* 12(7):889.

VDH (Virginia Department of Health). 2022a. Fish consumption advisory. https://www.vdh.virginia.gov/environmental-health/public-health-toxicology/fish-consumption-advisory (accessed July 8, 2022).

VDH. 2022b. Private well program. https://www.vdh.virginia.gov/environmental-health/onsite-sewage-water-services-updated/private-well-program (accessed June 17, 2022).

VDOT (Virginia Department of Transportation). 2020. Considerations of climate change and coastal storms—Chapter 33. https://www.virginiadot.org/business/resources/bridge/Manuals/Part2/Chapter33.pdf (accessed September 27, 2022).

VDOT. 2022. Resource identification and impact analysis methodologies: Interstate 495 southside express lanes study.

Veiga, M. M., and J. A. Meech. 1999. Reduction of mercury emissions from gold mining activities and remedial procedures for polluted sites. In Environmental impacts of mining activities. Berlin: Springer. Pp. 143–162.

Vergilio, C. d. S., D. Lacerda, B. C. V. d. Oliveira, E. Sartori, G. M. Campos, A. L. d. S. Pereira, D. B. d. Aguiar, T. d. S. Souza, M. G. d. Almeida, and F. Thompson. 2020. Metal concentrations and biological effects from one of the largest mining disasters in the world (Brumadinho, Minas Gerais, Brazil). *Scientific Reports* 10(1):1–12.

Verplanck, P. L., S. H. Mueller, R. J. Goldfarb, D. K. Nordstrom, and E. K. Youcha. 2008. Geochemical controls of elevated arsenic concentrations in groundwater, Ester Dome, Fairbanks District, Alaska. *Chemical Geology* 255(1–2):160–172.

Virginia DCR (Department of Conservation and Recreation). 2022a. Probable maximum precipitation study and evaluation tool. https://www.dcr.virginia.gov/dam-safety-and-floodplains/pmp-tool (accessed July 5, 2022).

Virgina DCR. 2022b. Virginia's protected lands. https://www.dcr.virginia.gov/land-conservation/protected-lands (accessed July 6, 2022).

Virginia DEQ (Department of Environmental Quality). 2000. Guidance memo No. 00-2011; guidance on preparing VPDES permit limits.

Virginia DEQ. 2019. Virginia Water Protection (VWP) program overview. https://www.deq.virginia.gov/home/showpublisheddocument/9440/637588239140500000 (accessed June 22, 2022).

Virginia DEQ. 2022a. Discharge to surface waters—Virginia pollutant discharge elimination system. https://www.deq.virginia.gov/permits-regulations/permits/water/surface-water-virginia-pollutant-discharge-elimination-system (accessed June 22, 2022).

Virginia DEQ. 2022b. Phase III WIP. https://www.deq.virginia.gov/water/chesapeake-bay/phase-iii-wip (accessed June 22, 2022).

Virginia DEQ. 2022c. Water quality assessments.

Virginia DMME (Department of Mines, Minerals, and Energy). 2007. State minerals management plan. https://townhall.virginia.gov/l/GetFile.cfm?File=C:%5CTownHall%5Cdocroot%5CGuidanceDocs%5C409%5CGDoc_DMME_1540_v2.pdf (accessed September 27, 2022).

Virginia DMME. 2011. Mineral mine operator's manual. https://www.energy.virginia.gov/mineral-mining/documents/PERMITTING/OperatorsManual.pdf (accessed September 27, 2022).

Virginia DWR (Department of Wildlife Resources). 2022. How fishing benefits Virginians. https://dwr.virginia.gov/fishing/how-fishing-benefits-virginians (accessed June 16, 2022).

Virginia Energy. 2015a. Procedure No. 2.6.00. Inspection frequency: Reclamation. In Enforcement policy and procedures manual.

Virginia Energy. 2015b. Procedure No. 2.12.00. Safety/Health Reclamation Complaints. In Enforcement policy and procedures manual.

Virginia Energy. 2015c. Procedure No. 2.8.00. Special orders/notices of violations. In Enforcement policy and procedures manual.

Virginia Energy. 2021. Gold. https://energy.virginia.gov/geology/gold.shtml (accessed June 16, 2022).

Virginia Energy. 2022a. About us. https://energy.virginia.gov/mineral-mining/aboutus.shtml (accessed June 17, 2022).

Virginia Energy. 2022b. Earthquakes. https://energy.virginia.gov/geology/earthquakes.shtml (accessed June 16, 2022).

Virginia Energy. 2022c. Inspectors report. Violation number V72882.

Virginia Energy. 2022d. Mineral mining. https://energy.virginia.gov/mineral-mining/mineralmining.shtml (accessed June 16, 2022).

Virginia Energy. 2022e. Mineral mining. Orphaned Land Sites: MRV183C_101,103,104. https://energy.virginia.gov/webmaps/MineralMining (accessed September 27, 2022).

Virginia Energy. 2022f. Public hearing informational brochure. https://energy.virginia.gov/mineral-mining/documents/BROCHURES/HEARINGS%20BROCHURE.pdf (accessed July 5, 2022).

Virginia Natural and Historic Resources. 2022. Virginia council on environmental justice. https://www.naturalresources.virginia.gov/initiatives/environmental-justice (accessed July 5, 2022).

Virginia Places. 2022. Native American reservations in Virginia. http://www.virginiaplaces.org/nativeamerican/reservation.html (accessed August 26, 2022).

Virginia Water Resources Research Center. 2022. Virginia stormwater BMP clearinghouse. https://swbmp.vwrrc.vt.edu (accessed June 17, 2022).

Virginia.gov. 2022. Virginia regulatory town hall. https://townhall.virginia.gov (accessed June 22, 2022).

Vogelsong, S. 2021. After opposition to gold mining in Buckingham, general assembly weighs temporary ban and study. Virginia Mercury. https://www.virginiamercury.com/2021/02/12/after-opposition-to-gold-mining-in-buckingham-general-assembly-weighs-temporary-ban-and-study (accessed September 27, 2022).

VTSO (Virginia Tech Seismological Observatory). 2021. Virginia earthquakes. http://www.magma.geos.vt.edu/vtso/va_quakes.html (accessed September 19, 2022).

Walton, R. 2016. Zinc cementation. In Gold ore processing, edited by M. D. Adams. Amsterdam: Elsevier.

Wang, H., F. Chen, C. Zhang, M. Wang, and J. Kan. 2021. Estuarine gradients dictate spatiotemporal variations of microbiome networks in the Chesapeake Bay. *Environmental Microbiome* 16(1):1–18.

Wang, J., X. Feng, C. W. Anderson, Y. Xing, and L. Shang. 2012. Remediation of mercury contaminated sites—a review. *Journal of Hazardous Materials* 221:1–18.

Wang, X., B. Mukherjee, and S. K. Park. 2018. Associations of cumulative exposure to heavy metal mixtures with obesity and its comorbidities among us adults in NHANES 2003–2014. *Environment International* 121:683–694.

Wang, X., C. A. Karvonen-Gutierrez, W. H. Herman, B. Mukherjee, S. D. Harlow, and S. K. Park. 2020a. Urinary metals and incident diabetes in midlife women: Study of women's health across the nation (SWAN). *BMJ Open Diabetes Research and Care* 8(1):e001233.

Wang, X., B. Mukherjee, C. A. Karvonen-Gutierrez, W. H. Herman, S. Batterman, S. D. Harlow, and S. K. Park. 2020b. Urinary metal mixtures and longitudinal changes in glucose homeostasis: The study of women's health across the nation (SWAN). *Environment International* 145:106109.

Wasana, H., D. Aluthpatabendi, W. Kularatne, P. Wijekoon, R. Weerasooriya, and J. Bandara. 2016. Drinking water quality and chronic kidney disease of unknown etiology (CKDU): Synergic effects of fluoride, cadmium and hardness of water. *Environmental Geochemistry and Health* 38(1):157–168.

Wheatley, B. 2020. Mining the sun: How Nevada and West Virginia are reclaiming former mine lands with solar panels. The Nature Conservancy.

White, W., III, and T. Jeffers. 1994. Chemical predictive modeling of acid mine drainage from metallic sulfide-bearing waste rock. In Environmental geochemistry of sulfide oxidation, edited by C. N. Alpers and D. W. Blowes. ACS Publications. Pp. 608–630.

WHO (World Health Organization). 2005. Mercury in drinking-water. WHO.

Williams, T. J., T. M. Brady, D. C. Bayer, M. J. Bren, R. T. Pakalnis, J. A. Marjerison, and R. B. Langston. 2007. Underhand cut and fill mining as practiced in three deep hard rock mines in the United States. Paper read at Proceedings of the CIM Conference and Exhibition, April 29–May 2, 2007, Montreal, Quebec.

The World Bank. 2015. World Bank board committee authorizes release of revised draft environmental and social framework. https://www.worldbank.org/en/news/press-release/2015/08/04/world-bank-board-committee-authorizes-release-of-revised-draft-environmental-and-social-framework (accessed July 5, 2022).

World Gold Council. 2022. How gold is mined: The lifecycle of a gold mine. https://www.gold.org/gold-supply/gold-mining-lifecycle (accessed September 27, 2022).

Wu, S., B. N. Duncan, D. J. Jacob, A. M. Fiore, and O. Wild. 2009. Chemical nonlinearities in relating intercontinental ozone pollution to anthropogenic emissions. *Geophysical Research Letters* 36(5).

Wurtsbaugh, W. A., and A. Horne. 1983. Iron in eutrophic clear lake, California: Its importance for algal nitrogen fixation and growth. *Canadian Journal of Fisheries and Aquatic Sciences* 40(9):1419–1429.

Xiao, L.-J., L.-M. Lei, L. Peng, Q.-Q. Lin, and L. Naselli-Flores. 2021a. Iron operates as an important factor promoting year-round diazotrophic cyanobacteria blooms in eutrophic reservoirs in the tropics. *Ecological Indicators* 125:107446.

Xiao, L., G. Zan, J. Qin, X. Wei, G. Lu, X. Li, H. Zhang, Y. Zou, L. Yang, and M. He. 2021b. Combined exposure to multiple metals and cognitive function in older adults. *Ecotoxicology and Environmental Safety* 222:112465.

Yasuno, M., S. Fukushima, F. Shioyama, J. Hasegawa, and S. Kasuga. 1981. Recovery processes of benthic flora and fauna in a stream after discharge of slag containing cyanide. *Internationale Vereinigung für theoretische und angewandte Limnologie: Verhandlungen* 21(2):1154–1164.

Yeates, W. S., S. W. McCallie, and F. P. King. 1896. A preliminary report on a part of the gold deposits of Georgia. Geological Survey of Georgia.

Young, R. S. 1956. Sulfides in Virginia. *Virginia minerals* 2(1):1–7.

Zhou, S., C. Di Paolo, X. Wu, Y. Shao, T.-B. Seiler, and H. Hollert. 2019. Optimization of screening-level risk assessment and priority selection of emerging pollutants–the case of pharmaceuticals in European surface waters. *Environment International* 128:1–10.

Zhuang, W., and J. Song. 2021. Thallium in aquatic environments and the factors controlling Tl behavior. *Environmental Science and Pollution Research* 28(27):35472–35487.

Glossary

Agglomerate: reform into lumps of suitable size (Encyclopedia Britannica, 2022b)

Base metal: a metal (e.g., zinc, lead) of comparatively low value compared to precious metals (e.g., gold, silver) (Merriam-Webster, 2022)

Breccia: sharp-angled fragmented rock (Merriam-Webster, 2022)

Cambrian: a geologic period that began approximately 542 million years ago and ended 488 million years ago (USGS, 2007a)

Commercial gold mining: higher-tech gold mining that occurs on a larger scale than small-scale gold mining

Comminution: the crushing and grinding of a material/ore to reduce it to smaller or finer particles (Multotec Group, 2019)

Criteria air pollutants: particulate matter, photochemical oxidants (including ozone), carbon monoxide, sulfur oxides, nitrogen oxides, and lead (EPA, 2022c)

Dip: the angle of inclination measured from horizontal of a planar geologic feature (Encyclopedia Britannica, 2022b)

Doré: recovered metal containing gold and silver (Encyclopedia Britannica, 2022b)

En echelon: parallel or subparallel, closely spaced, steplike minor structural features in rock (Davis et al., 2011)

Epithermal: deposited under conditions in the lower ranges of temperature and pressure (Merriam-Webster, 2022)

Fugitive dust: small particles emitted to the air from open air sources or opening that are not a stack, chimney, or vent. Fugitive dust may include particulate material of a range of sizes, including PM_{10} (generally 10 microns in diameter or smaller) and $PM_{2.5}$ (generally 2.5 microns in diameter or smaller) (9VAC5-50-70)

Gabions: rectangular baskets filled with stone or dirt for support (Merriam-Webster, 2022)

Gangue: the economically worthless material that surrounds, or is closely mixed with, a desirable mineral in an ore deposit (Mindat, 2022a)

Geosynthetic: a planar product made from polymeric material used with soil, rock, or earth for construction (Idaho Transportation Department, 2011)

Geotextiles: a permeable geosynthetic comprised solely of textiles (Idaho Transportation Department, 2011)

Greenschist-facies: low to medium metamorphism corresponding to temperatures of about 300°C to 500°C and pressures of 3 to 20 kbar, which is typical of continental collision tectonics (Arndt, 2011)

Hydrometallurgy: the treatment of ores by wet processes such as leaching (Merriam-Webster, 2022)

Igneous rock: rock formed by magma or lava (Merriam-Webster, 2022)

Indicated mineral resource: that part of a mineral resource for which quantity and grade or quality are estimated on the basis of adequate geological evidence and sampling (17 CFR § 229.1300)

Intrusive: igneous rock formed within the crust

Island arc: a curved chain of volcanic islands that are found along tectonic plate margins (Merriam-Webster, 2022)

Lixiviant: a liquid medium used in hydrometallurgy to selectively extract the desired metal from the ore or mineral (AIME, 1917)

Lode: an ore body found within rock (Encyclopedia Britannica, 2022b)

Mafic: a rock rich in magnesium and iron and relatively depleted in silica (Merriam-Webster, 2022)

Measured mineral resource: that part of a mineral resource for which quantity and grade or quality are estimated on the basis of conclusive geological evidence and sampling (17 CFR § 229.1300)

Mesozoic: a geologic era that began 251 million years ago and ended 66 million years ago (USGS, 2007a)

Metaclastics: metamorphosed clastic sedimentary rocks, which is composed of detrital rock and mineral fragments (Mindat, 2022b)

Metalloid: an element with properties that are intermediate between those of metals and nonmetals (Merriam-Webster, 2022)

Metamorphic rock: formed under extreme heat and temperature (Merriam-Webster, 2022)

Metasedimentary rocks: sedimentary rocks that have been metamorphosed

Mineral: a crystalline inorganic substance (Merriam-Webster, 2022)

Ordovician: a geologic period started 488 million years ago and ended 444 million years ago (USGS, 2007a)

Ore: a naturally occurring accumulation of one or more valuable mineral resources (Merriam-Webster, 2022)

Orogenic gold deposits: Deposits that are formed from the remobilization of gold scavenged during metamorphism and redeposited elsewhere (Sillitoe, 2020)

Paleozoic: a geologic era that began 542 billion years ago and ended 251 million years ago (USGS, 2007a)

Placer: deposit that has been moved following river, marine, or glacial action (Merriam-Webster, 2022)

Plutonic: igneous rock formed deep underground

Precambrian: a geologic period that began approximately 4,600 million years ago and ended 542 million years ago (USGS, 2007a)

Processing: in this report, processing indicates both the physical processing sometimes called "beneficiation" along with chemical processing that is principally hydrometallurgical for gold, although the last steps are pyrometallurgical

Proterozoic: a geologic eon that began approximately 2.5 billion years ago and ended 542 million years ago (USGS, 2007a)

Pulp: is a freely flowing mixture of powdered ore and water (911 Metallurgist, 2016)

Pyrometallurgy: chemical metallurgy depending on heat action (such as roasting and smelting) (Merriam-Webster, 2022)

Recreational gold mining: mining, often by a few individuals, primarily for recreation. This is often limited to panning for alluvial gold in streams

Sedimentary rock: formed from sediment deposited by water or air (Merriam-Webster, 2022)

Shotcrete: pneumatically applied concrete consisting of cement, aggregates, water, and additives such as accelerators, silica fume, and steel fibers (USACE, 1993)

Slag: the material that is left when rocks that contain metal are heated to get the metal out (Encyclopedia Britannica, 2022b)

Small-scale mining: low-tech, labor-intensive mineral extraction and processing carried out mostly by local people (Hilson and Maconachie, 2020)

Stopes: excavation areas underground that are formed as the ore is mined in successive layers (Merriam-Webster, 2022)

Strike: the orientation of an imaginary horizontal line across the plan of a geologic feature (Encyclopedia Britannica, 2022b)

Surfactant: a surface-active substance that lowers the surface tension (or interfacial tension) (Merriam-Webster, 2022)

Tailings: the remaining waste following ore processing (Merriam-Webster, 2022)

Terranes: coherent units of Earth's crust that have a distinct geologic history and that are bounded by faults (Encyclopedia Britannica, 2022b)

Triassic: a geologic period that began approximately 251 million years ago and ended 200 million years ago (USGS, 2007a)

Volcanic: igneous rock formed near the surface (Merriam-Webster, 2022)

Volcanogenic massive sulfide: accumulations of sulfide minerals that form due to hydrothermal action on the seafloor; ancient varieties may now be exposed on land (Jamieson et al., 2016)

Waste rock: bedrock that has been mined and transported out of the pit but does not have gold concentrations of economic interest

Appendix A

Committee Member and Staff Biographical Sketches

William A. Hopkins, *Chair*, is a professor in the Department of Fish and Wildlife Conservation at Virginia Tech. He is also the Associate Executive Director of the Fralin Life Sciences Institute, the founding Director of the Global Change Center at Virginia Tech, and the founding Director of one of the graduate school's largest interdisciplinary Ph.D. programs. Prior to joining the faculty at Virginia Tech, Hopkins was faculty at the University of Georgia's Savannah River Ecology Laboratory. Hopkins's research focuses on how anthropogenic disturbances such as climate change, pollution, and habitat loss affect wildlife. He has considerable experience evaluating how activities such as fossil fuel extraction, combustion, and accidental spills affect the environment. He regularly provides guidance to state and federal agencies, industry, and other stakeholders on issues related to environmental degradation and threats to biodiversity. He is an award-winning educator, researcher, and leader, including the highest awards offered to faculty at both Virginia Tech and in the Commonwealth of Virginia. He received a B.S. in biology from Mercer University, an M.S. in zoology from Auburn University, and a Ph.D. from the University of South Carolina. He has previously served on four National Academies committees, one of which he chaired, addressing issues related to freshwater resources, mining, management of wastes from fossil fuel combustion, and research data quality in federal agencies.

Kwame Awuah-Offei is currently the Union Pacific/Rocky Mountain Energy Professor in Mining Engineering and the Chair of the Department of Mining & Explosives Engineering at Missouri University of Science and Technology. He has served as a mining engineering academic fellow for the U.S. Securities & Exchange Commission and an alternate member of the U.S. Department of the Interior's Royalty Policy Committee. His research revolves around improving our understanding of the effects of mining on the environment and society in order to develop sustainable mining practices. He is a Fellow of the West African Institute of Mining, Metallurgy, and Petroleum; a two-time Henry Krumb Lecturer of the Society of Mining, Metallurgy and Exploration; and a past Carnegie African Diaspora Fellow. He holds a Ph.D. and a B.S. in mining engineering from the Missouri University of Science & Technology and the University of Mines & Technology, respectively. In the past 5 years, Awuah-Offei has consulted for Rio Tinto and received research funding from Komatsu Mining Corp.

Joel D. Blum (NAS) is the John D. MacArthur, Arthur F. Thurnau and Gerald J. Keeler Distinguished Professor of Earth and Environmental Sciences at the University of Michigan. His research expertise is in the sources, transport, and fate of toxic trace metals in the environment. He has experience in gold deposit exploration and the environmental consequences of gold and mercury mining and smelting of other metals. He was the recipient of the Patterson Medal given by the Geochemical Society for excellence in environmental geochemistry. He is a Fellow of the Geological Society of America, the American Geophysical Union, the American Association for the Advancement of Science, and the Geochemical Society, and is a member of the National Academy of Sciences. Blum has a B.A. in political science and geological science from Case Western Reserve University, an M.Sc. in geological science from the University of Alaska Fairbanks, and a Ph.D. in geochemistry from the California Institute of Technology.

Robert J. Bodnar is the C.C. Garvin Professor and University Distinguished Professor in the Department of Geosciences at Virginia Tech in Blacksburg, Virginia. Bodnar's research focuses on the role of geofluids in various geologic processes, including the formation of mineral deposits and extraction of energy and mineral resources. Bodnar's group has worked on gold, silver, copper, and other metal deposit types in various geological environments around the world, and recently led a multidisciplinary and multiyear study examining uranium deposits and environmental impacts of mining in Virginia's Piedmont region. Bodnar has been awarded the Society of Economic Geologists Lindgren Award and Silver Medal, the American Geophysical Union's N.L. Bowen Award, and the Thomas Jefferson Medal from the Virginia Museum of Natural History, and he was named Virginia's Outstanding Scientist in 2010. Bodnar has been an elected Fellow of the Geological Society of America, the Geochemical Society, the Society of Economic Geologists, the American Association for the Advancement of Science, the Mineralogical Society of America, the American Geophysical Union, the Geological Society of London, and was elected an honorary member of the Geological Society of India and the Italian Mineralogical Association. Bodnar earned a B.S. in chemistry from the University of Pittsburgh, an M.S. in geology from the University of Arizona, and a Ph.D. in geochemistry and mineralogy from The Pennsylvania State University.

Thomas Crafford recently retired from the U.S. Geological Survey, where he served as the Mineral Resources Program Coordinator. He previously worked for the State of Alaska as the Associate Director of State–Federal Relations in the Governor's Office and as the Director of the Office of Project Management and Permitting, the Mining Coordinator, and the Chief of the Mining Section in the Alaska Department of Natural Resources. Prior to his employment with the State of Alaska he worked as an independent minerals industry consultant; the Manager of Minerals and Coal for Cook Inlet Region, Inc. (CIRI), an Alaska Native corporation; as a mine geologist at the Greens Creek Mine; and an exploration geologist for multiple employers. He previously served as a Trustee of the Northwest Mining Association and the President of the Alaska Mining Association and remains a member of those associations. He is also a member of the Geological Society of America and the Society of Exploration Geologists. Crafford holds an M.S. in geology from Dartmouth College and a B.S. in geology from the University of Washington.

Fiona M. Doyle (NAE) is the Donald H. McLaughlin Professor Emerita in Mineral Engineering at the University of California, Berkeley (UC Berkeley). Previously, she has served as the Chair of the Department of Materials Science and Engineering and as the Executive Associate Dean of the College of Engineering at Berkeley. Doyle's research expertise includes the application of chemical thermodynamics, chemical and electrochemical kinetics, transport phenomena, and colloid and interfacial science to develop a fundamental mechanistic understanding of minerals and materials processing operations and materials–solution interactions, with a goal of developing a foundation for ensuring sustainability and economic competitiveness in the supply of resources and energy. She was the program leader of the Singapore Berkeley Research Initiative on Sustainable Energy from 2013 to 2015. Doyle served as the Vice Provost for Graduate Studies and the Dean of the Graduate Division at UC Berkeley from 2015 to 2019. Her honors include the Milton E. Wadsworth Award of the Society for Mining, Metallurgy, and Exploration; election as a Fellow of the Minerals, Metals, and Materials Society; and membership of the National Academy of Engineering. Doyle received her B.A. from the University of Cambridge and her M.Sc. and Ph.D. in hydrometallurgy from Imperial College, London.

Jami Dwyer is a licensed professional engineer with nearly 30 years of experience in the mining industry specializing in rock mechanics, blasting, operational efficiency, health and safety, maintenance strategies, mine design, and mine planning. She recently retired from Barr Engineering where she was responsible for business development for its Engineering and Design Business Unit. Previous to that, Dwyer worked for Barrick Gold Corporation for nearly 11 years where she served in a variety of roles including management of the engineering, maintenance, and mine operations departments. While at Barrick, she received a Corporate Environmental Excellence Award and a Corporate Social Responsibility Award. Dwyer spent 15 years employed by the National Institute for Occupational Safety and Health Office of Mine Safety and Health Research (formerly U.S. Bureau of Mines) in Spokane, Washington, where she led and developed several rock mechanics research projects related to innovative geotechnical monitoring technologies, blast damage assessments, and evaluation of ground support. She was also instrumental in developing early versions of software to locate and analyze mine seismicity and rock bursts in deep underground hard rock mine. Dwyer has served on the board of directors for the American Rock Mechanics Association, a past Chair of the Society of Mining, Metallurgy, and Exploration's Mining & Exploration Division's Executive Committee, and a member of the National Academies of Sciences, Engineering, and Medicine's Standing Committee on Geological and Geotechnical Engineering (COGGE). She holds a B.S. in applied computer science and a B.S. in mining engineering from Montana Technological University and an M.S. in mining engineering from the University of Missouri–Rolla. Dwyer was employed by Barrick Gold Corporation from 2007 to 2017 and Barr Engineering Company from 2018 to 2021.

Elizabeth Holley is an associate professor of mineral exploration and mining geology in the Department of Mining Engineering at the Colorado School of Mines, where she studies the processes responsible for ore deposit genesis, as well as the geologic characteristics that determine how ore bodies are developed, mined, and reclaimed. Her interdisciplinary work examines the intersections between technical and social risks in mining. She is a Fellow of the Payne Institute for Public Policy at the Colorado School of Mines, as well as the Site Director for a mining and mineral exploration-focused National Science Foundation (NSF) Industry–University Collaborative Research Center. Holley is an NSF Career Awardee, as well as the lead investigator for an NSF Growing Convergence Research project on responsible approaches to critical mineral supply. Her Mining Geology Research Group has been supported by NSF, the U.S. Centers for Disease Control and Prevention, the National Institute for Occupational Safety and Health, the U.S. Geological Survey, the Gates Environmental Fund, as well as major and mid-tier mining companies. Holley has worked in the industry on five continents, and she contributed to the discovery of the White Gold deposit in the Yukon. She is also a fellow of the Society of Economic Geologists (SEG) and has organized more than 175 professional development short courses as the SEG Education and Training Program coordinator. Holley holds a B.A. in geology from Pomona College, California; an M.Sc. in geochemistry from the University of Otago, New Zealand; and a Ph.D. in geology from the Colorado School Mines.

Paul A. Locke is an associate professor at the Johns Hopkins Bloomberg School of Public Health in the Department of Environmental Health and Engineering. In addition to his teaching and research, he co-directs the Bloomberg School's Dr.P.H. concentration in environmental health sciences and a certificate program in humane sciences and toxicology policy. Locke is an experienced environmental health professional with expertise in environmental health risk assessment, radiation risk communication, environmental law, and occupational health and toxicology policy. He is a National Associate of the National Academies of Sciences, Engineering, and Medicine's National Research Council and has chaired two National Academies committees—the Committee on Uranium Mining in Virginia and the Committee to Study the Potential Health Effects of Surface Coal Mining in Central Appalachia—and served as a committee member on at least seven other National Academies committees. Locke is admitted to practice law in the District of Columbia, the State of New York, and before the U.S. Supreme Court. Locke holds an M.P.H. and a Dr.P.H. from Yale University and Johns Hopkins, respectively, and a J.D. from Vanderbilt University.

Scott M. Olson is a professor and Faculty Excellence Scholar in the Civil and Environmental Engineering Department at the University of Illinois. Prior to joining Illinois, Olson worked in practice for more than 7 years for Woodward-Clyde Consultants and URS Corporation. For more than 25 years, Olson has been involved in dozens of research and consulting projects involving static and seismic liquefaction; geotechnical earthquake engineering; tailings dam engineering; in situ, laboratory, and centrifuge testing; soil–foundation–structure interaction; and paleoliquefaction and geohazards analysis. From these activities, Olson has published more than 150 journal papers, book chapters, conference articles, and reports, and has received numerous awards, including the American Society of Civil Engineers (ASCE) Walter L. Huber Civil Engineering Research Prize, the ASCE Arthur Casagrande Award, and the Canadian Geotechnical Society R.M. Quigley Award. Olson serves in various capacities for the Geo-Institute, the U.S. Universities Council on Geotechnical Education and Research, Earthquake Engineering Research Institute, and the Transportation Research Board. Recently, Olson became a Founding/Steering Committee member of the U.S.-based Tailings and Industrial Waste Engineering Center. Olson holds a B.S., an M.S., and a Ph.D. in civil and environmental engineering from the University of Illinois at Urbana-Champaign. In the past 5 years, Olson has consulted with Vale S.A., PolyMet Mining, and AECOM.

Brian S. Schwartz is a physician, environmental health scientist, and environmental and occupational epidemiologist. He joined the faculty at the Johns Hopkins Bloomberg School of Public Health in 1990 and has spent his entire career on the faculty there, where he is currently a professor of environmental health and engineering, epidemiology, and medicine within the Johns Hopkins University School of Medicine. He has done extensive epidemiologic research on the human health effects of metals exposures in occupational and environmental settings. He has also published recent studies on the human health effects of unconventional natural gas development, industrial farm animal production, and abandoned coal mine lands in Pennsylvania. He received an M.D. from the Feinberg School of Medicine at Northwestern University and an M.S. in clinical epidemiology from the University of Pennsylvania School of Medicine. He completed an internal medicine residency and general internal medicine fellowship at the Hospital of the University of Pennsylvania, and a fellowship in occupational and environmental medicine at the Johns Hopkins Bloomberg School of Public Health. He is board certified in internal medicine and occupational and environmental medicine.

M. Garrett Smith is currently the Geochemist with the Montana Department of Environmental Quality (DEQ), Hard Rock Mining Section. Having previously served as an assistant research professor with the Montana Bureau of Mines and Geology, his areas of interest include geochemical and hydrogeologic characterization of active and abandoned mines, stable isotope dynamics in riparian environments, tailings facility design and management, digital mapping and modeling, and the characterization and development of geothermal systems. This technical background supports the permitting and regulatory duties of his current role at DEQ, which include implementing state regulations to analyze operation plans, reclamation plans, and bonding for hard rock mines and mills; producing interdisciplinary environmental impact review documents; engaging with mine operators through site inspections and evaluating compliance with state and federal regulations; and conducting meetings with the public and other stakeholders. Smith's education and training include an M.S. in geoscience-geochemistry and a B.S. in chemistry from Montana Tech of the University of Montana, the U.S. Department of the Interior and Office of Surface Mining: Acid-Forming Materials Technical Training, Montana Environmental Policy Act (MEPA) Training, 40-Hour OSHA-HAZWOPER Certification, 20-Hour MSHA Certification, and repair and performance training for field and laboratory analytical instruments.

Shiliang Wu is a professor at Michigan Technological University with a joint appointment in the Department of Geological and Mining Engineering and Sciences and the Department of Civil, Environmental and Geospatial Engineering. His primary research areas include the interactions between anthropogenic emissions, land use/land cover, climate and atmospheric chemistry; the impacts of extreme events (such as heat waves, temperature inversion, etc.) on atmospheric chemistry and air quality; and the transport and cycling of emerging pollutants (such as arsenic, mercury, and selenium) in the global environment. He has published extensively in these areas. Wu holds a Ph.D. in atmospheric chemistry from Harvard University.

STAFF

Stephanie Johnson is a Senior Program Officer with the Water Science and Technology Board. Since joining the National Academies of Sciences, Engineering, and Medicine in 2002, she has worked on a wide range of water-related studies, on topics such as desalination, wastewater reuse, contaminant source remediation, coal and uranium mining, coastal risk reduction, and ecosystem restoration. Johnson received her B.A. from Vanderbilt University in chemistry and geology and her M.S. and Ph.D. in environmental sciences from the University of Virginia.

Margo Regier is a Program Officer with the Board on Earth Sciences and Resources and the Water Science and Technology Board at the National Academies. Regier received her B.S. from Beloit College, M.S. from Arizona State University, and Ph.D. in geology from the University of Alberta.

Miles Lansing is a Program Assistant with the Board on Earth Sciences and Resources and the Water Science and Technology Board at the National Academies. Lansing received his B.A. in political science from the University of Pittsburgh.

Clara Phipps was a Senior Program Assistant with the Board on Earth Sciences and Resources and the Water Science and Technology Board at the National Academies. Phipps is a graduate from the University of Puget Sound in Tacoma, Washington, with a B.S. in geology.

Chioma Onwumelu is a Ph.D. candidate in geology at the University of North Dakota and was a Fellow of the Christine Mirzayan Science & Technology Policy Graduate Fellowship Program.

Appendix B

Disclosure of Unavoidable Conflicts of Interest

The conflict-of-interest policy of the National Academies of Sciences, Engineering, and Medicine (https://www.nationalacademies.org/about/institutional-policies-and-procedures/conflict-of-interest-policies-and-procedures) prohibits the appointment of an individual to a committee like the one that authored this Consensus Study Report if the individual has a conflict of interest that is relevant to the task to be performed. An exception to this prohibition is permitted only if the National Academies determine that the conflict is unavoidable and the conflict is promptly and publicly disclosed.

When the committee that authored this report was established a determination of whether there was a conflict of interest was made for each committee member given the individual's circumstances and the task being undertaken by the committee. A determination that an individual has a conflict of interest is not an assessment of that individual's actual behavior or character or ability to act objectively despite the conflicting interest.

Kwame Awuah-Offei was determined to have a conflict of interest in relation to his service on the Committee on Potential Impacts of Goldmining in Virginia because he owns Sphinx Mining Systems, a company that consults with companies in the mining industry, including Barr Engineering, Weir International, and Vale International. The National Academies have determined that the experience and expertise of Awuah-Offei are needed for the committee to accomplish the task for which it has been established. The National Academies could not find another available individual with the equivalent experience and expertise who does not have a conflict of interest. Therefore, the National Academies have concluded that the conflict is unavoidable.

Jami Dwyer was determined to have a conflict of interest because her spouse holds stock in SSR Mining, a mining company focused on the operation, development, exploration, and acquisition of precious metal projects. The National Academies concluded that for this committee to accomplish the tasks for which it was established, its membership must include at least one person who has detailed first-hand knowledge of gold mine management and operations. As described in her biographical summary, as a General Supervisor for the Goldstrike Mine operated by Barrick Gold Corporation, and as a senior industry expert in gold and precious metals mining at Barr Engineering Company, Dwyer has extensive experience in managing gold mine operations, including operations in mine safety and health, rock mechanics, blasting, mine design, and technological innovation.

Scott M. Olson was determined to have a conflict of interest because he currently consults with companies with some interests in the gold mining industry (Statum and Intertechne) and is on the Steering Committee of the Tailings and Industrial Waste Engineering Center that conducts research related to mine tailings and tailings storage facilities, which is supported through donations from numerous firms involved in the mining industry. The National Academies have concluded that for this committee to accomplish the tasks for which it was established, its membership must include at least one person who has current and detailed knowledge of environmental engineering as it pertains to tailings management. The task requires that the committee evaluate the impacts of potential gold mining and processing operations on public health, safety, and welfare. Tailings dam failure and contamination of the surface and groundwater by improperly managed tailings are arguably the greatest risks of gold mining and processing to public welfare and health. As described in his biographical summary, Olson has extensive experience as a consultant for numerous clients providing advice on tailings dam engineering. Olson also has specialized expertise in research related to static liquefaction and geotechnical earthquake engineering, which are two of the top design issues related to tailings dam failure and ground and surface water contamination.

In each case, the National Academies determined that the experience and expertise of the individual were needed for the committee to accomplish the task for which it was established. The National Academies could not find another available individual with the equivalent experience and expertise who did not have a conflict of interest. Therefore, the National Academies concluded that the conflict was unavoidable and publicly disclosed it on its website (www.nationalacademies.org).

Appendix C

Metal Distribution and Potential for Mobilization

TABLE C-1 Distribution of Some Metals in Virginia Gold Deposits and Potential for Mobilization in Waters

Antimony	Distribution	Antimony can be a trace to minor constituent in sulfides, like pyrite (USGS, 2017). Tetrahedrite $((Cu,Fe)_{12}Sb_4S_{13})$ is reported in massive sulfides in Virginia and the Eldridge deposit (Mangan et al., 1984).
	Mobilization	Pyrite and other sulfides, including tetrahedrite, are unstable in the oxidized, weathering environment, and any antimony hosted in these phases would likely be released during mining, processing, and long-term storage of waste and could potentially make its way into the local groundwater/surface water system.
Arsenic	Distribution	*Distribution in deposit:* The main arsenic-bearing mineral phase associated with gold deposits is arsenopyrite (FeAsS). Pardee and Park (1948) report that arsenopyrite is rare in most gold deposits in Virginia. Park (1936), in referring to deposits in the gold-pyrite belt, notes that "arsenopyrite has previously been reported as occurring in the ores but has not been observed in the recent work." Rimstidt et al. (1994) report that arsenopyrite is among the most reactive of all sulfide minerals studied, indicating that any arsenopyrite in the ores would be oxidized relatively quickly. Arsenic also occurs as a trace to minor component in pyrite, and arsenian pyrite is common in many gold deposits. Arsenic contents of gold ores in the gold-pyrite belt are expected to be in concentrations of a few hundred to perhaps a few thousand parts per million, with essentially all of the arsenic hosted in trace arsenopyrite and as a trace element in pyrite.
	Mobilization	Both pyrite and arsenopyrite are unstable in the oxidized, weathering environment, and any arsenic hosted in these phases would likely be released during mining, processing, and long-term storage of waste and could potentially make its way into the local groundwater/ surface water system.
Cadmium	Distribution	Minerals in which cadmium is a major component, including the cadmium sulfide phase greenockite (CdS), are not reported in any gold deposits in Virginia. Most cadmium that occurs in the gold deposits of Virginia is thought to occur as a trace element in sphalerite (ZnS) and, to a lesser extent, in other sulfide minerals. Schwartz (2000) reports concentrations of cadmium in sphalerite ranging up to 5 wt.% with chalcopyrite (600–1,200 mg/kg), galena (10–500 mg/kg), and pyrite (1–200 mg/kg) also serving as potential hosts for cadmium. Hammarstrom et al. (2006) report cadmium concentrations in pyrite from the Valzinco Mine of 0.18 wt.% (1,800 mg/kg).
	Mobilization	During mining, processing, and long-term storage of waste rock and tailings, all (or most) cadmium contained in sphalerite and other sulfide phases could be released into the local environment and potentially make its way into the local groundwater/surface water system.
Copper	Distribution	Chalcopyrite $(CuFeS_2)$ is a common trace sulfide phase in many gold deposits in the gold-pyrite belt, and in the Virgilina district the main copper-bearing phase is bornite (Cu_5FeS_4). Other copper-bearing phases that have been reported include native copper, chalcocite (Cu_2S), cuprite (Cu_2O), malachite $[CuCO_3 \bullet Cu(OH)_2]$, and azurite $[2CuCO_3 \bullet Cu(OH)_2]$.
	Mobilization	Copper in chalcopyrite and bornite are likely to break down during exposure to humid, oxidizing conditions and release copper to the local environment, but Rimstidt et al. (1994) note that the rate of oxidation of chalcopyrite is about 30 times slower than that of pyrite under similar conditions. Copper hosted in native copper, chalcocite, cuprite, malachite, and azurite are more stable at near surface conditions. Copper contained in these mineral phases may remain sequestered and not release into the environment.
Lead	Distribution	Trace to minor amounts of galena (PbS) are reported in most mines in the gold-pyrite belt.
	Mobilization	Galena reacts (oxidizes) about 300 times faster than sphalerite at pH 2 (but still slower than pyrite) (Rimstidt et al., 1994). Products of weathering of lead-bearing sulfides, including pyromorphite $[Pb_5(PO_4)_3Cl]$, vanadinite $[Pb_5(VO_4)_3Cl]$, and cerrusite $(PbCO_3)$, have also been reported at some locations in the gold-pyrite belt. While the overall lead content of the ores is low, it is expected that all or most of the lead in a given deposit could be mobilized as a result of mining activities.

TABLE C-1 Continued

Selenium	Distribution	No minerals in which selenium is a major component have been reported in any deposits from the gold-pyrite belt. Stillings (2017) reports that while selenium-bearing minerals are common, in hydrothermal sulfide ore deposits most of the selenium occurs as a trace element in sulfides such as bornite (Cu_5FeS_4), chalcocite (Cu_2S), chalcopyrite ($CuFeS_2$), galena (PbS), and pyrite (FeS_2), all of which have been reported in gold deposits in the gold-pyrite belt. In the Virgilina district, the main ore in the copper deposits is bornite. Babedi et al. (2022) report selenium concentrations in pyrite from orogenic gold deposits ranging from below detection to 200 mg/kg, with a mean value of 25 mg/kg, and report values ranging from 1,500 to 1,900 mg/kg in pyrites from low-sulfidation gold deposits. Chinnasamy et al. (2021) report selenium contents of pyrites from different stages in the Jonnagiri shear-zone hosted gold deposit in India that range from below detection to 80 mg/kg.
	Mobilization	While the concentrations of selenium appear to be low in ores from the gold-pyrite belt, if most of the selenium is contained as a trace component in pyrite (and other sulfide minerals), the selenium would be released during mining, processing, and long-term storage of waste.
Thallium	Distribution	Thallium minerals are rare, and none have been reported in the gold deposits of the gold-pyrite belt. In hydrothermal sulfide ores, thallium dominantly occurs as a trace component in pyrite, galena, and sphalerite. Bojakowska and Paulo (2013) report average thallium concentrations of 52.1 mg/kg in lead-zinc ores from ore deposits in Poland, and 1.4 mg/kg thallium in copper ores. Maximum values reported are 17.9 mg/kg for copper ores and 547 mg/kg for zinc-lead ores, with most of the thallium in lead-zinc ores contained in sphalerite. Chinnasamy et al. (2021) analyzed numerous pyrites from different stages in the Jonnagiri shear-zone hosted gold deposit in India and reported thallium concentrations ranging from below detection to 0.388 mg/kg. Majumdar et al. (2019) studied pyrites from shear-zone hosted gold occurrences in the South Purulia shear zone, India, and report thallium values ranging from below detection to 0.75 mg/kg. These workers also report 0.64 and 1.25 mg/kg thallium in chalcopyrite from these same deposits. Babedi et al. (2022) report thallium concentrations in pyrite from orogenic gold deposits ranging from below detection to 4,244 mg/kg, with a mean value of 98 mg/kg, and report values ranging from 10 to 2,700 mg/kg in pyrites from low-sulfidation gold deposits, with a mean value of 262 mg/kg.
	Mobilization	Most or all of the thallium in gold deposits in the gold-pyrite belt is likely hosted in pyrite and other sulfide phases and would be released during mining, processing, and long-term storage of waste.
Zinc	Distribution	Trace to minor amounts of sphalerite (ZnS) are reported in most mines in the gold-pyrite belt, and trace amounts of the alteration product smithsonite ($ZnCO_3$) have been rarely reported.
	Mobilization	Most zinc in sphalerite will likely be released to the environment as a result of alteration and decomposition during mining activities. The insignificant amount of zinc hosted by smithsonite will remain sequestered in the mineral, which is stable at surface conditions.

Appendix D

Valzinco Samples from Puddle

TABLE D-1 Metal Concentrations and pH in Water Samples Collected in Stagnant Puddles on Top of Tailings Piles at the Valzinco Mine (Spotsylvania County)

Sample	Comments (sample name)	pH	Al (mg/L)	As (mg/L)	Cd (mg/L)	Cu (mg/L)	Pb (mg/L)	Sb (mg/L)	Se (mg/L)	Tl (mg/L)	Zn (mg/L)
Valzinco filtered water at mine	Collected from a small puddle on the tailings (VLZN-10-2 FA)	1.1	1038.40	<0.02	3	59	2.1	0.01	<0.04	<0.005	2,300
Valzinco unfiltered water at mine	Collected from a small puddle on the tailings (VLZN-10-2 RA)	1.1	1090.32	0.28	3.1	62	1.5	0.01	0.07	<0.005	2,400

NOTE: Valzinco data samples are described as "an anomalous geochemical setting."
SOURCES: Data are from Seal and Hammarstrom (2002) and Seal et al. (2002).

Appendix E

Comparison to Other Deposits

The Statement of Task requests that the committee include a discussion of current gold mining operations at sites with comparable geologic, mineralogical, hydrologic, and climatic characteristics to those found in the Commonwealth (see Box 1-3). Below, we summarize those features that are common to all (or most) deposits in Virginia and discuss a few individual non-Virginia deposits that show many features that are similar to Virginia deposits.

All of the gold deposits in Virginia's gold-pyrite belt and in the Virgilina district are classified as orogenic deposits (Goldfarb et al., 2005). This class of gold deposit is associated with mountain building, metamorphism, deformation, and regional shearing. Orogenic gold deposits are further subdivided based on the intensity, or temperature and pressure, of the metamorphic and deformation events, and include deposits hosted in subgreenschist to lower greenschist (~<1–1.5 kbar; <~5 km) rocks, greenschist grade (~1.5–3 kbar; <~5–10 km) rocks, and amphibolite grade (~3–5 kbar; <~10–18 km) rocks. Deposits in the gold-pyrite belt and the Virgilina district are associated with greenschist grade metamorphic rocks, although some approach amphibolite grade and some in the Virgilina district are lower greenschist grade. The tectonic, structural, and alteration characteristics of gold-pyrite belt deposits match the greenschist class of orogenic deposits well, but differ from many greenschist-hosted orogenic deposits in that a larger proportion of gold in this class occurs as "free" gold, whereas most gold in the gold-pyrite belt is encapsulated in pyrite. The gold-pyrite belt deposits are also lower in arsenic than the typical greenschist grade orogenic deposit, and veins/lenses in gold-pyrite belt deposits tend to be smaller than the average reported for this class of deposit (Goldfarb et al., 2005).

COMPARISON WITH OTHER NON-VIRGINIA DEPOSITS

Below, we summarize the characteristics from non-Virginia deposits that have comparable geologic, mineralogical, hydrologic, or climatic characteristics to Virginia gold deposits.

Haile Gold Mine, South Carolina

The Haile Gold Mine is comparable to gold deposits in Virginia in that it is found in the Piedmont physiographic province, which has a humid, subtropical climate, and aquifers hosted in fractured bedrock (SCDHEC, 2022). Similar to Virginia gold mines, Haile Gold Mine is an orogenic deposit that occurs in greenschist facies rocks (although of lower metamorphic grade than in Virginia) that has quartz-sericite (carbonate) alteration and low pyritic ores that are associated with shearing and occur along a linear trend. However, Haile differs from

Virginia deposits in that much of the mineralization is disseminated in a siltstone at Haile rather than occurring in distinct quartz veins, the mineralized zones are much wider with barren or low-grade rock between mineralized zones (which requires open pit mining), and the ores are more arsenic rich and contain much fewer base metals (Foley and Ayuso, 2012; Foley et al., 2001; Mobley et al., 2014; SRK Consulting, 2020). As a result, even though the original formation of gold deposits in Virginia may have occurred in similar tectonic environments as gold deposits in South Carolina, the gold deposits in these two states have experienced different postformation histories that have led to different styles of mineralization.

Another important difference between the Haile mine and gold deposits in Virginia is the scale of the deposits. The Haile mine extracts much larger volumes of rock compared to mining that would occur in Virginia, generating much more waste and tailings, and requiring a much larger surface footprint to accommodate the infrastructure and waste and tailings storage. The potential environmental impact of a tailings dam failure or release of water from the site at Haile is commensurately larger because of these various factors.

Dahlonega District, Georgia

The gold deposits of the Dahlonega district, Georgia, are very comparable to the Virginia deposits. Similar to Virginia, the Dahlonega district deposits are associated with shear zones in greenschist wallrock, have gold hosted in pyrite that is found in steeply dipping veins, and have relatively low arsenic content. Graton and Lindgren (1906) note that the mode of origin was similar to that of the Californian and Australian deposits, which are classified as orogenic gold deposits today. The main differences between the Dahlonega deposits and those in Virginia are that there are no massive sulfide deposits spatially associated with the Dahlonega deposits, the gangue mineral assemblage at Dahlonega contains less carbonate (and no reported ankerite) as is common in the Virginia deposits, and pyrrhotite appears to be more common in the Dahlonega deposits. Most other characteristics of the Dahlonega deposits are in good agreement with the Virginia gold deposits (Graton and Lindgren, 1906; Jones, 1909; Yeates et al., 1896).

Most of the mining in the Dahlonega district occurred in the saprolite using hydraulic mining methods. In some cases, such as at the Findley Mine, the quartz veins in saprolite are identifiable and competent and mined using conventional methods (not hydraulic mining) (Jones, 1909).

The Kensington Deposit, Alaska

The Kensington deposit is an orogenic gold deposit that consists of the Kensington, Raven, Eureka, Jualin, and Elmira deposits/ore bodies. Vein mineralization in the Raven ore body is characterized by gold and gold–silver telluride minerals, where most of the gold is contained in calaverite ($AuTe_2$) that occurs in association in and interstitial to pyrite grains and in microfractures in pyrite. In the other ore bodies, mineralization occurs primarily as disseminated pyrite or auriferous pyrite seams and blebs. The Kensington deposits are comparable to deposits in Virginia in terms of genetic classification as orogenic deposits—both are associated with shearing, have steeply dipping veins, show similar alteration assemblages, are low in arsenic, and have similar gold grades. The Kensington deposits differ from gold-pyrite belt deposits in that schists are not associated with the Kensington deposits, and mineralization occurs completely within one rock type (Jualin diorite). Additionally, there are no associated massive sulfide deposits at the Kensington deposits (Cox and Bagbey, 1992; Pascoe et al., 2022; USGS, 2017c).

The scale of the mining operation and the processing methods used at Kensington are similar to what might occur in Virginia. Kensington has a total of 983,000 indicated and measured ounces of gold (Pascoe et al., 2022), which is approximately one to two orders of magnitude larger than deposits in Virginia. The mining operations use underground mining, crushing, and flotation to produce a concentrate that is then shipped off-site for further processing (Pascoe et al., 2022).

Bousquet Group, Canada

The Mic Mac, Mooshla A, Mooshla B, and parts of the Mouska deposit at the Bousquet Group of orogenic gold deposits in Quebec, Canada, are comparable to the Virginia deposits in that there is a clear genetic association between their shear zone–hosted gold mineralization and nearby volcanogenic massive sulfide deposits. In other respects, the Bousquet deposits are not comparable to Virginia gold deposits. For example, their upper greenschist to amphibolite metamorphic grade is distinct from the lower to mid greenschist metamorphic grade in Virginia. Additionally, Bousquet has a much higher pyrite content than Virginia, different alteration assemblages, and gold associated with chalcopyrite, bornite, and pyrrhotite, instead of pyrite (Groves et al., 2003; Mercier-Langevin et al., 2007; Tourigny et al., 1993).

Mikado Deposit, Alaska

The Mikado deposit is an orogenic deposit that is most comparable to the Virginia deposits in terms of size, geometry, and the scale of the mining operation required. In other ways, the Mikado deposit is not similar to Virginia deposits. For example, the Mikado deposit is much higher in arsenopyrite than Virginia ores, veins that represent open space filling rather than replacement, wall rocks that are highly graphitic, and more free gold than gold encapsulated in sulfides (Ashworth, 1983).

Eagle Shawmut Mine, Mother Lode District, California

The Eagle Shawmut Mine is an orogenic deposit in the Mother Lode of California is comparable to Virginia deposits in many respects. For example, the Eagle Shawmut deposit is of a similar size and is associated with steeply dipping veins in greenschist-hosted shear zones. However, the deposit differs from Virginia in that it has higher arsenic and contains more free gold (Knopf, 1929).